Computers as Components
Principles of Embedded Computing System Design

About the Author

Wayne Wolf is Professor, Rhesea "Ray" P. Farmer Distinguished Chair in Embedded Computing, and Georgia Research Alliance Eminent Scholar at the Georgia Institute of Technology. Before joining Georgia Tech, he was with Princeton University and AT&T Bell Laboratories in Murray Hill, New Jersey. He received his B.S., M.S., and Ph.D. in electrical engineering from Stanford University. He is well known for his research in the areas of hardware/software co-design, embedded computing, VLSI CAD, and multimedia computing systems. He is a fellow of the IEEE and ACM. He co-founded several conferences in the area, including CODES, MPSoC, and Embedded Systems Week. He was founding co-editor-in-chief of *Design Automation for Embedded Systems* and founding editor-in-chief of *ACM Transactions on Embedded Computing Systems*. He has received the ASEE Frederick E. Terman Award and the IEEE Circuits and Society Education Award. He is also co-series editor of the Morgan Kaufmann Series in Systems on Silicon.

Computers as Components
Principles of Embedded Computing System Design

Second Edition

Wayne Wolf

AMSTERDAM • BOSTON • HEIDELBERG • LONDON
NEW YORK • OXFORD • PARIS • SAN DIEGO
SAN FRANCISCO • SINGAPORE • SYDNEY • TOKYO

Morgan Kaufmann Publishers is an imprint of Elsevier

ELSEVIER

Morgan Kaufmann Publishers is an imprint of Elsevier.
30 Corporate Drive, Suite 400, Burlington, MA 01803, USA

This book is printed on acid-free paper. ⊗

Library of Congress Cataloging-in-Publication Data
Wolf, Wayne Hendrix.
 Computers as components: principles of embedded computing system design / by Wayne Wolf – 2nd ed.
 p. cm.
 Includes bibliographical references and index.
 ISBN 978-0-12-374397-8 (pbk. : alk. paper)
1. System design. 2. Embedded computer systems. I. Title.
 QA76.9.S88W64 2001
 004.16–dc22

 2008012300

ISBN: 978-0-12-374397-8

For information on all Morgan Kaufmann publications,
visit our website at *www.mkp.com* or *www.books.elsevier.com*

Printed in the United States of America
08 09 10 11 12 5 4 3 2 1

To Nancy and Alec

Disclaimer

Designations used by companies to distinguish their products are often claimed as trademarks or registered trademarks. In all instances where Morgan Kaufmann Publishers is aware of a claim, the product names appear in initial capital or all capital letters. Readers, however, should contact the appropriate companies for more complete information regarding trademarks and registration.

ARM, the ARM Powered logo, StrongARM, Thumb and ARM7TDMI are registered trademarks of ARM Ltd. ARM Powered, ARM7, ARM7TDMI-S, ARM710T, ARM740T, ARM9, ARM9TDMI, ARM940T, ARM920T, EmbeddedICE, ARM7T-S, Embedded-ICE-RT, ARM9E, ARM946E, ARM966E, ARM10, AMBA, and Multi-ICE are trademarks of ARM Limited. All other brand names or product names are the property of their respective holders. "ARM" is used to represent ARM Holdings plc (LSE: ARM and NASDAQ: ARMHY); its operating company, ARM Ltd; and the regional subsidiaries: ARM, INC., ARM KK; ARM Korea, Ltd.

Microsoft and Windows are registered trademarks and Windows NT is a trademark of Microsoft Corporation. Pentium is a trademark of Intel Corporation. All other trademarks and logos are property of their respective holders. TMS320C55x, C55x, and Code Composer Studio are trademarks of Texas Instruments Incorporated.

Foreword to The First Edition

Digital system design has entered a new era. At a time when the design of microprocessors has shifted into a classical optimization exercise, the design of embedded computing systems in which microprocessors are merely components has become a wide-open frontier. Wireless systems, wearable systems, networked systems, smart appliances, industrial process systems, advanced automotive systems, and biologically interfaced systems provide a few examples from across this new frontier.

Driven by advances in sensors, transducers, microelectronics, processor performance, operating systems, communications technology, user interfaces, and packaging technology on the one hand, and by a deeper understanding of human needs and market possibilities on the other, a vast new range of systems and applications is opening up. It is now up to the architects and designers of embedded systems to make these possibilities a reality.

However, embedded system design is practiced as a craft at the present time. Although knowledge about the component hardware and software subsystems is clear, there are no system design methodologies in common use for orchestrating the overall design process, and embedded system design is still run in an *ad hoc* manner in most projects.

Some of the challenges in embedded system design come from changes in underlying technology and the subtleties of how it can all be correctly mingled and integrated. Other challenges come from new and often unfamiliar types of system requirements. Then too, improvements in infrastructure and technology for communication and collaboration have opened up unprecedented possibilities for fast design response to market needs. However, effective design methodologies and associated design tools have not been available for rapid follow-up of these opportunities.

At the beginning of the VLSI era, transistors and wires were the fundamental components, and the rapid design of computers on a chip was the dream. Today the CPU and various specialized processors and subsystems are merely basic components, and the rapid, effective design of very complex embedded systems is the dream. Not only are system specifications now much more complex, but they must also meet real-time deadlines, consume little power, effectively support complex real-time user interfaces, be very cost-competitive, and be designed to be upgradable.

Wayne Wolf has created the first textbook to systematically deal with this array of new system design requirements and challenges. He presents formalisms and a methodology for embedded system design that can be employed by the new type of "tall-thin" system architect who really understands the foundations of system design across a very wide range of its component technologies.

Moving from the basics of each technology dimension, Wolf presents formalisms for specifying and modeling system structures and behaviors and then clarifies these

ideas through a series of design examples. He explores the complexities involved and how to systematically deal with them. You will emerge with a sense of clarity about the nature of the design challenges ahead and with knowledge of key methods and tools for tackling those challenges.

As the first textbook on embedded system design, this book will prove invaluable as a means for acquiring knowledge in this important and newly emerging field. It will also serve as a reference in actual design practice and will be a trusted companion in the design adventures ahead. I recommend it to you highly.

Lynn Conway
Professor Emerita, Electrical Engineering and
Computer Science University of Michigan

Contents

CHAPTER 5 Program Design and Analysis 209

List of Examples

Preface to The Second Edition

Embedded computing is more important today than it was in 2000, when the first edition of this book appeared. Embedded processors are in even more products, ranging from toys to airplanes. Systems-on-chips now use up to hundreds of CPUs. The cell phone is on its way to becoming the new standard computing platform. As my column in *IEEE Computer* in September 2006 indicated, there are at least a half-million embedded systems programmers in the world today, probably closer to 800,000.

In this edition I have tried to both update and revamp. One major change is that the book now uses the TI TMS320C55x™ (C55x) DSP. I seriously rewrote the discussion of real-time scheduling. I have tried to expand on performance analysis as a theme at as many levels of abstraction as possible. Given the importance of multiprocessors in even the most mundane embedded systems, this edition also talks more generally about hardware/software co-design and multiprocessors.

One of the changes in the field is that this material is taught at lower and lower levels of the curriculum. What used to be graduate material is now upper-division undergraduate; some of this material will percolate down to the sophomore level in the foreseeable future. I think that you can use subsets of this book to cover both more advanced and more basic courses. Some advanced students may not need the background material of the earlier chapters and you can spend more time on software performance analysis, scheduling, and multiprocessors. When teaching introductory courses, software performance analysis is an alternative path to exploring microprocessor architectures as well as software; such courses can concentrate on the first few chapters.

The new Web site for this book and my other books is http://www.waynewolf.us. On this site, you can find overheads for the material in this book, suggestions for labs, and links to more information on embedded systems.

ACKNOWLEDGMENTS

I would like to thank a number of people who helped me with this second edition. Cathy Wicks and Naser Salameh of Texas Instruments gave me invaluable help in figuring out the C55x. Richard Barry of freeRTOS.org not only graciously allowed me to quote from the source code of his operating system but he also helped clarify the explanation of that code. My editor at Morgan Kaufmann, Chuck Glaser, knew when to be patient, when to be encouraging, and when to be cajoling. (He also has great taste in sushi restaurants.) And of course, Nancy and Alec patiently let me type away. Any problems, small or large, with this book are, of course, solely my responsibility.

<div style="text-align: right">

Wayne Wolf
Atlanta, GA, USA

</div>

xix

Preface to The First Edition

Microprocessors have long been a part of our lives. However, microprocessors have become powerful enough to take on truly sophisticated functions only in the past few years. The result of this explosion in microprocessor power, driven by Moore's Law, is the emergence of embedded computing as a discipline. In the early days of microprocessors, when all the components were relatively small and simple, it was necessary and desirable to concentrate on individual instructions and logic gates. Today, when systems contain tens of millions of transistors and tens of thousands of lines of high-level language code, we must use design techniques that help us deal with complexity.

This book tries to capture some of the basic principles and techniques of this new discipline of embedded computing. Some of the challenges of embedded computing are well known in the desktop computing world. For example, getting the highest performance out of pipelined, cached architectures often requires careful analysis of program traces. Similarly, the techniques developed in software engineering for specifying complex systems have become important with the growing complexity of embedded systems. Another example is the design of systems with multiple processes. The requirements on a desktop general-purpose operating system and a real-time operating system are very different; the real-time techniques developed over the past 30 years for larger real-time systems are now finding common use in microprocessor-based embedded systems.

Other challenges are new to embedded computing. One good example is power consumption. While power consumption has not been a major consideration in traditional computer systems, it is an essential concern for battery-operated embedded computers and is important in many situations in which power supply capacity is limited by weight, cost, or noise. Another challenge is deadline-driven programming. Embedded computers often impose hard deadlines on completion times for programs; this type of constraint is rare in the desktop world. As embedded processors become faster, caches and other CPU elements also make execution times less predictable. However, by careful analysis and clever programming, we can design embedded programs that have predictable execution times even in the face of unpredictable system components such as caches.

Luckily, there are many tools for dealing with the challenges presented by complex embedded systems: high-level languages, program performance analysis tools, processes and real-time operating systems, and more. But understanding how all these tools work together is itself a complex task. This book takes a bottom-up approach to understanding embedded system design techniques. By first understanding the fundamentals of microprocessor hardware and software, we can build powerful abstractions that help us create complex systems.

A NOTE TO EMBEDDED SYSTEM PROFESSIONALS

This book is not a manual for understanding a particular microprocessor. Why should the techniques presented here be of interest to you? There are two reasons. First, techniques such as high-level language programming and real-time operating systems are very important in making large, complex embedded systems that actually work. The industry is littered with failed system designs that didn't work because their designers tried to hack their way out of problems rather than stepping back and taking a wider view of the problem. Second, the components used to build embedded systems are constantly changing, but the principles remain constant. Once you understand the basic principles involved in creating complex embedded systems, you can quickly learn a new microprocessor (or even programming language) and apply the same fundamental principles to your new components.

A NOTE TO TEACHERS

The traditional microprocessor system design class originated in the 1970s when microprocessors were exotic yet relatively limited. That traditional class emphasizes breadboarding hardware and software to build a complete system. As a result, it concentrates on the characteristics of a particular microprocessor, including its instruction set, bus interface, and so on.

This book takes a more abstract approach to embedded systems. While I have taken every opportunity to discuss real components and applications, this book is fundamentally not a microprocessor data book. As a result, its approach may seem initially unfamiliar. Rather than concentrating on particulars, the book tries to study more generic examples to come up with more generally applicable principles. However, I think that this approach is both fundamentally easier to teach and in the long run more useful to students. It is easier because one can rely less on complex lab setups and spend more time on pencil-and-paper exercises, simulations, and programming exercises. It is more useful to the students because their eventual work in this area will almost certainly use different components and facilities than those used at your school. Once students learn fundamentals, it is much easier for them to learn the details of new components.

Hands-on experience is essential in gaining physical intuition about embedded systems. Some hardware building experience is very valuable; I believe that every student should know the smell of burning plastic integrated circuit packages. But I urge you to avoid the tyranny of hardware building. If you spend too much time building a hardware platform, you will not have enough time to write interesting programs for it. And as a practical matter, most classes do not have the time to let students build sophisticated hardware platforms with high-performance I/O devices and possibly multiple processors. A lot can be learned about hardware by measuring and evaluating an existing hardware platform. The experience of programming

complex embedded systems will teach students quite a bit about hardware as well—debugging interrupt-driven code is an experience that few students are likely to forget.

A home page for the book (www.mkp.com/embed) includes overheads, instructor's manual, lab materials, links to related Web sites, and a link to a password-protected ftp site that contains solutions to the exercises.

ACKNOWLEDGMENTS

I owe a word of thanks to many people who helped me in the preparation of this book. Several people gave me advice about various aspects of the book: Steve Johnson (Indiana University) about specification, Louise Trevillyan and Mark Charney (both IBM Research) on program tracing, Margaret Martonosi (Princeton University) on cache miss equations, Randy Harr (Synopsys) on low power, Phil Koopman (Carnegie Mellon University) on distributed systems, Joerg Henkel (NEC C&C Labs) on low-power computing and accelerators, Lui Sha (University of Illinois) on real-time operating systems, John Rayfield (ARM) on the ARM architecture, David Levine (Analog Devices) on compilers and SHARC, and Con Korikis (Analog Devices) on the SHARC. Many people acted as reviewers at various stages: David Harris (Harvey Mudd College); Jan Rabaey (University of California at Berkeley); David Nagle (Carnegie Mellon University); Randy Harr (Synopsys); Rajesh Gupta, Nikil Dutt, Frederic Doucet, and Vivek Sinha (University of California at Irvine); Ronald D. Williams (University of Virginia); Steve Sapiro (SC Associates); Paul Chow (University of Toronto); Bernd G. Wenzel (Eurostep); Steve Johnson (Indiana University); H. Alan Mantooth (University of Arkansas); Margarida Jacome (University of Texas at Austin); John Rayfield (ARM); David Levine (Analog Devices); Ardsher Ahmed (University of Massachusetts/Dartmouth University); and Vijay Madisetti (Georgia Institute of Technology). I also owe a big word of thanks to my editor, Denise Penrose. Denise put in a great deal of effort finding and talking to potential users of this book to help us understand what readers wanted to learn. This book owes a great deal to her insight and persistence. Cheri Palmer and her production team did an excellent job on an impossibly tight schedule. The mistakes and miscues are, of course, all mine.

Embedded Computing

- Why we embed microprocessors in systems.
- What is difficult and unique about embedding computing.
- Design methodologies.
- System specification.
- A guided tour of this book.

INTRODUCTION

In this chapter we set the stage for our study of embedded computing system design. In order to understand the design processes, we first need to understand how and why microprocessors are used for control, user interface, signal processing, and many other tasks. The ***microprocessor*** has become so common that it is easy to forget how hard some things are to do without it.

We first review the various uses of microprocessors and then review the major reasons why microprocessors are used in system design–delivering complex behaviors, fast design turnaround, and so on. Next, in Section 1.2, we walk through the design of an example system to understand the major steps in designing a system. Section 1.3 includes an in-depth look at techniques for specifying embedded systems—we use these specification techniques throughout the book. In Section 1.4, we use a model train controller as an example for applying the specification techniques introduced in Section1.3 that we use throughout the rest of the book. Section 1.5 provides a chapter-by-chapter tour of the book.

1.1 COMPLEX SYSTEMS AND MICROPROCESSORS

What is an ***embedded computer system***? Loosely defined, it is any device that includes a programmable computer but is not itself intended to be a general-purpose computer. Thus, a PC is not itself an embedded computing system, although PCs are often used to build embedded computing systems. But a fax machine or a clock built from a microprocessor is an embedded computing system.

This means that embedded computing system design is a useful skill for many types of product design. Automobiles, cell phones, and even household appliances make extensive use of microprocessors. Designers in many fields must be able to identify where microprocessors can be used, design a hardware platform with I/O devices that can support the required tasks, and implement software that performs the required processing. Computer engineering, like mechanical design or thermo-dynamics, is a fundamental discipline that can be applied in many different domains. But of course, embedded computing system design does not stand alone. Many of the challenges encountered in the design of an embedded computing system are not computer engineering—for example, they may be mechanical or analog electri-cal problems. In this book we are primarily interested in the embedded computer itself, so we will concentrate on the hardware and software that enable the desired functions in the final product.

1.1.1 Embedding Computers

Computers have been embedded into applications since the earliest days of com-puting. One example is the Whirlwind, a computer designed at MIT in the late 1940s and early 1950s. Whirlwind was also the first computer designed to support *real-time* operation and was originally conceived as a mechanism for controlling an aircraft simulator. Even though it was extremely large physically compared to today's computers (e.g., it contained over 4,000 vacuum tubes), its complete design from components to system was attuned to the needs of real-time embedded com-puting. The utility of computers in replacing mechanical or human controllers was evident from the very beginning of the computer era—for example, computers were proposed to control chemical processes in the late 1940s [Sto95].

A microprocessor is a single-chip CPU. Very large scale integration (VLSI) stet—the acronym is the name technology has allowed us to put a complete CPU on a single chip since 1970s, but those CPUs were very simple. The first microproces-sor, the Intel 4004, was designed for an embedded application, namely, a calculator. The calculator was not a general-purpose computer—it merely provided basic arithmetic functions. However, Ted Hoff of Intel realized that a general-purpose computer programmed properly could implement the required function, and that the computer-on-a-chip could then be reprogrammed for use in other products as well. Since integrated circuit design was (and still is) an expensive and time-consuming process, the ability to reuse the hardware design by changing the software was a key breakthrough. The HP-35 was the first handheld calculator to perform transcendental functions [Whi72]. It was introduced in 1972, so it used several chips to implement the CPU, rather than a single-chip microprocessor. How-ever, the ability to write programs to perform math rather than having to design digital circuits to perform operations like trigonometric functions was critical to the successful design of the calculator.

Automobile designers started making use of the microprocessor soon after single-chip CPUs became available. The most important and sophisticated use of

microprocessors in automobiles was to control the engine: determining when spark plugs fire, controlling the fuel/air mixture, and so on. There was a trend toward electronics in automobiles in general—electronic devices could be used to replace the mechanical distributor. But the big push toward microprocessor-based engine control came from two nearly simultaneous developments: The oil shock of the 1970s caused consumers to place much higher value on fuel economy, and fears of pollution resulted in laws restricting automobile engine emissions. The combination of low fuel consumption and low emissions is very difficult to achieve; to meet these goals without compromising engine performance, automobile manufacturers turned to sophisticated control algorithms that could be implemented only with microprocessors.

Microprocessors come in many different levels of sophistication; they are usually classified by their word size. An 8-bit *microcontroller* is designed for low-cost applications and includes on-board memory and I/O devices; a 16-bit microcontroller is often used for more sophisticated applications that may require either longer word lengths or off-chip I/O and memory; and a 32-bit *RISC* microprocessor offers very high performance for computation-intensive applications.

Given the wide variety of microprocessor types available, it should be no surprise that microprocessors are used in many ways. There are many household uses of microprocessors. The typical microwave oven has at least one microprocessor to control oven operation. Many houses have advanced thermostat systems, which change the temperature level at various times during the day. The modern camera is a prime example of the powerful features that can be added under microprocessor control.

Digital television makes extensive use of embedded processors. In some cases, specialized CPUs are designed to execute important algorithms—an example is the CPU designed for audio processing in the SGS Thomson chip set for DirecTV [Lie98]. This processor is designed to efficiently implement programs for digital audio decoding. A programmable CPU was used rather than a hardwired unit for two reasons: First, it made the system easier to design and debug; and second, it allowed the possibility of upgrades and using the CPU for other purposes.

A high-end automobile may have 100 microprocessors, but even inexpensive cars today use 40 microprocessors. Some of these microprocessors do very simple things such as detect whether seat belts are in use. Others control critical functions such as the ignition and braking systems.

Application Example 1.1 describes some of the microprocessors used in the BMW 850i.

Application Example 1.1

BMW 850i brake and stability control system

The BMW 850i was introduced with a sophisticated system for controlling the wheels of the car. An antilock brake system (ABS) reduces skidding by pumping the brakes. An automatic

stability control (ASC + T) system intervenes with the engine during maneuvering to improve the car's stability. These systems actively control critical systems of the car; as control systems, they require inputs from and output to the automobile.

Let's first look at the ABS. The purpose of an ABS is to temporarily release the brake on a wheel when it rotates too slowly—when a wheel stops turning, the car starts skidding and becomes hard to control. It sits between the hydraulic pump, which provides power to the brakes, and the brakes themselves as seen in the following diagram. This hookup allows the ABS system to modulate the brakes in order to keep the wheels from locking. The ABS system uses sensors on each wheel to measure the speed of the wheel. The wheel speeds are used by the ABS system to determine how to vary the hydraulic fluid pressure to prevent the wheels from skidding.

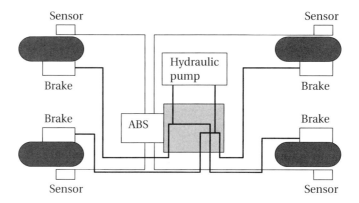

The ASC + T system's job is to control the engine power and the brake to improve the car's stability during maneuvers. The ASC + T controls four different systems: throttle, ignition timing, differential brake, and (on automatic transmission cars) gear shifting. The ASC + T can be turned off by the driver, which can be important when operating with tire snow chains.

The ABS and ASC + T must clearly communicate because the ASC + T interacts with the brake system. Since the ABS was introduced several years earlier than the ASC + T, it was important to be able to interface ASC + T to the existing ABS module, as well as to other existing electronic modules. The engine and control management units include the electronically controlled throttle, digital engine management, and electronic transmission control. The ASC + T control unit has two microprocessors on two printed circuit boards, one of which concentrates on logic-relevant components and the other on performance-specific components.

1.1.2 Characteristics of Embedded Computing Applications

Embedded computing is in many ways much more demanding than the sort of programs that you may have written for PCs or workstations. Functionality is

important in both general-purpose computing and embedded computing, but embedded applications must meet many other constraints as well.

On the one hand, embedded computing systems have to provide sophisticated functionality:

- *Complex algorithms:* The operations performed by the microprocessor may be very sophisticated. For example, the microprocessor that controls an automobile engine must perform complicated filtering functions to optimize the performance of the car while minimizing pollution and fuel utilization.

- *User interface:* Microprocessors are frequently used to control complex user interfaces that may include multiple menus and many options. The moving maps in Global Positioning System (GPS) navigation are good examples of sophisticated user interfaces.

To make things more difficult, embedded computing operations must often be performed to meet deadlines:

- *Real time:* Many embedded computing systems have to perform in real time— if the data is not ready by a certain deadline, the system breaks. In some cases, failure to meet a deadline is unsafe and can even endanger lives. In other cases, missing a deadline does not create safety problems but does create unhappy customers—missed deadlines in printers, for example, can result in scrambled pages.

- *Multirate:* Not only must operations be completed by deadlines, but many embedded computing systems have several real-time activities going on at the same time. They may simultaneously control some operations that run at slow rates and others that run at high rates. Multimedia applications are prime examples of **multirate** behavior. The audio and video portions of a multimedia stream run at very different rates, but they must remain closely synchronized. Failure to meet a deadline on either the audio or video portions spoils the perception of the entire presentation.

Costs of various sorts are also very important:

- *Manufacturing cost:* The total cost of building the system is very important in many cases. Manufacturing cost is determined by many factors, including the type of microprocessor used, the amount of memory required, and the types of I/O devices.

- *Power and energy:* Power consumption directly affects the cost of the hardware, since a larger power supply may be necessary. Energy consumption affects battery life, which is important in many applications, as well as heat consumption, which can be important even in desktop applications.

Finally, most embedded computing systems are designed by small teams on tight deadlines. The use of small design teams for microprocessor-based systems is a self-fulfilling prophecy—the fact that systems can be built with microprocessors by only a few people invariably encourages management to assume that all microprocessor-based systems can be built by small teams. Tight deadlines are facts of life in today's internationally competitive environment. However, building a product using embedded software makes a lot of sense: Hardware and software can be debugged somewhat independently and design revisions can be made much more quickly.

1.1.3 Why Use Microprocessors?

There are many ways to design a digital system: custom logic, field-programmable gate arrays (FPGAs), and so on. Why use microprocessors? There are two answers:

- Microprocessors are a very efficient way to implement digital systems.

- Microprocessors make it easier to design families of products that can be built to provide various feature sets at different price points and can be extended to provide new features to keep up with rapidly changing markets.

The paradox of digital design is that using a predesigned instruction set processor may in fact result in faster implementation of your application than designing your own custom logic. It is tempting to think that the overhead of fetching, decoding, and executing instructions is so high that it cannot be recouped.

But there are two factors that work together to make microprocessor-based designs fast. First, microprocessors execute programs very efficiently. Modern RISC processors can execute one instruction per clock cycle most of the time, and high-performance processors can execute several instructions per cycle. While there is overhead that must be paid for interpreting instructions, it can often be hidden by clever utilization of parallelism within the CPU.

Second, microprocessor manufacturers spend a great deal of money to make their CPUs run very fast. They hire large teams of designers to tweak every aspect of the microprocessor to make it run at the highest possible speed. Few products can justify the dozens or hundreds of computer architects and VLSI designers customarily employed in the design of a single microprocessor; chips designed by small design teams are less likely to be as highly optimized for speed (or power) as are microprocessors. They also utilize the latest manufacturing technology. Just the use of the latest generation of VLSI fabrication technology, rather than one-generation-old technology, can make a huge difference in performance. Microprocessors generally dominate new fabrication lines because they can be manufactured in large volume and are guaranteed to command high prices. Customers who wish to fabricate their own logic must often wait to make use of VLSI technology from the latest generation of microprocessors. Thus, even if logic you design avoids all the overhead of executing instructions, the fact that it is built from slower circuits often means that its performance advantage is small and perhaps nonexistent.

It is also surprising but true that microprocessors are very efficient utilizers of logic. The generality of a microprocessor and the need for a separate memory may suggest that microprocessor-based designs are inherently much larger than custom logic designs. However, in many cases the microprocessor is smaller when size is measured in units of logic gates. When special-purpose logic is designed for a particular function, it cannot be used for other functions. A microprocessor, on the other hand, can be used for many different algorithms simply by changing the program it executes. Since so many modern systems make use of complex algorithms and user interfaces, we would generally have to design many different custom logic blocks to implement all the required functionality. Many of those blocks will often sit idle—for example, the processing logic may sit idle when user interface functions are performed. Implementing several functions on a single processor often makes much better use of the available hardware budget.

Given the small or nonexistent gains that can be had by avoiding the use of microprocessors, the fact that microprocessors provide substantial advantages makes them the best choice in a wide variety of systems. The programmability of microprocessors can be a substantial benefit during the design process. It allows program design to be separated (at least to some extent) from design of the hardware on which programs will be run. While one team is designing the board that contains the microprocessor, I/O devices, memory, and so on, others can be writing programs at the same time. Equally important, programmability makes it easier to design families of products. In many cases, high-end products can be created simply by adding code without changing the hardware. This practice substantially reduces manufacturing costs. Even when hardware must be redesigned for next-generation products, it may be possible to reuse software, reducing development time and cost.

Why not use PCs for all embedded computing? Put another way, how many different hardware ***platforms*** do we need for embedded computing systems? PCs are widely used and provide a very flexible programming environment. Components of PCs are, in fact, used in many embedded computing systems. But several factors keep us from using the stock PC as the universal embedded computing platform.

First, real-time performance requirements often drive us to different architectures. As we will see later in the book, real-time performance is often best achieved by multiprocessors.

Second, low power and low cost also drive us away from PC architectures and toward multiprocessors. Personal computers are designed to satisfy a broad mix of computing requirements and to be very flexible. Those features increase the complexity and price of the components. They also cause the processor and other components to use more energy to perform a given function. Custom embedded systems that are designed for an application, such as a cell phone, burn several orders of magnitude less power than do PCs with equivalent computational performance, and they are considerably less expensive as well.

The cell phone may, in fact, be the next computing platform. Since over one billion cell phones are sold each year, a great deal of effort is put into designing them. Cell phones operate on batteries, so they must be very power efficient. They

must also perform huge amounts of computation in real time. Not only are cell phones taking over some PC-oriented tasks, such as e-mail and Web browsing, but the components of the cell phone can also be used to build non-cell-phone systems that are very energy efficient for certain classes of applications.

1.1.4 The Physics of Software

Computing is a physical act. Although PCs have trained us to think about computers as purveyors of abstract information, those computers in fact do their work by moving electrons and doing work. This is the fundamental reason why programs take time to finish, why they consume energy, etc.

A prime subject of this book is what we might think of as the ***physics of software***. Software performance and energy consumption are very important properties when we are connecting our embedded computers to the real world. We need to understand the sources of performance and power consumption if we are to be able to design programs that meet our application's goals. Luckily, we don't have to optimize our programs by pushing around electrons. In many cases, we can make very high-level decisions about the structure of our programs to greatly improve their real-time performance and power consumption. As much as possible, we want to make computing abstractions work for us as we work on the physics of our software systems.

1.1.5 Challenges in Embedded Computing System Design

External constraints are one important source of difficulty in embedded system design. Let's consider some important problems that must be taken into account in embedded system design.

How much hardware do we need?
We have a great deal of control over the amount of computing power we apply to our problem. We cannot only select the type of microprocessor used, but also select the amount of memory, the peripheral devices, and more. Since we often must meet both performance deadlines and manufacturing cost constraints, the choice of hardware is important—too little hardware and the system fails to meet its deadlines, too much hardware and it becomes too expensive.

How do we meet deadlines?
The brute force way of meeting a deadline is to speed up the hardware so that the program runs faster. Of course, that makes the system more expensive. It is also entirely possible that increasing the CPU clock rate may not make enough difference to execution time, since the program's speed may be limited by the memory system.

How do we minimize power consumption?
In battery-powered applications, power consumption is extremely important. Even in nonbattery applications, excessive power consumption can increase heat dissipation. One way to make a digital system consume less power is to make it

run more slowly, but naively slowing down the system can obviously lead to missed deadlines. Careful design is required to slow down the noncritical parts of the machine for power consumption while still meeting necessary performance goals.

How do we design for upgradability?

The hardware platform may be used over several product generations, or for several different versions of a product in the same generation, with few or no changes. However, we want to be able to add features by changing software. How can we design a machine that will provide the required performance for software that we haven't yet written?

Does it really work?

Reliability is always important when selling products—customers rightly expect that products they buy will work. Reliability is especially important in some applications, such as safety-critical systems. If we wait until we have a running system and try to eliminate the bugs, we will be too late—we won't find enough bugs, it will be too expensive to fix them, and it will take too long as well. Another set of challenges comes from the characteristics of the components and systems themselves. If workstation programming is like assembling a machine on a bench, then embedded system design is often more like working on a car—cramped, delicate, and difficult. Let's consider some ways in which the nature of embedded computing machines makes their design more difficult.

- *Complex testing:* Exercising an embedded system is generally more difficult than typing in some data. We may have to run a real machine in order to generate the proper data. The timing of data is often important, meaning that we cannot separate the testing of an embedded computer from the machine in which it is embedded.

- *Limited observability and controllability:* Embedded computing systems usually do not come with keyboards and screens. This makes it more difficult to see what is going on and to affect the system's operation. We may be forced to watch the values of electrical signals on the microprocessor bus, for example, to know what is going on inside the system. Moreover, in real-time applications we may not be able to easily stop the system to see what is going on inside.

- *Restricted development environments:* The development environments for embedded systems (the tools used to develop software and hardware) are often much more limited than those available for PCs and workstations. We generally compile code on one type of machine, such as a PC, and download it onto the embedded system. To debug the code, we must usually rely on programs that run on the PC or workstation and then look inside the embedded system.

1.1.6 Performance in Embedded Computing

When we talk about performance when writing programs for our PC, what do we really mean? Most programmers have a fairly vague notion of performance—they want their program to run "fast enough" and they may be worried about the asympototic complexity of their program. Most general-purpose programmers use no tools that are designed to help them improve the performance of their programs.

Embedded system designers, in contrast, have a very clear performance goal in mind—their program must meet its ***deadline***. At the heart of embedded computing is ***real-time computing***, which is the science and art of programming to deadlines. The program receives its input data; the deadline is the time at which a computation must be finished. If the program does not produce the required output by the deadline, then the program does not work, even if the output that it eventually produces is functionally correct.

This notion of deadline-driven programming is at once simple and demanding. It is not easy to determine whether a large, complex program running on a sophisticated microprocessor will meet its deadline. We need tools to help us analyze the real-time performance of embedded systems; we also need to adopt programming disciplines and styles that make it possible to analyze these programs.

In order to understand the real-time behavior of an embedded computing system, we have to analyze the system at several different levels of abstraction. As we move through this book, we will work our way up from the lowest layers that describe components of the system up through the highest layers that describe the complete system. Those layers include:

- *CPU:* The CPU clearly influences the behavior of the program, particularly when the CPU is a pipelined processor with a cache.

- *Platform:* The platform includes the bus and I/O devices. The platform components that surround the CPU are responsible for feeding the CPU and can dramatically affect its performance.

- *Program:* Programs are very large and the CPU sees only a small window of the program at a time. We must consider the structure of the entire program to determine its overall behavior.

- *Task:* We generally run several programs simultaneously on a CPU, creating a ***multitasking system***. The tasks interact with each other in ways that have profound implications for performance.

- *Multiprocessor:* Many embedded systems have more than one processor—they may include multiple programmable CPUs as well as accelerators. Once again, the interaction between these processors adds yet more complexity to the analysis of overall system performance.

1.2 **THE EMBEDDED SYSTEM DESIGN PROCESS**

This section provides an overview of the embedded system design process aimed at two objectives. First, it will give us an introduction to the various steps in embedded system design before we delve into them in more detail. Second, it will allow us to consider the design **methodology** itself. A design methodology is important for three reasons. First, it allows us to keep a scorecard on a design to ensure that we have done everything we need to do, such as optimizing **performance** or performing functional tests. Second, it allows us to develop computer-aided design tools. Developing a single program that takes in a concept for an embedded system and emits a completed design would be a daunting task, but by first breaking the process into manageable steps, we can work on automating (or at least semiautomating) the steps one at a time. Third, a design methodology makes it much easier for members of a design team to communicate. By defining the overall process, team members can more easily understand what they are supposed to do, what they should receive from other team members at certain times, and what they are to hand off when they complete their assigned steps. Since most embedded systems are designed by teams, coordination is perhaps the most important role of a well-defined design methodology.

Figure 1.1 summarizes the major steps in the embedded system design process. In this top–down view, we start with the system **requirements**. In the next step,

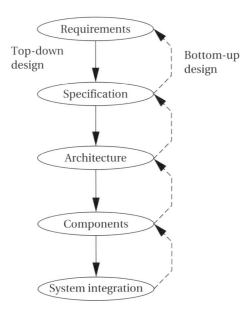

FIGURE 1.1

Major levels of abstraction in the design process.

specification, we create a more detailed description of what we want. But the specification states only how the system behaves, not how it is built. The details of the system's internals begin to take shape when we develop the architecture, which gives the system structure in terms of large components. Once we know the components we need, we can design those components, including both software modules and any specialized hardware we need. Based on those components, we can finally build a complete system.

In this section we will consider design from the *top–down*—we will begin with the most abstract description of the system and conclude with concrete details. The alternative is a **bottom–up** view in which we start with components to build a system. Bottom-up design steps are shown in the figure as dashed-line arrows. We need bottom–up design because we do not have perfect insight into how later stages of the design process will turn out. Decisions at one stage of design are based upon estimates of what will happen later: How fast can we make a particular function run? How much memory will we need? How much system bus capacity do we need? If our estimates are inadequate, we may have to backtrack and amend our original decisions to take the new facts into account. In general, the less experience we have with the design of similar systems, the more we will have to rely on bottom-up design information to help us refine the system.

But the steps in the design process are only one axis along which we can view embedded system design. We also need to consider the major goals of the design:

- manufacturing cost;

- performance (both overall speed and deadlines); and

- power consumption.

We must also consider the tasks we need to perform at every step in the design process. At each step in the design, we add detail:

- We must *analyze* the design at each step to determine how we can meet the specifications.

- We must then *refine* the design to add detail.

- And we must verify the design to ensure that it still meets all system goals, such as cost, speed, and so on.

1.2.1 Requirements

Clearly, before we design a system, we must know what we are designing. The initial stages of the design process capture this information for use in creating the architecture and components. We generally proceed in two phases: First, we gather an informal description from the customers known as requirements, and we refine the requirements into a specification that contains enough information to begin designing the system architecture.

Separating out requirements analysis and specification is often necessary because of the large gap between what the customers can describe about the system they want and what the architects need to design the system. Consumers of embedded systems are usually not themselves embedded system designers or even product designers. Their understanding of the system is based on how they envision users' interactions with the system. They may have unrealistic expectations as to what can be done within their budgets; and they may also express their desires in a language very different from system architects' jargon. Capturing a consistent set of requirements from the customer and then massaging those requirements into a more formal specification is a structured way to manage the process of translating from the consumer's language to the designer's.

Requirements may be *functional* or *nonfunctional*. We must of course capture the basic functions of the embedded system, but functional description is often not sufficient. Typical nonfunctional requirements include:

- *Performance:* The speed of the system is often a major consideration both for the usability of the system and for its ultimate cost. As we have noted, performance may be a combination of soft performance metrics such as approximate time to perform a user-level function and hard deadlines by which a particular operation must be completed.

- *Cost:* The target cost or purchase price for the system is almost always a consideration. Cost typically has two major components: *manufacturing cost* includes the cost of components and assembly; *nonrecurring engineering* **(NRE)** costs include the personnel and other costs of designing the system.

- *Physical size and weight:* The physical aspects of the final system can vary greatly depending upon the application. An industrial control system for an assembly line may be designed to fit into a standard-size rack with no strict limitations on weight. A handheld device typically has tight requirements on both size and weight that can ripple through the entire system design.

- *Power consumption:* Power, of course, is important in battery-powered systems and is often important in other applications as well. Power can be specified in the requirements stage in terms of battery life—the customer is unlikely to be able to describe the allowable wattage.

Validating a set of requirements is ultimately a psychological task since it requires understanding both what people want and how they communicate those needs. One good way to refine at least the user interface portion of a system's requirements is to build a *mock-up*. The mock-up may use canned data to simulate functionality in a restricted demonstration, and it may be executed on a PC or a workstation. But it should give the customer a good idea of how the system will be used and how the user can react to it. Physical, nonfunctional models of devices can also give customers a better idea of characteristics such as size and weight.

Name

Purpose

Inputs

Outputs

Functions

Performance

Manufacturing cost

Power

Physical size and weight

FIGURE 1.2

Sample requirements form.

Requirements analysis for big systems can be complex and time consuming. However, capturing a relatively small amount of information in a clear, simple format is a good start toward understanding system requirements. To introduce the discipline of requirements analysis as part of system design, we will use a simple requirements methodology.

Figure 1.2 shows a sample ***requirements form*** that can be filled out at the start of the project. We can use the form as a checklist in considering the basic characteristics of the system. Let's consider the entries in the form:

- *Name:* This is simple but helpful. Giving a name to the project not only simplifies talking about it to other people but can also crystallize the purpose of the machine.

- *Purpose:* This should be a brief one- or two-line description of what the system is supposed to do. If you can't describe the essence of your system in one or two lines, chances are that you don't understand it well enough.

- *Inputs and outputs:* These two entries are more complex than they seem. The inputs and outputs to the system encompass a wealth of detail:

 - *Types of data:* Analog electronic signals? Digital data? Mechanical inputs?
 - *Data characteristics:* Periodically arriving data, such as digital audio samples? Occasional user inputs? How many bits per data element?
 - *Types of I/O devices:* Buttons? Analog/digital converters? Video displays?

- *Functions:* This is a more detailed description of what the system does. A good way to approach this is to work from the inputs to the outputs: When the system receives an input, what does it do? How do user interface inputs affect these functions? How do different functions interact?

- *Performance:* Many embedded computing systems spend at least some time controlling physical devices or processing data coming from the physical world. In most of these cases, the computations must be performed within a certain time frame. It is essential that the performance requirements be identified early since they must be carefully measured during implementation to ensure that the system works properly.

- *Manufacturing cost:* This includes primarily the cost of the hardware components. Even if you don't know exactly how much you can afford to spend on system components, you should have some idea of the eventual cost range. Cost has a substantial influence on architecture: A machine that is meant to sell at $10 most likely has a very different internal structure than a $100 system.

- *Power:* Similarly, you may have only a rough idea of how much power the system can consume, but a little information can go a long way. Typically, the most important decision is whether the machine will be battery powered or plugged into the wall. Battery-powered machines must be much more careful about how they spend energy.

- *Physical size and weight:* You should give some indication of the physical size of the system to help guide certain architectural decisions. A desktop machine has much more flexibility in the components used than, for example, a lapel-mounted voice recorder.

A more thorough requirements analysis for a large system might use a form similar to Figure 1.2 as a summary of the longer requirements document. After an introductory section containing this form, a longer requirements document could include details on each of the items mentioned in the introduction. For example, each individual feature described in the introduction in a single sentence may be described in detail in a section of the specification.

After writing the requirements, you should check them for internal consistency: Did you forget to assign a function to an input or output? Did you consider all the modes in which you want the system to operate? Did you place an unrealistic number of features into a battery-powered, low-cost machine?

To practice the capture of system requirements, Example 1.1 creates the requirements for a GPS moving map system.

Example 1.1

Requirements analysis of a GPS moving map

The moving map is a handheld device that displays for the user a map of the terrain around the user's current position; the map display changes as the user and the map device change position. The moving map obtains its position from the GPS, a satellite-based navigation system. The moving map display might look something like the following figure.

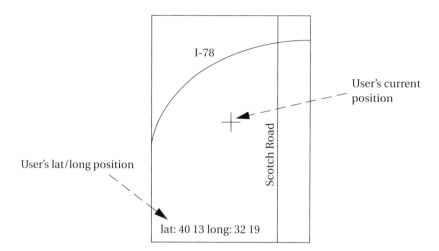

What requirements might we have for our GPS moving map? Here is an initial list:

- *Functionality:* This system is designed for highway driving and similar uses, not nautical or aviation uses that require more specialized databases and functions. The system should show major roads and other landmarks available in standard topographic databases.

- *User interface:* The screen should have at least 400 × 600 pixel resolution. The device should be controlled by no more than three buttons. A menu system should pop up on the screen when buttons are pressed to allow the user to make selections to control the system.

- *Performance:* The map should scroll smoothly. Upon power-up, a display should take no more than one second to appear, and the system should be able to verify its position and display the current map within 15 s.

- *Cost:* The selling cost (street price) of the unit should be no more than $100.

- *Physical size and weight:* The device should fit comfortably in the palm of the hand.

- *Power consumption:* The device should run for at least eight hours on four AA batteries.

Note that many of these requirements are not specified in engineering units—for example, physical size is measured relative to a hand, not in centimeters. Although these requirements must ultimately be translated into something that can be used by the designers, keeping a record of what the customer wants can help to resolve questions about the specification that may crop up later during design.

Based on this discussion, let's write a requirements chart for our moving map system:

Name	GPS moving map
Purpose	Consumer-grade moving map for driving use
Inputs	Power button, two control buttons
Outputs	Back-lit LCD display 400 × 600
Functions	Uses 5-receiver GPS system; three user-selectable resolutions; always displays current latitude and longitude
Performance	Updates screen within 0.25 seconds upon movement
Manufacturing cost	$30
Power	100 mW
Physical size and weight	No more than 2" × 6, " 12 ounces

This chart adds some requirements in engineering terms that will be of use to the designers. For example, it provides actual dimensions of the device. The manufacturing cost was derived from the selling price by using a simple rule of thumb: The selling price is four to five times the **cost of goods sold** (the total of all the component costs).

1.2.2 Specification

The specification is more precise—it serves as the contract between the customer and the architects. As such, the specification must be carefully written so that it accurately reflects the customer's requirements and does so in a way that can be clearly followed during design.

Specification is probably the least familiar phase of this methodology for neophyte designers, but it is essential to creating working systems with a minimum of designer effort. Designers who lack a clear idea of what they want to build when they begin typically make faulty assumptions early in the process that aren't obvious until they have a working system. At that point, the only solution is to take the machine apart, throw away some of it, and start again. Not only does this take a lot of extra time, the resulting system is also very likely to be inelegant, kludgey, and bug-ridden.

The specification should be understandable enough so that someone can verify that it meets system requirements and overall expectations of the customer. It should also be unambiguous enough that designers know what they need to build. Designers can run into several different types of problems caused by unclear specifications. If the behavior of some feature in a particular situation is unclear from the specification, the designer may implement the wrong functionality. If global characteristics of the specification are wrong or incomplete, the overall system architecture derived from the specification may be inadequate to meet the needs of implementation.

A specification of the GPS system would include several components:

- Data received from the GPS satellite constellation.

- Map data.

- User interface.

- Operations that must be performed to satisfy customer requests.

- Background actions required to keep the system running, such as operating the GPS receiver.

UML, a language for describing specifications, will be introduced in Section 1.3, and we will use it to write a specification in Section 1.4. We will practice writing specifications in each chapter as we work through example system designs. We will also study specification techniques in more detail in Chapter 9.

1.2.3 Architecture Design

The specification does not say how the system does things, only what the system does. Describing how the system implements those functions is the purpose of the architecture. The architecture is a plan for the overall structure of the system that will be used later to design the components that make up the architecture. The creation of the architecture is the first phase of what many designers think of as design.

To understand what an architectural description is, let's look at a sample architecture for the moving map of Example 1.1. Figure 1.3 shows a sample system architecture in the form of a **block diagram** that shows major operations and data flows among them.

This block diagram is still quite abstract—we have not yet specified which operations will be performed by software running on a CPU, what will be done by special-purpose hardware, and so on. The diagram does, however, go a long way toward describing how to implement the functions described in the specification. We clearly see, for example, that we need to search the topographic database and to render (i.e., draw) the results for the display. We have chosen to separate those functions so that we can potentially do them in parallel—performing rendering separately from searching the database may help us update the screen more fluidly.

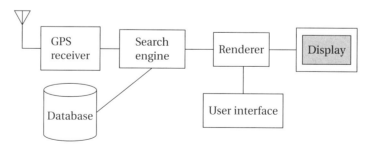

FIGURE 1.3

Block diagram for the moving map.

Only after we have designed an initial architecture that is not biased toward too many implementation details should we refine that system block diagram into two block diagrams: one for hardware and another for software. These two more refined block diagrams are shown in Figure 1.4. The hardware block diagram clearly shows that we have one central CPU surrounded by memory and I/O devices. In particular, we have chosen to use two memories: a frame buffer for the pixels to be displayed and a separate program/data memory for general use by the CPU. The software block diagram fairly closely follows the system block diagram, but we have added a timer to control when we read the buttons on the user interface and render data onto the screen. To have a truly complete architectural description, we require more detail, such as where units in the software block diagram will be executed in the hardware block diagram and when operations will be performed in time.

Architectural descriptions must be designed to satisfy both functional and non-functional requirements. Not only must all the required functions be present, but we must meet cost, speed, power, and other nonfunctional constraints. Starting out with a system architecture and refining that to hardware and software architectures

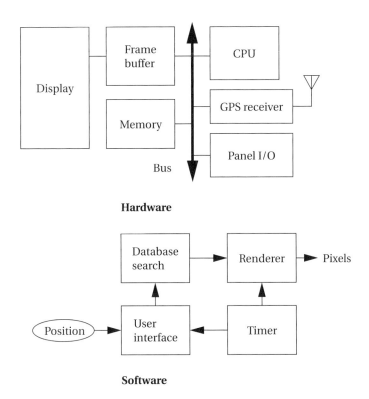

FIGURE 1.4

Hardware and software architectures for the moving map.

is one good way to ensure that we meet all specifications: We can concentrate on the functional elements in the system block diagram, and then consider the nonfunctional constraints when creating the hardware and software architectures.

How do we know that our hardware and software architectures in fact meet constraints on speed, cost, and so on? We must somehow be able to estimate the properties of the components of the block diagrams, such as the search and rendering functions in the moving map system. Accurate estimation derives in part from experience, both general design experience and particular experience with similar systems. However, we can sometimes create simplified models to help us make more accurate estimates. Sound estimates of all nonfunctional constraints during the architecture phase are crucial, since decisions based on bad data will show up during the final phases of design, indicating that we did not, in fact, meet the specification.

1.2.4 Designing Hardware and Software Components

The architectural description tells us what components we need. The component design effort builds those components in conformance to the architecture and specification. The components will in general include both hardware—FPGAs, boards, and so on—and software modules.

Some of the components will be ready-made. The CPU, for example, will be a standard component in almost all cases, as will memory chips and many other components. In the moving map, the GPS receiver is a good example of a specialized component that will nonetheless be a predesigned, standard component. We can also make use of standard software modules. One good example is the topographic database. Standard topographic databases exist, and you probably want to use standard routines to access the database—not only is the data in a predefined format, but it is highly compressed to save storage. Using standard software for these access functions not only saves us design time, but it may give us a faster implementation for specialized functions such as the data decompression phase.

You will have to design some components yourself. Even if you are using only standard integrated circuits, you may have to design the printed circuit board that connects them. You will probably have to do a lot of custom programming as well. When creating these embedded software modules, you must of course make use of your expertise to ensure that the system runs properly in real time and that it does not take up more memory space than is allowed. The power consumption of the moving map software example is particularly important. You may need to be very careful about how you read and write memory to minimize power—for example, since memory accesses are a major source of power consumption, memory transactions must be carefully planned to avoid reading the same data several times.

1.2.5 System Integration

Only after the components are built do we have the satisfaction of putting them together and seeing a working system. Of course, this phase usually consists of

a lot more than just plugging everything together and standing back. Bugs are typically found during system integration, and good planning can help us find the bugs quickly. By building up the system in phases and running properly chosen tests, we can often find bugs more easily. If we debug only a few modules at a time, we are more likely to uncover the simple bugs and able to easily recognize them. Only by fixing the simple bugs early will we be able to uncover the more complex or obscure bugs that can be identified only by giving the system a hard workout. We need to ensure during the architectural and component design phases that we make it as easy as possible to assemble the system in phases and test functions relatively independently.

System integration is difficult because it usually uncovers problems. It is often hard to observe the system in sufficient detail to determine exactly what is wrong—the debugging facilities for embedded systems are usually much more limited than what you would find on desktop systems. As a result, determining why things do not stet work correctly and how they can be fixed is a challenge in itself. Careful attention to inserting appropriate debugging facilities during design can help ease system integration problems, but the nature of embedded computing means that this phase will always be a challenge.

1.3 FORMALISMS FOR SYSTEM DESIGN

As mentioned in the last section, we perform a number of different design tasks at different levels of abstraction throughout this book: creating requirements and specifications, architecting the system, designing code, and designing tests. It is often helpful to conceptualize these tasks in diagrams. Luckily, there is a visual language that can be used to capture all these design tasks: the *Unified Modeling Language (UML)* [Boo99, Pil05]. UML was designed to be useful at many levels of abstraction in the design process. UML is useful because it encourages design by successive refinement and progressively adding detail to the design, rather than rethinking the design at each new level of abstraction.

UML is an *object-oriented* modeling language. We will see precisely what we mean by an object in just a moment, but object-oriented design emphasizes two concepts of importance:

- It encourages the design to be described as a number of interacting objects, rather than a few large monolithic blocks of code.

- At least some of those objects will correspond to real pieces of software or hardware in the system. We can also use UML to model the outside world that interacts with our system, in which case the objects may correspond to people or other machines. It is sometimes important to implement something we think of at a high level as a single object using several distinct pieces of code or to otherwise break up the object correspondence in the implementation.

However, thinking of the design in terms of actual objects helps us understand the natural structure of the system.

Object-oriented (often abbreviated OO) specification can be seen in two complementary ways:

- Object-oriented specification allows a system to be described in a way that closely models real-world objects and their interactions.

- Object-oriented specification provides a basic set of primitives that can be used to describe systems with particular attributes, irrespective of the relationships of those systems' components to real-world objects.

Both views are useful. At a minimum, object-oriented specification is a set of linguistic mechanisms. In many cases, it is useful to describe a system in terms of real-world analogs. However, performance, cost, and so on may dictate that we change the specification to be different in some ways from the real-world elements we are trying to model and implement. In this case, the object-oriented specification mechanisms are still useful.

What is the relationship between an object-oriented specification and an object-oriented programming language (such as C++ [Str97])? A specification language may not be executable. But both object-oriented specification and programming languages provide similar basic methods for structuring large systems.

Unified Modeling Language (UML)—the acronym is the name is a large language, and covering all of it is beyond the scope of this book. In this section, we introduce only a few basic concepts. In later chapters, as we need a few more UML concepts, we introduce them to the basic modeling elements introduced here. Because UML is so rich, there are many graphical elements in a UML diagram. It is important to be careful to use the correct drawing to describe something—for instance, UML distinguishes between arrows with open and filled-in arrowheads, and solid and broken lines. As you become more familiar with the language, uses of the graphical primitives will become more natural to you.

We also won't take a strict object-oriented approach. We may not always use objects for certain elements of a design—in some cases, such as when taking particular aspects of the implementation into account, it may make sense to use another design style. However, object-oriented design is widely applicable, and no designer can consider himself or herself design literate without understanding it.

1.3.1 Structural Description

By ***structural description***, we mean the basic components of the system; we will learn how to describe how these components act in the next section. The principal component of an object-oriented design is, naturally enough, the ***object***. An object includes a set of ***attributes*** that define its internal state. When implemented in a programming language, these attributes usually become variables or constants held in a data structure. In some cases, we will add the type of the attribute after

the attribute name for clarity, but we do not always have to specify a type for an attribute. An object describing a display (such as a CRT screen) is shown in UML notation in Figure 1.5. The text in the folded-corner page icon is a ***note***; it does not correspond to an object in the system and only serves as a comment. The attribute is, in this case, an array of pixels that holds the contents of the display. The object is identified in two ways: It has a unique name, and it is a member of a ***class***. The name is underlined to show that this is a description of an object and not of a class.

A class is a form of type definition—all objects derived from the same class have the same characteristics, although their attributes may have different values. A class defines the attributes that an object may have. It also defines the ***operations*** that determine how the object interacts with the rest of the world. In a programming language, the operations would become pieces of code used to manipulate the object. The UML description of the *Display* class is shown in Figure 1.6. The class has the name that we saw used in the *d*1 object since *d*1 is an instance of class *Display*. The *Display* class defines the *pixels* attribute seen in the object; remember that when we instantiate the class an object, that object will have its own memory so that different objects of the same class have their own values for the attributes. Other classes can examine and modify class attributes; if we have to do something more complex than use the attribute directly, we define a behavior to perform that function.

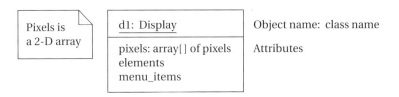

FIGURE 1.5

An object in UML notation.

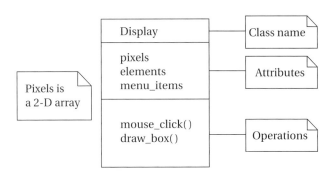

FIGURE 1.6

A class in UML notation.

A class defines both the ***interface*** for a particular type of object and that object's ***implementation***. When we use an object, we do not directly manipulate its attributes—we can only read or modify the object's state through the operations that define the interface to the object. (The implementation includes both the attributes and whatever code is used to implement the operations.) As long as we do not change the behavior of the object seen at the interface, we can change the implementation as much as we want. This lets us improve the system by, for example, speeding up an operation or reducing the amount of memory required without requiring changes to anything else that uses the object.

Clearly, the choice of an interface is a very important decision in object-oriented design. The proper interface must provide ways to access the object's state (since we cannot directly see the attributes) as well as ways to update the state. We need to make the object's interface general enough so that we can make full use of its capabilities. However, excessive generality often makes the object large and slow. Big, complex interfaces also make the class definition difficult for designers to understand and use properly.

There are several types of ***relationships*** that can exist between objects and classes:

- ***Association*** occurs between objects that communicate with each other but have no ownership relationship between them.

- ***Aggregation*** describes a complex object made of smaller objects.

- ***Composition*** is a type of aggregation in which the owner does not allow access to the component objects.

- ***Generalization*** allows us to define one class in terms of another.

The elements of a UML class or object do not necessarily directly correspond to statements in a programming language—if the UML is intended to describe something more abstract than a program, there may be a significant gap between the contents of the UML and a program implementing it. The attributes of an object do not necessarily reflect variables in the object. An attribute is some value that reflects the current state of the object. In the program implementation, that value could be computed from some other internal variables. The behaviors of the object would, in a higher-level specification, reflect the basic things that can be done with an object. Implementing all these features may require breaking up a behavior into several smaller behaviors—for example, initialize the object before you start to change its internal state-derived classes.

Unified Modeling Language, like most object-oriented languages, allows us to define one class in terms of another. An example is shown in Figure 1.7, where we ***derive*** two particular types of displays. The first, *BW_display*, describes a black-and-white display. This does not require us to add new attributes or operations, but we can specialize both to work on one-bit pixels. The second, *Color_map_display*, uses a graphic device known as a color map to allow the user to select from a

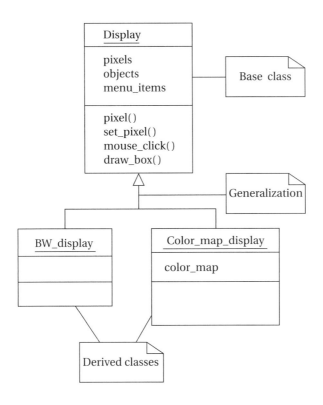

FIGURE 1.7

Derived classes as a form of generalization in UML.

large number of available colors even with a small number of bits per pixel. This class defines a *color_map* attribute that determines how pixel values are mapped onto display colors. A ***derived class*** inherits all the attributes and operations from its ***base class***. In this class, *Display* is the base class for the two derived classes. A derived class is defined to include all the attributes of its base class. This relation is transitive—if *Display* were derived from another class, both *BW_display* and *Color_map_display* would inherit all the attributes and operations of *Display's* base class as well. Inheritance has two purposes. It of course allows us to succinctly describe one class that shares some characteristics with another class. Even more important, it captures those relationships between classes and documents them. If we ever need to change any of the classes, knowledge of the class structure helps us determine the reach of changes—for example, should the change affect only *Color_map_display* objects or should it change all Display objects?

 Unified Modeling Language considers inheritance to be one form of generalization. A generalization relationship is shown in a UML diagram as an arrow with an open (unfilled) arrowhead. Both *BW_display* and *Color_map_display* are specific

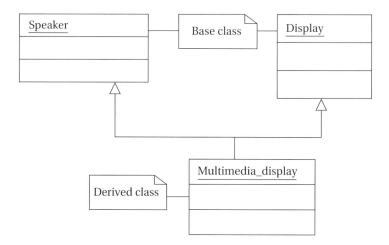

FIGURE 1.8

Multiple inheritance in UML.

versions of *Display*, so *Display* generalizes both of them. UML also allows us to define *multiple inheritance*, in which a class is derived from more than one base class. (Most object-oriented programming languages support multiple inheritance as well.) An example of multiple inheritance is shown in Figure 1.8; we have omitted the details of the classes' attributes and operations for simplicity. In this case, we have created a *Multimedia_display* class by combining the *Display* class with a *Speaker* class for sound. The derived class inherits all the attributes and operations of both its base classes, *Display* and *Speaker*. Because multiple inheritance causes the sizes of the attribute set and operations to expand so quickly, it should be used with care.

A *link* describes a relationship between objects; association is to link as class is to object. We need links because objects often do not stand alone; associations let us capture type information about these links. Figure 1.9 shows examples of links and an association. When we consider the actual objects in the system, there is a set of messages that keeps track of the current number of active messages (two in this example) and points to the active messages. In this case, the link defines the *contains* relation. When generalized into classes, we define an association between the message set class and the message class. The association is drawn as a line between the two labeled with the name of the association, namely, *contains*. The ball and the number at the message class end indicate that the message set may include zero or more message objects. Sometimes we may want to attach data to the links themselves; we can specify this in the association by attaching a class-like box to the association's edge, which holds the association's data.

Typically, we find that we use a certain combination of elements in an object or class many times. We can give these patterns names, which are called *stereotypes*

Links between objects

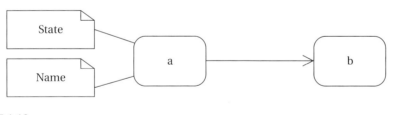

Association between classes

FIGURE 1.9

Links and association.

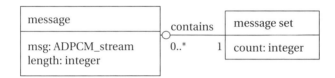

FIGURE 1.10

A state and transition in UML.

in UML. A stereotype name is written in the form <<signal>>. Figure 1.11 shows a stereotype for a signal, which is a communication mechanism.

1.3.2 Behavioral Description

We have to specify the behavior of the system as well as its structure. One way to specify the behavior of an operation is a *state machine*. Figure 1.10 shows UML states; the transition between two states is shown by a skeleton arrow.

These state machines will not rely on the operation of a clock, as in hardware; rather, changes from one state to another are triggered by the occurrence of *events*.

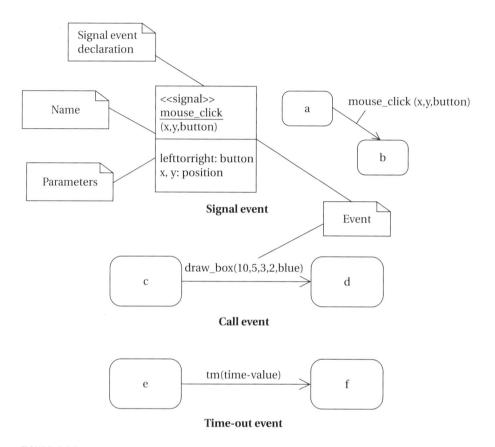

FIGURE 1.11

Signal, call, and time-out events in UML.

An event is some type of action. The event may originate outside the system, such as a user pressing a button. It may also originate inside, such as when one routine finishes its computation and passes the result on to another routine. We will concentrate on the following three types of events defined by UML, as illustrated in Figure 1.11:

- A **signal** is an asynchronous occurrence. It is defined in UML by an object that is labeled as a <<*signal*>>. The object in the diagram serves as a declaration of the event's existence. Because it is an object, a signal may have parameters that are passed to the signal's receiver.

- A **call event** follows the model of a procedure call in a programming language.

- A **time-out event** causes the machine to leave a state after a certain amount of time. The label *tm(time-value)* on the edge gives the amount of time after which the transition occurs. A time-out is generally implemented with an

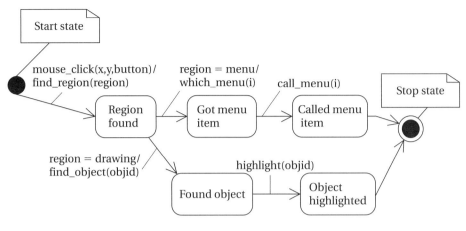

FIGURE 1.12

A state machine specification in UML.

external timer. This notation simplifies the specification and allows us to defer implementation details about the time-out mechanism.

We show the occurrence of all types of signals in a UML diagram in the same way—as a label on a transition.

Let's consider a simple state machine specification to understand the semantics of UML state machines. A state machine for an operation of the display is shown in Figure 1.12. The start and stop states are special states that help us to organize the flow of the state machine. The states in the state machine represent different conceptual operations. In some cases, we take conditional transitions out of states based on inputs or the results of some computation done in the state. In other cases, we make an unconditional transition to the next state. Both the unconditional and conditional transitions make use of the call event. Splitting a complex operation into several states helps document the required steps, much as subroutines can be used to structure code.

It is sometimes useful to show the sequence of operations over time, particularly when several objects are involved. In this case, we can create a sequence diagram, like the one for a mouse click scenario shown in Figure 1.13. A ***sequence diagram*** is somewhat similar to a hardware timing diagram, although the time flows vertically in a sequence diagram, whereas time typically flows horizontally in a timing diagram. The sequence diagram is designed to show a particular scenario or choice of events—it is not convenient for showing a number of mutually exclusive possibilities. In this case, the sequence shows what happens when a mouse click is on the menu region. Processing includes three objects shown at the top of the diagram. Extending below each object is its *lifeline*, a dashed line that shows how long the object is alive. In this case, all the objects remain alive for the entire sequence, but in other cases objects may be created or destroyed during processing. The boxes

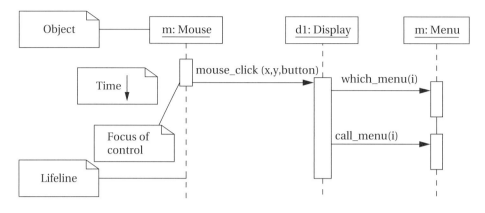

FIGURE 1.13

A sequence diagram in UML.

along the lifelines show the *focus of control* in the sequence, that is, when the object is actively processing. In this case, the mouse object is active only long enough to create the *mouse_click* event. The display object remains in play longer; it in turn uses call events to invoke the menu object twice: once to determine which menu item was selected and again to actually execute the menu call. The find_region() call is internal to the display object, so it does not appear as an event in the diagram.

1.4 MODEL TRAIN CONTROLLER

In order to learn how to use UML to model systems, we will specify a simple system, a model train controller, which is illustrated in Figure 1.14. The user sends messages to the train with a control box attached to the tracks. The control box may have familiar controls such as a throttle, emergency stop button, and so on. Since the train receives its electrical power from the two rails of the track, the control box can send signals to the train over the tracks by modulating the power supply voltage. As shown in the figure, the control panel sends packets over the tracks to the receiver on the train. The train includes analog electronics to sense the bits being transmitted and a control system to set the train motor's speed and direction based on those commands. Each packet includes an address so that the console can control several trains on the same track; the packet also includes an error correction code (ECC) to guard against transmission errors. This is a one-way communication system—the model train cannot send commands back to the user.

We start by analyzing the requirements for the train control system. We will base our system on a real standard developed for model trains. We then develop two specifications: a simple, high-level specification and then a more detailed specification.

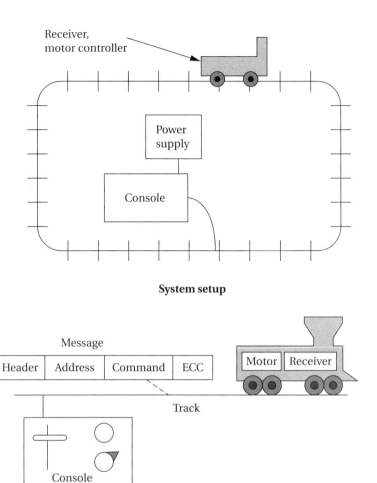

System setup

Signaling the train

FIGURE 1.14

A model train control system.

1.4.1 **Requirements**

Before we can create a system specification, we have to understand the require-
ments. Here is a basic set of requirements for the system:

■ The console shall be able to control up to eight trains on a single track.

■ The speed of each train shall be controllable by a throttle to at least 63 different
levels in each direction (forward and reverse).

- There shall be an inertia control that shall allow the user to adjust the responsiveness of the train to commanded changes in speed. Higher inertia means that the train responds more slowly to a change in the throttle, simulating the inertia of a large train. The inertia control will provide at least eight different levels.

- There shall be an emergency stop button.

- An error detection scheme will be used to transmit messages.

We can put the requirements into our chart format:

Name	Model train controller
Purpose	Control speed of up to eight model trains
Inputs	Throttle, inertia setting, emergency stop, train number
Outputs	Train control signals
Functions	Set engine speed based upon inertia settings; respond to emergency stop
Performance	Can update train speed at least 10 times per second
Manufacturing cost	$50
Power	10W (plugs into wall)
Physical size and weight	Console should be comfortable for two hands, approximate size of standard keyboard; weight <2 pounds

We will develop our system using a widely used standard for model train control. We could develop our own train control system from scratch, but basing our system upon a standard has several advantages in this case: It reduces the amount of work we have to do and it allows us to use a wide variety of existing trains and other pieces of equipment.

1.4.2 DCC

The **Digital Command Control (DCC)** standard (http://www.nmra.org/standards/DCC/standards_rps/DCCStds.html) was created by the National Model Railroad Association to support interoperable digitally-controlled model trains. Hobbyists started building homebrew digital control systems in the 1970s and Marklin developed its own digital control system in the 1980s. DCC was created to provide a standard that could be built by any manufacturer so that hobbyists could mix and match components from multiple vendors.

The DCC standard is given in two documents:

- Standard S-9.1, the DCC Electrical Standard, defines how bits are encoded on the rails for transmission.

- Standard S-9.2, the DCC Communication Standard, defines the packets that carry information.

Any DCC-conforming device must meet these specifications. DCC also provides several recommended practices. These are not strictly required but they provide some hints to manufacturers and users as to how to best use DCC.

The DCC standard does not specify many aspects of a DCC train system. It doesn't define the control panel, the type of microprocessor used, the programming language to be used, or many other aspects of a real model train system. The standard concentrates on those aspects of system design that are necessary for interoperability. Overstandardization, or specifying elements that do not really need to be standardized, only makes the standard less attractive and harder to implement.

The Electrical Standard deals with voltages and currents on the track. While the electrical engineering aspects of this part of the specification are beyond the scope of the book, we will briefly discuss the data encoding here. The standard must be carefully designed because the main function of the track is to carry power to the locomotives. The signal encoding system should not interfere with power transmission either to DCC or non-DCC locomotives. A key requirement is that the data signal should not change the DC value of the rails.

The data signal swings between two voltages around the power supply voltage. As shown in Figure 1.15, bits are encoded in the time between transitions, not by voltage levels. A 0 is at least 100 μs while a 1 is nominally 58 μs. The durations of the high (above nominal voltage) and low (below nominal voltage) parts of a bit are equal to keep the DC value constant. The specification also gives the allowable variations in bit times that a conforming DCC receiver must be able to tolerate.

The standard also describes other electrical properties of the system, such as allowable transition times for signals.

The DCC Communication Standard describes how bits are combined into packets and the meaning of some important packets. Some packet types are left undefined in the standard but typical uses are given in Recommended Practices documents.

We can write the basic packet format as a regular expression:

$$PSA(sD) + E \qquad (1.1)$$

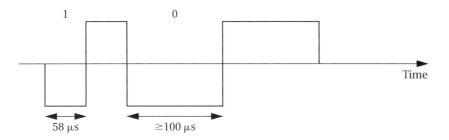

FIGURE 1.15

Bit encoding in DCC.

In this regular expression:

- P is the preamble, which is a sequence of at least 10 1 bits. The command station should send at least 14 of these 1 bits, some of which may be corrupted during transmission.

- S is the packet start bit. It is a 0 bit.

- A is an address data byte that gives the address of the unit, with the most significant bit of the address transmitted first. An address is eight bits long. The addresses 00000000, 11111110, and 11111111 are reserved.

- s is the data byte start bit, which, like the packet start bit, is a 0.

- D is the data byte, which includes eight bits. A data byte may contain an address, instruction, data, or error correction information.

- E is a packet end bit, which is a 1 bit.

A packet includes one or more data byte start bit/data byte combinations. Note that the address data byte is a specific type of data byte.

A ***baseline packet*** is the minimum packet that must be accepted by all DCC implementations. More complex packets are given in a Recommended Practice document. A baseline packet has three data bytes: an address data byte that gives the intended receiver of the packet; the instruction data byte provides a basic instruction; and an error correction data byte is used to detect and correct transmission errors.

The instruction data byte carries several pieces of information. Bits 0–3 provide a 4-bit speed value. Bit 4 has an additional speed bit, which is interpreted as the least significant speed bit. Bit 5 gives direction, with 1 for forward and 0 for reverse. Bits 7–8 are set at 01 to indicate that this instruction provides speed and direction.

The error correction databyte is the bitwise exclusive OR of the address and instruction data bytes.

The standard says that the command unit should send packets frequently since a packet may be corrupted. Packets should be separated by at least 5 ms.

1.4.3 Conceptual Specification

Digital Command Control specifies some important aspects of the system, particularly those that allow equipment to interoperate. But DCC deliberately does not specify everything about a model train control system. We need to round out our specification with details that complement the DCC spec. A ***conceptual specification*** allows us to understand the system a little better. We will use the experience gained by writing the conceptual specification to help us write a detailed specification to be given to a system architect. This specification does not correspond to what any commercial DCC controllers do, but it is simple enough to allow us to cover some basic concepts in system design.

A train control system turns ***commands*** into ***packets***. A command comes from the command unit while a packet is transmitted over the rails. Commands and packets may not be generated in a 1-to-1 ratio. In fact, the DCC standard says that command units should resend packets in case a packet is dropped during transmission.

We now need to model the train control system itself. There are clearly two major subsystems: the command unit and the train-board component as shown in Figure 1.16. Each of these subsystems has its own internal structure. The basic relationship between them is illustrated in Figure 1.17. This figure shows a UML ***collaboration diagram***; we could have used another type of figure, such as a class or object diagram, but we wanted to emphasize the transmit/receive relationship between these major subsystems. The command unit and receiver are each represented by objects; the command unit sends a sequence of packets to the train's receiver, as illustrated by the arrow. The notation on the arrow provides both the type of message sent and its sequence in a flow of messages; since the console sends all the messages, we have numbered the arrow's messages as $1..n$. Those messages are of course carried over the track. Since the track is not a computer component and is purely passive, it does not appear in the diagram. However, it would be perfectly legitimate to model the track in the collaboration diagram, and in some situations it may be wise to model such nontraditional components in the specification diagrams. For example, if we are worried about what happens when the track breaks,

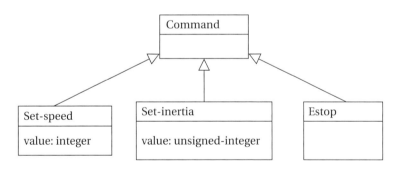

FIGURE 1.16

Class diagram for the train controller messages.

FIGURE 1.17

UML collaboration diagram for major subsystems of the train controller system.

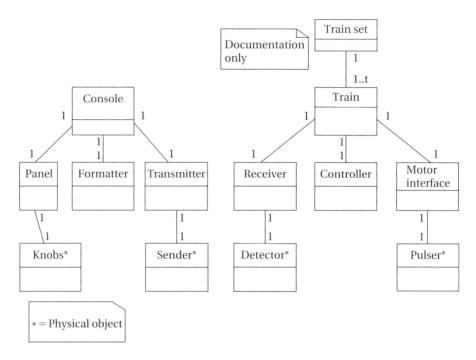

FIGURE 1.18

A UML class diagram for the train controller showing the composition of the subsystems.

modeling the tracks would help us identify failure modes and possible recovery mechanisms.

Let's break down the command unit and receiver into their major components. The console needs to perform three functions: read the state of the front panel on the command unit, format messages, and transmit messages. The train receiver must also perform three major functions: receive the message, interpret the message (taking into account the current speed, inertia setting, etc.), and actually control the motor. In this case, let's use a class diagram to represent the design; we could also use an object diagram if we wished. The UML class diagram is shown in Figure 1.18. It shows the console class using three classes, one for each of its major components. These classes must define some behaviors, but for the moment we will concentrate on the basic characteristics of these classes:

■ The *Console* class describes the command unit's front panel, which contains the analog knobs and hardware to interface to the digital parts of the system.

■ The *Formatter* class includes behaviors that know how to read the panel knobs and creates a bit stream for the required message.

■ The *Transmitter* class interfaces to analog electronics to send the message along the track.

There will be one instance of the *Console* class and one instance of each of the component classes, as shown by the numeric values at each end of the relationship links. We have also shown some special classes that represent analog components, ending the name of each with an asterisk:

- *Knobs** describes the actual analog knobs, buttons, and levers on the control panel.

- *Sender** describes the analog electronics that send bits along the track.

Likewise, the Train makes use of three other classes that define its components:

- The *Receiver* class knows how to turn the analog signals on the track into digital form.

- The *Controller* class includes behaviors that interpret the commands and figures out how to control the motor.

- The *Motor interface* class defines how to generate the analog signals required to control the motor.

We define two classes to represent analog components:

- *Detector** detects analog signals on the track and converts them into digital form.

- *Pulser** turns digital commands into the analog signals required to control the motor speed.

We have also defined a special class, *Train set*, to help us remember that the system can handle multiple trains. The values on the relationship edge show that one train set can have *t* trains. We would not actually implement the train set class, but it does serve as useful documentation of the existence of multiple receivers.

1.4.4 Detailed Specification

Now that we have a conceptual specification that defines the basic classes, let's refine it to create a more detailed specification. We won't make a complete specification, but we will add detail to the classes and look at some of the major decisions in the specification process to get a better handle on how to write good specifications.

At this point, we need to define the analog components in a little more detail because their characteristics will strongly influence the *Formatter* and *Controller*. Figure 1.19 shows a class diagram for these classes; this diagram shows a little more detail than Figure 1.18 since it includes attributes and behaviors of these classes. The *Panel* has three knobs: *train* number (which train is currently being controlled), *speed* (which can be positive or negative), and *inertia*. It also has one button for *emergency-stop*. When we change the train number setting, we also want to reset the other controls to the proper values for that train so that the previous train's control settings are not used to change the current train's settings. To do this, *Knobs** must

provide a *set-knobs* behavior that allows the rest of the system to modify the knob settings. (If we wanted or needed to model the user, we would expand on this class definition to provide methods that a user object would call to specify these parameters.) The motor system takes its motor commands in two parts. The *Sender* and *Detector* classes are relatively simple: They simply put out and pick up a bit, respectively.

To understand the *Pulser* class, let's consider how we actually control the train motor's speed. As shown in Figure 1.20, the speed of electric motors is commonly controlled using pulse-width modulation: Power is applied in a pulse for a fraction of some fixed interval, with the fraction of the time that power is applied determining the speed. The digital interface to the motor system specifies that pulse width as an integer, with the maximum value being maximum engine speed. A separate binary value controls direction. Note that the motor control takes an unsigned speed with a

Knobs*
train-knob: integer speed-knob: integer inertia-knob: unsigned-integer emergency-stop: boolean
set-knobs()

Pulser*
pulse-width: unsigned-integer direction: boolean

Sender*
send-bit()

Detector*
<integer> read-bit(): integer

FIGURE 1.19

Classes describing analog physical objects in the train control system.

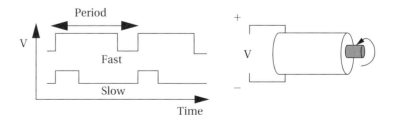

FIGURE 1.20

Controlling motor speed by pulse-width modulation.

separate direction, while the panel specifies speed as a signed integer, with negative speeds corresponding to reverse.

Figure 1.21 shows the classes for the panel and motor interfaces. These classes form the software interfaces to their respective physical devices. The *Panel* class defines a behavior for each of the controls on the panel; we have chosen not to define an internal variable for each control since their values can be read directly from the physical device, but a given implementation may choose to use internal variables. The *new-settings* behavior uses the *set-knobs* behavior of the *Knobs** class to change the knobs settings whenever the train number setting is changed. The *Motor-interface* defines an attribute for speed that can be set by other classes. As we will see in a moment, the controller's job is to incrementally adjust the motor's speed to provide smooth acceleration and deceleration.

The *Transmitter* and *Receiver* classes are shown in Figure 1.22. They provide the software interface to the physical devices that send and receive bits along the track.

Panel
panel-active(): boolean train-number(): integer speed(): integer inertia(): integer estop(): boolean new-settings()

Motor-interface
speed: integer

FIGURE 1.21

Class diagram for the Panel and Motor interface.

Transmitter
send-speed(adrs: integer, speed: integer) send-inertia(adrs: integer, val: integer) send-estop(adrs: integer)

Receiver
current: command new: boolean
read-cmd() new-cmd(): boolean rcv-type(msg-type: command) rcv-speed(val: integer) rcv-inertia(val: integer)

FIGURE 1.22

Class diagram for the Transmitter and Receiver.

The *Transmitter* provides a distinct behavior for each type of message that can be sent; it internally takes care of formatting the message. The *Receiver* class provides a *read-cmd* behavior to read a message off the tracks. We can assume for now that the receiver object allows this behavior to run continuously to monitor the tracks and intercept the next command. (We consider how to model such continuously running behavior as processes in Chapter 6.) We use an internal variable to hold the current command. Another variable holds a flag showing when the command has been processed. Separate behaviors let us read out the parameters for each type of command; these messages also reset the new flag to show that the command has been processed. We do not need a separate behavior for an *Estop* message since it has no parameters—knowing the type of message is sufficient.

Now that we have specified the subsystems around the formatter and controller, it is easier to see what sorts of interfaces these two subsystems may need.

The *Formatter* class is shown in Figure 1.23. The formatter holds the current control settings for all of the trains. The *send-command* method is a utility function that serves as the interface to the transmitter. The *operate* function performs the basic actions for the object. At this point, we only need a simple specification, which states that the formatter repeatedly reads the panel, determines whether any settings have changed, and sends out the appropriate messages. The *panel-active* behavior returns true whenever the panel's values do not correspond to the current values.

The role of the formatter during the panel's operation is illustrated by the sequence diagram of Figure 1.24. The figure shows two changes to the knob settings: first to the throttle, inertia, or emergency stop; then to the train number. The panel is called periodically by the formatter to determine if any control settings have changed. If a setting has changed for the current train, the formatter decides to send a command, issuing a *send-command* behavior to cause the transmitter to send the bits. Because transmission is serial, it takes a noticeable amount of time for the transmitter to finish a command; in the meantime, the formatter continues to

Formatter
current-train: integer current-speed[ntrains]: integer current-inertia[ntrains]: unsigned-integer current-estop[ntrains]: boolean
send-command() panel-active(): boolean operate()

FIGURE 1.23

Class diagram for the Formatter class.

check the panel's control settings. If the train number has changed, the formatter must cause the knob settings to be reset to the proper values for the new train.

We have not yet specified the operation of any of the behaviors. We define what a behavior does by writing a state diagram. The state diagram for a very simple version of the *operate* behavior of the *Formatter* class is shown in Figure 1.25. This behavior watches the panel for activity: If the train number changes, it updates

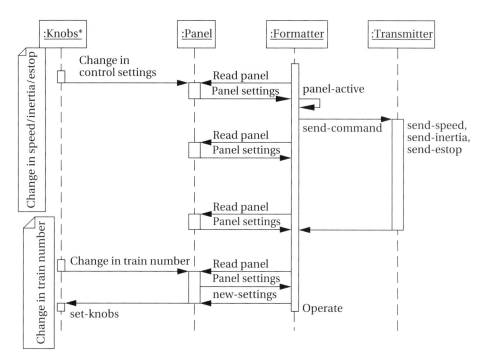

FIGURE 1.24

Sequences diagram for transmitting a control input.

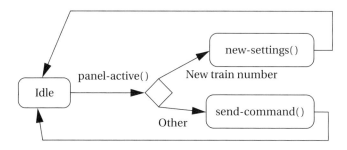

FIGURE 1.25

State diagram for the formatter operate behavior.

the panel display; otherwise, it causes the required message to be sent. Figure 1.26 shows a state diagram for the *panel-active* behavior.

The definition of the train's *Controller* class is shown in Figure 1.27. The *operate* behavior is called by the receiver when it gets a new command; *operate* looks at the contents of the message and uses the *issue-command* behavior to change the speed, direction, and inertia settings as necessary. A specification for *operate* is shown in Figure 1.28.

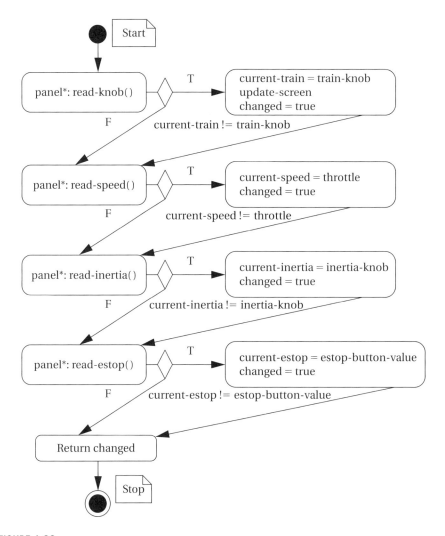

FIGURE 1.26

State diagram for the panel-active behavior.

The operation of the *Controller* class during the reception of a *set-speed* command is illustrated in Figure 1.29. The *Controller's operate* behavior must execute several behaviors to determine the nature of the message. Once the speed command has been parsed, it must send a sequence of commands to the motor to smoothly change the train's speed.

It is also a good idea to refine our notion of a command. These changes result from the need to build a potentially upward-compatible system. If the messages were entirely internal, we would have more freedom in specifying messages that we could use during architectural design. But since these messages must work with a variety of trains and we may want to add more commands in a later version of the system, we need to specify the basic features of messages for compatibility. There are three important issues. First, we need to specify the number of bits used to determine the message type. We choose three bits, since that gives us five unused message codes. Second, we need to include information about the length of the

FIGURE 1.27

Class diagram for the Controller class.

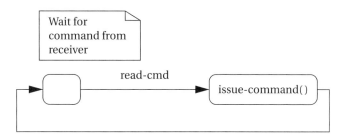

FIGURE 1.28

State diagram for the Controller operate behavior.

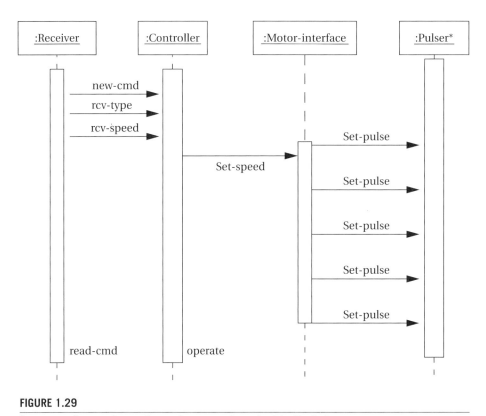

FIGURE 1.29

Sequence diagram for a set-speed command received by the train.

data fields, which is determined by the resolution for speeds and inertia set by the requirements. Third, we need to specify the error correction mechanism; we choose to use a single-parity bit. We can update the classes to provide this extra information as shown in Figure 1.30.

1.4.5 Lessons Learned

We have learned a couple of things in this exercise beyond gaining experience with UML notation. First, standards are important. We often can't avoid working with standards but standards often save us work and allow us to make use of components designed by others. Second, specifying a system is not easy. You often learn a lot about the system you are trying to build by writing a specification. Third, specification invariably requires making some choices that may influence the implementation. Good system designers use their experience and intuition to guide them when these kinds of choices must be made.

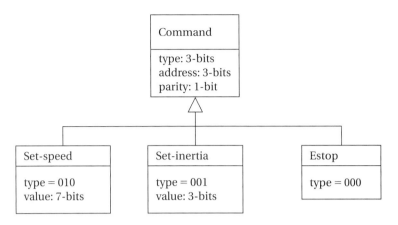

FIGURE 1.30

Refined class diagram for the train controller commands.

1.5 A GUIDED TOUR OF THIS BOOK

The most efficient way to learn all the necessary concepts is to move from the bottom–up. This book is arranged so that you learn about the properties of components and build toward more complex systems and a more complete view of the system design process. Veteran designers have learned enough bottom–up knowledge from experience to know how to use a top–down approach to designing a system, but when learning things for the first time, the bottom–up approach allows you to build more sophisticated concepts on the basis of lower-level ideas.

We will use several organizational devices throughout the book to help you. Application Examples focus on a particular end-use application and how it relates to embedded system design. We will also make use of Programming Examples to describe software designs. In addition to these examples, each chapter will use a significant system design example to demonstrate the major concepts of the chapter.

Each chapter includes questions that are intended to be answered on paper as homework assignments. The chapters also include lab exercises. These are more open ended and are intended to suggest activities that can be performed in the lab to help illuminate various concepts in the chapter.

Throughout the book, we will use two CPUs as examples: the ARM RISC processor and the Texas Instruments TI TMS320C55x™ (C55x) digital signal processor (DSP). Both are well-known microprocessors used in many embedded applications. Using real microprocessors helps make concepts more concrete. However, our aim is to learn concepts that can be applied to many different microprocessors, not only ARM and the C55x. While microprocessors will evolve over time (Warhol's Law of

Computer Architecture [Wol92] states that every microprocessor architecture will be the price/performance leader for 15 min), the concepts of embedded system design are fundamental and long term.

1.5.1 Chapter 2: Instruction Sets

In Chapter 2, we begin our study of microprocessors by concentrating on **instruction sets**. The chapter covers the instruction sets of the ARM and C55x microprocessors in separate sections. These two microprocessors are very different. Understanding all details of both is not strictly necessary to the design of embedded systems. However, comparing the two does provide some interesting lessons in instruction set architectures.

Understanding details of the instruction set is important both for concreteness and for seeing how architectural features can affect performance and other system attributes. But many mechanisms, such as caches and memory management, can be understood in general before we go on to details of how they are implemented in ARM and C55x.

We do not introduce a design example in this chapter—it is difficult to build even a simple working system without understanding other aspects of the CPU that will be introduced in Chapter 3. However, understanding instruction sets is critical to understanding problems such as execution speed and code size that we study throughout the book.

1.5.2 Chapter 3: CPUs

Chapter 3 rounds out our discussion of microprocessors by focusing on the following important mechanisms that are not part of the instruction set itself:

- We will introduce the fundamental mechanisms of *input* and *output*, including interrupts.

- We also study the *cache* and *memory management unit*.

We also begin to consider how the CPU hardware affects important characteristics of program execution. Program performance and power consumption are very important parameters in embedded system design. An understanding of how architectural aspects such as pipelining and caching affect these system characteristics is a foundation for analyzing and optimizing programs in later chapters.

Our study of program performance will begin with instruction-level performance. The basics of pipeline and cache timing will serve as the foundation for our studies of larger program units.

We use as an example a simple data compression unit, concentrating on the programming of the core compression algorithm.

1.5.3 Chapter 4: Bus-Based Computer Systems

Chapter 4 looks at the basic hardware and software platform for embedded computing. The microprocessor is very important, but only part of a system that

includes memory, I/O devices, and low-level software. We need to understand the basic characteristics of the platform before we move on to build sophisticated systems.

The basic embedded computing platform includes a microprocessor, I/O hardware, I/O driver software, and memory. Application-specific software and hardware can be added to this platform to turn it into an embedded computing platform. The microprocessor is at the center of both the hardware and software structure of the embedded computing system. The CPU controls the bus that connects to memory and I/O devices; the CPU also runs software that talks to the devices. In particular, I/O is central to embedded computing. Many aspects of I/O are not typically studied in modern computer architecture courses, so we need to master the basic concepts of input and output before we can design embedded systems.

Chapter 4 covers several important aspects of the platform:

- We study in detail how the CPU talks to memory and devices using the microprocessor *bus*.

- Based on our knowledge of bus operation, we study the structure of the *memory system* and types of *memory components*.

- We survey some important types of *I/O devices* to understand how to implement various types of real-world interfaces.

- We look at basic techniques for embedded system *design* and *debugging*.

System performance includes the bus and memory system, too. We will see how bus and memory transactions affect the execution time of systems.

We use an alarm clock as a design example. The clock does relatively little computation but a lot of I/O: It uses a timer to tell the CPU when to update the time, it reads buttons on the clock to respond to the user, and it continually updates the clock display.

1.5.4 Chapter 5: Program Design and Analysis

Chapter 5 looks inside the CPU to understand how instructions are executed as programs. Given the challenges of embedded programming—meeting strict performance goals, minimizing program size, reducing power consumption—this is an especially important topic. We build upon the fundamentals of computer architecture to understand how to design embedded programs.

- As a part of our study of the relationship between programs and instructions, we introduce a model for high-level language programs known as the *control/data flow graph (CDFG)*. We use this model extensively to help us analyze and optimize programs.

- Because embedded programs are largely written in higher-level languages, we will look at the processes for compiling, assembling, and linking to understand how high-level language programs are translated into instructions and data.

Some of the discussion surveys basic techniques for translating high-level language programs, but we also spend time on compilation techniques designed specifically to meet embedded system challenges.

- We develop techniques for the ***performance analysis*** of programs. It is difficult to determine the speed of a program simply by examining its source code. We learn how to use a combination of the source code, its assembly language implementation, and expected data inputs to analyze program execution time. We also study some basic techniques for optimizing program performance.

- An important topic related to performance analysis is ***power analysis***. We build on performance analysis methods to learn how to estimate the power consumption of programs.

- It is critical that the programs that we design function correctly. The control/data flow graph and techniques we have learned for performance analysis are related to techniques for ***testing programs***. We develop techniques that can methodically develop a set of tests for a program in order to exercise likely bugs.

At this point, we can consider the performance of a complete program. We will introduce the concept of worst-case execution time as a basic measure of program execution time.

Our design example for Chapter 5 is a software modem. A modem translates between the digital world of the microprocessor and the analog transmission scheme of the telephone network. Rather than use analog electronics to build a modem, we can use a microprocessor and special-purpose software. Because the modem has strict real-time deadlines, this example lets us exercise our knowledge of the microprocessor and of program analysis.

1.5.5 Chapter 6: Processes and Operating Systems

Chapter 6 builds on our knowledge of programs to study a special type of software component, the ***process***, and operating systems that use processes to create systems. A process is an execution of a program; an embedded system may have several processes running concurrently. A separate ***real-time operating system (RTOS)*** controls when the processes run on the CPU. Processes are important to embedded system design because they help us juggle multiple events happening at the same time. A real-time embedded system that is designed without processes usually ends up as a mess of spaghetti code that does not operate properly.

We will study the basic concepts of processes and process-based design in this chapter:

- We begin by introducing the ***process abstraction***. A process is defined by a combination of the program being executed and the current state of the program. We will learn how to switch contexts between processes.

- We cover the fundamentals of ***interprocess communication***, including the various styles of communication and how they can be implemented.

- In order to make use of processes, we must be able to ***schedule*** them. We discuss process priorities and how they can be used to guide scheduling.

- The real-time operating system is the software component that implements the process abstraction and scheduling. We study how RTOSs implement schedules, how programs interface to the operating system, and how we can evaluate the performance of systems built from RTOSs.

Tasks introduce a new level of complexity to performance analysis. Our study of real-time scheduling provides an important foundation for the study of multi-tasking systems.

Chapter 6 uses as a design example a digital telephone answering machine. Not only does an answering machine require real-time operation—telephone data are regularly sampled and stored to memory—but it must juggle several tasks at once. The answering machine must be able to operate the user interface simultaneously with recording voice data. In the most complex version of the answering machine, we must also simultaneously compress voice data during recording and uncompress it during playback. To emphasize the role of processes in structuring real-time computation, we compare the answering machine design with and without processes. It becomes apparent that the implementation that does not use processes will be considerably harder to design and debug.

1.5.6 Chapter 7: Multiprocessors

Many embedded systems are multiprocessors—computer systems with more than one processing element. The multiprocessor may use CPUs and DSPs; it may also include non-programmable elements known as ***accelerators***. Multiprocessors are often more energy-efficient and less expensive than platforms that try to do all the required computing on one big CPU.

Chapter 7 studies the design of multiprocessor embedded systems. We will spend a good amount of time on ***hardware/software co-design*** and the design of accelerated systems. Designing an accelerated system requires more than just building the accelerator itself. We have to determine how to connect the accelerator into the hardware and software so that we make best use of its capabilities. For example, the data transfers between the CPU and accelerator can consume all of the time savings created by the accelerator if we are not careful. We can also introduce added parallelism into the system if we have the CPU working on something else while the accelerator does its job.

Understanding the performance of accelerators requires a basic understanding of multiprocessor performance. We also need to extend our knowledge of bus and memory system performance. We will look at the architecture of several consumer electronics devices. A surprising number of devices make use of multiple processors under the hood.

We use as our example a video accelerator. Digital video requires performing a huge number of operations in real time; video also requires large volumes of data transfers. As such, it provides a good way to study not only the design of the accelerator itself but also how it fits into the overall system.

1.5.7 Chapter 8: Networks

Chapter 8 studies how we can build more complex embedded systems by letting several components communicate on a ***network***. The network may include several microprocessors, I/O devices, and special-purpose acceleration units. Embedded systems that are built from multiple microprocessors are called ***distributed embedded systems***. The automobile is a prime example of a distributed embedded system: Microprocessors are distributed all over the automobile performing distributed computations and coordinating the operation of the vehicle using networks.

This chapter builds on our knowledge of processes in particular to understand networks and their use in system design as follows:

- We start by discussing the fundamentals of network protocols and how networks differ from simple buses.

- Based on our knowledge of interprocess communication, we see how to allow processes to communicate over networks. We see how real-time operating systems can be extended to support multiple microprocessors whose processes communicate over a network.

- We study how to break a design into multiple components that communicate over a network. In particular, we need to know how to factor the communication delay of the network into our performance analysis.

We will also look at the networks used in automobiles and airplanes, which are prime examples of networked embedded systems. Chapter 8 uses as a design example a simple elevator system. An elevator is necessarily a distributed system operating over a network: We must have control in each elevator, but we must also coordinate the elevators to respond to user requests. And because the elevator includes some real-time control requirements—we must be able to stop the elevator at the door to the right floor—it provides a very good example to show how to properly distribute computations over the network to maximize responsiveness.

1.5.8 Chapter 9: System Design Techniques

Chapter 9 is our capstone chapter. This chapter studies the design of large, complex embedded systems. We introduce important concepts that are essential for the successful completion of large embedded system projects, and we use those techniques to help us integrate the knowledge obtained throughout the book.

This chapter delves into several topics related to large-scale embedded system design:

- We revisit the topic of ***design methodologies***. Based on our more detailed knowledge of embedded system design, we can better understand the role of methodology and the possible variations in methodologies.

- We study system ***specification methods***. Proper specifications become increasingly important as system complexity grows. More formal specification techniques help us capture intent clearly, consistently, and unambiguously.

- We look at ***quality assurance*** techniques. The program testing techniques covered in Chapter 5 are a good foundation but may not scale easily to complex systems. Additional methods are required to ensure that we exercise complex systems to shake out bugs.

SUMMARY

Embedded microprocessors are everywhere. Microprocessors allow sophisticated algorithms and user interfaces to be added relatively inexpensively to an amazing variety of products. Microprocessors also help reduce design complexity and time by separating out hardware and software design. Embedded system design is much more complex than programming PCs because we must meet multiple design constraints, including performance, cost, and so on. In the remainder of this book, we will build a set of techniques from the bottom up that will allow us to conceive, design, and implement sophisticated microprocessor-based systems.

What We Learned

- Embedded computing can be fun. It can also be difficult.

- Trying to hack together a complex embedded system probably won't work. You need to master a number of skills and understand the design process.

- Your system must meet certain functional requirements, such as features. It may also have to perform tasks to meet deadlines, limit its power consumption, be of a certain size, or meet other nonfunctional requirements.

- A hierarchical design process takes the design through several different levels of abstraction. You may need to do both top–down and bottom–up design.

- We use UML to describe designs at several levels of abstraction.

- This book takes a bottom–up view of embedded system design.

FURTHER READING

Spasov [Spa99] describes how 68HC11 microcontrollers are used in Canon EOS cameras. Douglass [Dou98] gives a good introduction to UML for embedded

systems. Other foundational books on object-oriented design include Rumbaugh et al. [Rum91], Booch [Boo91], Shlaer and Mellor [Shl92], and Selic et al. [Sel94].

QUESTIONS

Q1-1 Briefly describe the distinction between requirements and specification.

Q1-2 Briefly describe the distinction between specification and architecture.

Q1-3 At what stage of the design methodology would we determine what type of CPU to use (8-bit vs. 16-bit vs. 32-bit, which model of a particular type of CPU, etc.)?

Q1-4 At what stage of the design methodology would we choose a programming language?

Q1-5 At what stage of the design methodology would we test our design for functional correctness?

Q1-6 Compare and contrast top–down and bottom–up design.

Q1-7 Provide a concrete example of how bottom–up information from the software programming phase of design may be useful in refining the architectural design.

Q1-8 Give a concrete example of how bottom–up information from I/O device hardware design may be useful in refining the architectural design.

Q1-9 Create a UML state diagram for the *issue-command()* behavior of the *Controller* class of Figure 1.27.

Q1-10 Show how a *Set-speed* command flows through the refined class structure described in Figure 1.18, moving from a change on the front panel to the required changes on the train:

a. Show it in the form of a collaboration diagram.
b. Show it in the form of a sequence diagram.

Q1-11 Show how a *Set-inertia* command flows through the refined class structure described in Figure 1.18, moving from a change on the front panel to the required changes on the train:

a. Show it in the form of a collaboration diagram.
b. Show it in the form of a sequence diagram.

Q1-12 Show how an *Estop* command flows through the refined class structure described in Figure 1.18, moving from a change on the front panel to the required changes on the train:

 a. Show it in the form of a collaboration diagram.

 b. Show it in the form of a sequence diagram.

Q1-13 Draw a state diagram for a behavior that sends the command bits on the track. The machine should generate the address, generate the correct message type, include the parameters, and generate the ECC.

Q1-14 Draw a state diagram for a behavior that parses the received bits. The machine should check the address, determine the message type, read the parameters, and check the ECC.

Q1-15 Draw a class diagram for the classes required in a basic microwave oven. The system should be able to set the microwave power level between 1 and 9 and time a cooking run up to 59 min and 59 s in 1-s increments. Include * classes for the physical interfaces to the telephone line, microphone, speaker, and buttons.

Q1-16 Draw a collaboration diagram for the microwave oven of question Q1-15. The diagram should show the flow of messages when the user first sets the power level to 7, then sets the timer to 2:30, and then runs the oven.

LAB EXERCISES

L1-1 How would you measure the execution speed of a program running on a microprocessor? You may not always have a system clock available to measure time. To experiment, write a piece of code that performs some function that takes a small but measurable amount of time, such as a matrix algebra function. Compile and load the code onto a microprocessor, and then try to observe the behavior of the code on the microprocessor's pins.

L1-2 Complete the detailed specification of the train controller that was started in Section 1.4.4. Show all the required classes. Specify the behaviors for those classes. Use object diagrams to show the instantiated objects in the complete system. Develop at least one sequence diagram to show system operation.

L1-3 Develop a requirements description for an interesting device. The device may be a household appliance, a computer peripheral, or whatever you wish.

L1-4 Write a specification for an interesting device in UML. Try to use a variety of UML diagrams, including class diagrams, object diagrams, sequence diagrams, and so on.

Instruction Sets

2

- A brief review of computer architecture taxonomy and assembly language.

- Two very different architectures: ARM and TI C55x.

INTRODUCTION

In this chapter, we begin our study of microprocessors by studying *instruction sets*—the programmer's interface to the hardware. Although we hope to do as much programming as possible in high-level languages, the instruction set is the key to analyzing the performance of programs. By understanding the types of instructions that the CPU provides, we gain insight into alternative ways to implement a particular function.

We use two CPUs as examples. The ARM processor [Fur96, Jag95] is widely used in cell phones and many other systems. (The ARM architecture comes in several versions; we will concentrate on ARM version 7.) The Texas Instruments C55x is a family of *digital signal processors (DSPs)* [Tex01, Tex02].

We will start with a brief introduction to the terminology of computer architectures and instruction sets, followed by detailed descriptions of the ARM and C55x instruction sets.

2.1 PRELIMINARIES

In this section, we will look at some general concepts in computer architecture, including the different styles of computer architecture and the nature of assembly language.

2.1.1 Computer Architecture Taxonomy

Before we delve into the details of microprocessor instruction sets, it is helpful to develop some basic terminology. We do so by reviewing a taxonomy of the basic ways we can organize a computer.

A block diagram for one type of computer is shown in Figure 2.1. The computing system consists of a *central processing unit (CPU)* and a *memory*.

55

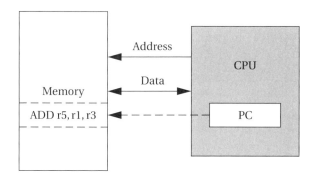

FIGURE 2.1

A von Neumann architecture computer.

FIGURE 2.2

A Harvard architecture.

The memory holds both data and instructions, and can be read or written when given an address. A computer whose memory holds both data and instructions is known as a *von Neumann* machine.

The CPU has several internal *registers* that store values used internally. One of those registers is the *program counter (PC)*, which holds the address in memory of an instruction. The CPU fetches the instruction from memory, decodes the instruction, and executes it. The program counter does not directly determine what the machine does next, but only indirectly by pointing to an instruction in memory. By changing only the instructions, we can change what the CPU does. It is this separation of the instruction memory from the CPU that distinguishes a stored-program computer from a general finite-state machine.

An alternative to the von Neumann style of organizing computers is the **Harvard architecture**, which is nearly as old as the von Neumann architecture. As shown in Figure 2.2, a Harvard machine has separate memories for data and program. The program counter points to program memory, not data memory. As a result, it is harder to write self-modifying programs (programs that write data values, then use those values as instructions) on Harvard machines.

Harvard architectures are widely used today for one very simple reason—the separation of program and data memories provides higher performance for digital signal processing. Processing signals in real-time places great strains on the data access system in two ways: First, large amounts of data flow through the CPU; and second, that data must be processed at precise intervals, not just when the CPU gets around to it. Data sets that arrive continuously and periodically are called **streaming data**. Having two memories with separate ports provides higher memory bandwidth; not making data and memory compete for the same port also makes it easier to move the data at the proper times. DSPs constitute a large fraction of all microprocessors sold today, and most of them are Harvard architectures. A single example shows the importance of DSP: Most of the telephone calls in the world go through at least two DSPs, one at each end of the phone call.

Another axis along which we can organize computer architectures relates to their instructions and how they are executed. Many early computer architectures were what is known today as **complex instruction set computers (CISC)**. These machines provided a variety of instructions that may perform very complex tasks, such as string searching; they also generally used a number of different instruction formats of varying lengths. One of the advances in the development of high-performance microprocessors was the concept of **reduced instruction set computers (RISC)**. These computers tended to provide somewhat fewer and simpler instructions. The instructions were also chosen so that they could be efficiently executed in **pipelined** processors. Early RISC designs substantially outperformed CISC designs of the period. As it turns out, we can use RISC techniques to efficiently execute at least a common subset of CISC instruction sets, so the performance gap between RISC-like and CISC-like instruction sets has narrowed somewhat.

Beyond the basic RISC/CISC characterization, we can classify computers by several characteristics of their instruction sets. The instruction set of the computer defines the interface between software modules and the underlying hardware; the instructions define what the hardware will do under certain circumstances. Instructions can have a variety of characteristics, including:

- Fixed versus variable length.

- Addressing modes.

- Numbers of operands.

- Types of operations supported.

The set of registers available for use by programs is called the **programming model**, also known as the **programmer model**. (The CPU has many other registers that are used for internal operations and are unavailable to programmers.)

There may be several different implementations of an architecture. In fact, the architecture definition serves to define those characteristics that must be true of all implementations and what may vary from implementation to implementation. Different CPUs may offer different clock speeds, different cache configurations,

changes to the bus or interrupt lines, and many other changes that can make one model of CPU more attractive than another for any given application.

2.1.2 Assembly Language

Figure 2.3 shows a fragment of ARM assembly code to remind us of the basic features of assembly languages. Assembly languages usually share the same basic features:

- One instruction appears per line.

- *Labels*, which give names to memory locations, start in the first column.

- Instructions must start in the second column or after to distinguish them from labels.

- Comments run from some designated comment character (; in the case of ARM) to the end of the line.

Assembly language follows this relatively structured form to make it easy for the **assembler** to parse the program and to consider most aspects of the program line by line. (It should be remembered that early assemblers were written in assembly language to fit in a very small amount of memory. Those early restrictions have carried into modern assembly languages by tradition.) Figure 2.4 shows the format of an ARM data processing instruction such as an ADD. For the instruction

```
ADDGT r0,r3,#5
```

the *cond* field would be set according to the GT condition (1100), the *opcode* field would be set to the binary code for the ADD instruction (0100), the first *operand* register *Rn* would be set to 3 to represent r3, the destination register *Rd* would be set to 0 for r0, and the *operand 2* field would be set to the immediate value of 5.

Assemblers must also provide some **pseudo-ops** to help programmers create complete assembly language programs. An example of a pseudo-op is one that allows data values to be loaded into memory locations. These allow constants, for example, to be set into memory. An example of a memory allocation pseudo-op for ARM is shown in Figure 2.5. The ARM % pseudo-op allocates a block of memory of the size specified by the operand and initializes those locations to zero.

```
label1    ADR r4,c
          LDR r0,[r4]        ; a comment
          ADR r4,d
          LDR r1,[r4]
          SUB r0,r0,r1       ; another comment
```

FIGURE 2.3

An example of ARM assembly language.

FIGURE 2.4

Format of ARM data processing instructions.

```
BIGBLOCK        % 10
```

FIGURE 2.5

Pseudo-ops for allocating memory.

2.2 ARM PROCESSOR

In this section, we concentrate on the ARM processor. ARM is actually a family of RISC architectures that have been developed over many years. ARM does not manufacture its own VLSI devices; rather, it licenses its architecture to companies who either manufacture the CPU itself or integrate the ARM processor into a larger system.

The textual description of instructions, as opposed to their binary representation, is called an assembly language. ARM instructions are written one per line, starting after the first column. Comments begin with a semicolon and continue to the end of the line. A label, which gives a name to a memory location, comes at the beginning of the line, starting in the first column. Here is an example:

```
      LDR r0,[r8]; a comment
label ADD r4,r0,r1
```

2.2.1 **Processor and Memory Organization**

Different versions of the ARM architecture are identified by different numbers. ARM7 is a von Neumann architecture machine, while ARM9 uses a Harvard architecture. However, this difference is invisible to the assembly language programmer, except for possible performance differences.

The ARM architecture supports two basic types of data:

- The standard ARM word is 32 bits long.

- The word may be divided into four 8-bit bytes.

ARM7 allows addresses up to 32 bits long. An address refers to a byte, not a word. Therefore, the word 0 in the ARM address space is at location 0, the word 1 is at 4, the word 2 is at 8, and so on. (As a result, the PC is incremented by 4 in the absence of a branch.) The ARM processor can be configured at power-up to address the bytes in a word in either ***little-endian*** mode (with the lowest-order byte residing in the low-order bits of the word) or ***big-endian*** mode (the lowest-order byte stored in the highest bits of the word), as illustrated in Figure 2.6 [Coh81]. General-purpose computers have sophisticated instruction sets. Some of this sophistication is required simply to provide the functionality of a general computer, while other aspects of instruction sets may be provided to increase performance, reduce code size, or otherwise improve program characteristics. In this section, we concentrate on the functionality of the ARM instruction set and will defer performance and other aspects of the CPU to Section 5.6.

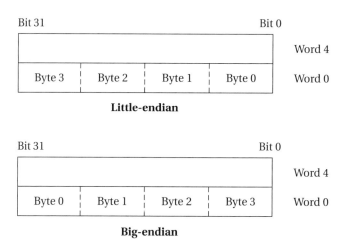

FIGURE 2.6

Byte organizations within an ARM word.

2.2.2 Data Operations

Arithmetic and logical operations in C are performed in variables. Variables are implemented as memory locations. Therefore, to be able to write instructions to perform C expressions and assignments, we must consider both arithmetic and logical instructions as well as instructions for reading and writing memory.

Figure 2.7 shows a sample fragment of C code with data declarations and several assignment statements. The variables a, b, c, x, y, and z all become data locations in memory. In most cases data are kept relatively separate from instructions in the program's memory image.

In the ARM processor, arithmetic and logical operations cannot be performed directly on memory locations. While some processors allow such operations to directly reference main memory, ARM is a *load-store architecture*—data operands must first be loaded into the CPU and then stored back to main memory to save the results. Figure 2.8 shows the registers in the basic ARM programming model. ARM has 16 general-purpose registers, r0 through r15. Except for r15, they are identical—any operation that can be done on one of them can be done on the other one also. The r15 register has the same capabilities as the other registers, but it is also used as the program counter. The program counter should of course not be overwritten for use in data operations. However, giving the PC the properties of a general-purpose register allows the program counter value to be used as an operand in computations, which can make certain programming tasks easier.

The other important basic register in the programming model is the *current program status register (CPSR)*. This register is set automatically during every arithmetic, logical, or shifting operation. The top four bits of the CPSR hold the following useful information about the results of that arithmetic/logical operation:

- The negative (N) bit is set when the result is negative in two's-complement arithmetic.

- The zero (Z) bit is set when every bit of the result is zero.

- The carry (C) bit is set when there is a carry out of the operation.

- The overflow (V) bit is set when an arithmetic operation results in an overflow.

```
int a, b, c, x, y, z;
x = (a + b) - c;
y = a*(b + c);
z = (a << 2) | (b & 15);
```

FIGURE 2.7

A C fragment with data operations.

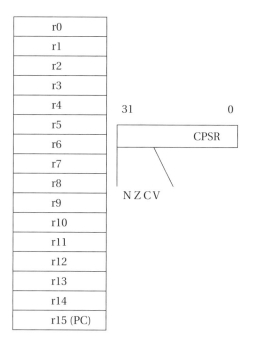

FIGURE 2.8

The basic ARM programming model.

These bits can be used to check easily the results of an arithmetic operation. However, if a chain of arithmetic or logical operations is performed and the intermediate states of the CPSR bits are important, then they must be checked at each step since the next operation changes the CPSR values. Example 2.1 illustrates the computation of CPSR bits.

Example 2.1

Status bit computation in the ARM

An ARM word is 32 bits. In C notation, a hexadecimal number starts with 0x, such as 0xffffffff, which is a two's-complement representation of -1 in a 32-bit word.
Here are some sample calculations:

- $-1 + 1 = 0$: Written in 32-bit format, this becomes 0xffffffff + 0x1 = 0x0, giving the CPSR value of NZCV = 1001.

- $0 - 1 = -1$: 0x0 − 0x1 = 0xffffffff, with NZCV = 1000.

- $2^{31} - 1 + 1 = -2^{31}$: 0x7fffffff + 0x1 = 0x80000000, with NZCV = 1001.

The basic form of a data instruction is simple:

```
ADD r0,r1,r2
```

This instruction sets register r0 to the sum of the values stored in r1 and r2. In addition to specifying registers as sources for operands, instructions may also provide *immediate operands*, which encode a constant value directly in the instruction. For example,

```
ADD r0,r1,#2
```

sets r0 to r1 + 2.

The major data operations are summarized in Figure 2.9. The arithmetic operations perform addition and subtraction; the with-carry versions include the current value of the carry bit in the computation. RSB performs a subtraction with the order of the two operands reversed, so that RSB r0,r1,r2 sets r0 to be r2 − r1. The bit-wise logical operations perform logical AND, OR, and XOR operations (the exclusive or is called EOR). The BIC instruction stands for bit clear: BIC r0, r1, r2 sets r0 to r1 and not r2. This instruction uses the second source operand as a mask: Where a bit in the mask is 1, the corresponding bit in the first source operand is cleared. The MUL instruction multiplies two values, but with some restrictions: No operand may be an immediate, and the two source operands must be different registers. The MLA instruction performs a multiply-accumulate operation, particularly useful in matrix operations and signal processing. The instruction

```
MLA r0,r1,r2,r3
```

sets r0 to the value r1 × r2 + r3.

The shift operations are not separate instructions—rather, shifts can be applied to arithmetic and logical instructions. The shift modifier is always applied to the second source operand. A left shift moves bits up toward the most-significant bits, while a right shift moves bits down to the least-significant bit in the word. The LSL and LSR modifiers perform left and right logical shifts, filling the least-significant bits of the operand with zeroes. The arithmetic shift left is equivalent to an LSL, but the ASR copies the sign bit—if the sign is 0, a 0 is copied, while if the sign is 1, a 1 is copied. The rotate modifiers always rotate right, moving the bits that fall off the least-significant bit up to the most-significant bit in the word. The RRX modifier performs a 33-bit rotate, with the CPSR's C bit being inserted above the sign bit of the word; this allows the carry bit to be included in the rotation.

The instructions in Figure 2.10 are comparison operations—they do not modify general-purpose registers but only set the values of the NZCV bits of the CPSR register. The compare instruction CMP r0,r1 computes r0 − r1, sets the status bits, and throws away the result of the subtraction. CMN uses an addition to set the status bits. TST performs a bit-wise AND on the operands, while TEQ performs an exclusive-or.

Figure 2.11 summarizes the ARM move instructions. The instruction MOV r0,r1 sets the value of r0 to the current value of r1. The MVN instruction complements the operand bits (one's complement) during the move.

ADD	Add
ADC	Add with carry
SUB	Subtract
SBC	Subtract with carry
RSB	Reverse subtract
RSC	Reverse subtract with carry
MUL	Multiply
MLA	Multiply and accumulate

Arithmetic

AND	Bit-wise and
ORR	Bit-wise or
EOR	Bit-wise exclusive-or
BIC	Bit clear

Logical

LSL	Logical shift left (zero fill)
LSR	Logical shift right (zero fill)
ASL	Arithmetic shift left
ASR	Arithmetic shift right
ROR	Rotate right
RRX	Rotate right extended with C

Shift/rotate

FIGURE 2.9

ARM data instructions.

Values are transferred between registers and memory using the load-store instructions summarized in Figure 2.12. LDRB and STRB load and store bytes rather than whole words, while LDRH and SDRH operate on half-words and LDRSH extends the sign bit on loading. An ARM address may be 32 bits long. The ARM load and store instructions do not directly refer to main memory addresses, since a 32-bit address would not fit into an instruction that included an opcode and operands. Instead, the ARM uses *register-indirect addressing*. In register-indirect addressing, the value

CMP	Compare
CMN	Negated compare
TST	Bit-wise test
TEQ	Bit-wise negated test

FIGURE 2.10

ARM comparison instructions.

MOV	Move
MVN	Move negated

FIGURE 2.11

ARM move instructions.

LDR	Load
STR	Store
LDRH	Load half-word
STRH	Store half-word
LDRSH	Load half-word signed
LDRB	Load byte
STRB	Store byte
ADR	Set register to address

FIGURE 2.12

ARM load-store instructions and pseudo-operations.

stored in the register is used as the address to be fetched from memory; the result of that fetch is the desired operand value. Thus, as illustrated in Figure 2.13, if we set $r1 = 0 \times 100$, the instruction

```
LDR r0,[r1]
```

sets r0 to the value of memory location 0x100. Similarly, STR r0,[r1] would store the contents of r0 in the memory location whose address is given in r1. There are several possible variations:

```
LDR r0,[r1, - r2]
```

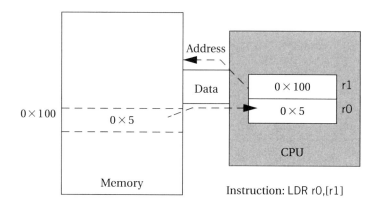

FIGURE 2.13

Register-indirect addressing in the ARM.

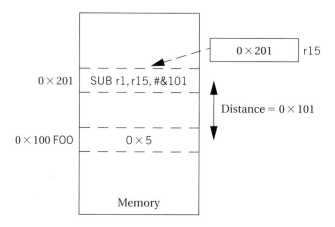

FIGURE 2.14

Computing an absolute address using the PC.

loads r0 from the address given by r1 − r2, while

```
LDR r0,[r1, #4]
```

loads r0 from the address r1 + 4.

This begs the question of how we get an address into a register—we need to be able to set a register to an arbitrary 32-bit value. In the ARM, the standard way to set a register to an address is by performing arithmetic on the program counter, which is stored in r15. By adding or subtracting to the PC a constant equal to the distance between the current instruction (i.e., the instruction that is computing the address) and the desired location, we can generate the desired address without performing a load. The ARM programming system provides an ADR pseudo-operation to simplify

this step. Thus, as shown in Figure 2.14, if we give location 0x100 the name FOO, we can use the pseudo-operation

```
ADR r1,FOO
```

to perform the same function of loading r1 with the address 0x100.

Example 2.2 illustrates how to implement C assignments in ARM instruction.

Example 2.2

C assignments in ARM instructions

We will use the assignments of Figure 2.7. The semicolon (;) begins a comment after an instruction, which continues to the end of that line. The statement

$$x = (a + b) - c;$$

can be implemented by using r0 for a, r1 for b, r2 for c, and r3 for x. We also need registers for indirect addressing. In this case, we will reuse the same indirect addressing register, r4, for each variable load. The code must load the values of a, b, and c into these registers before performing the arithmetic, and it must store the value of x back to memory when it is done. This code performs the following necessary steps:

```
ADR r4,a      ; get address for a
LDR r0,[r4]   ; get value of a
ADR r4,b      ; get address for b, reusing r4
LDR r1,[r4]   ; load value of b
ADD r3,r0,r1  ; set intermediate result for x to a + b
ADR r4,c      ; get address for c
LDR r2,[r4]   ; get value of c
SUB r3,r3,r2  ; complete computation of x
ADR r4,x      ; get address for x
STR r3,[r4]   ; store x at proper location
```

The operation

$$y = a * (b + c);$$

can be coded similarly, but in this case we will reuse more registers by using r0 for both a and b, r1 for c, and r2 for y. Once again, we will use r4 to store addresses for indirect addressing. The resulting code is

```
ADR r4,b      ; get address for b
LDR r0,[r4]   ; get value of b
ADR r4,c      ; get address for c
LDR r1,[r4]   ; get value of c
ADD r2,r0,r1  ; compute partial result of y
ADR r4,a      ; get address for a
```

```
LDR r0,[r4]   ; get value of a
MUL r2,r2,r0  ; compute final value of y
ADR r4,y      ; get address for y
STR r2,[r4]   ; store value of y at proper location
```

The C statement

$$z = (a \ll 2) \,|\, (b \,\&\, 15);$$

can be coded using r0 for *a* and *z*, r1 for *b*, and r4 for addresses as follows:

```
ADR r4,a         ; get address for a
LDR r0,[r4]      ; get value of a
MOV r0,r0,LSL 2  ; perform shift
ADR r4,b         ; get address for b
LDR r1,[r4]      ; get value of b
AND r1,r1,#15    ; perform logical AND
ORR r1,r0,r1     ; compute final value of z
ADR r4,z         ; get address for z
STR r1,[r4]      ; store value of z
```

We have already seen three addressing modes: register, immediate, and indirect. The ARM also supports several forms of ***base-plus-offset addressing***, which is related to indirect addressing. But rather than using a register value directly as an address, the register value is added to another value to form the address. For instance,

```
LDR r0,[r1,#16]
```

loads r0 with the value stored at location r1 + 16. Here, r1 is referred to as the ***base*** and the immediate value the ***offset***. When the offset is an immediate, it may have any value up to 4,096; another register may also be used as the offset. This addressing mode has two other variations: ***auto-indexing*** and ***post-indexing***. Auto-indexing updates the base register, such that

```
LDR r0,[r1,#16]!
```

first adds 16 to the value of r1, and then uses that new value as the address. The ! operator causes the base register to be updated with the computed address so that it can be used again later. Our examples of base-plus-offset and auto-indexing instructions will fetch from the same memory location, but auto-indexing will also modify the value of the base register r1. Post-indexing does not perform the offset calculation until after the fetch has been performed. Consequently,

```
LDR r0,[r1],#16
```

will load r0 with the value stored at the memory location whose address is given by r1, and then add 16 to r1 and set r1 to the new value. In this case, the post-indexed

mode fetches a different value than the other two examples, but ends up with the same final value for r1 as does auto-indexing.

We have used the ADR pseudo-op to load addresses into registers to access variables because this leads to simple, easy-to-read code (at least by assembly language standards). Compilers tend to use other techniques to generate addresses, because they must deal with global variables and automatic variables.

2.2.3 Flow of Control

The B (branch) instruction is the basic mechanism in ARM for changing the flow of control. The address that is the destination of the branch is often called the **branch target**. Branches are **PC-relative**—the branch specifies the offset from the current PC value to the branch target. The offset is in words, but because the ARM is byte-addressable, the offset is multiplied by four (shifted left two bits, actually) to form a byte address. Thus, the instruction

```
B #100
```

will add 400 to the current PC value.

We often wish to branch conditionally, based on the result of a given computation. The if statement is a common example. The ARM allows *any* instruction, including branches, to be executed conditionally. This allows branches to be conditional, as well as data operations. Figure 2.15 summarizes the condition codes.

EQ	Equals zero	$Z = 1$
NE	Not equal to zero	$Z = 0$
CS	Carry set	$C = 1$
CC	Carry clear	$C = 0$
MI	Minus	$N = 1$
PL	Nonnegative (plus)	$N = 0$
VS	Overflow	$V = 1$
VC	No overflow	$V = 0$
HI	Unsigned higher	$C = 1$ and $Z = 0$
LS	Unsigned lower or same	$C = 0$ or $Z = 1$
GE	Signed greater than or equal	$N = V$
LT	Signed less than	$N \ V$
GT	Signed greater than	$Z = 0$ and $N = V$
LE	Signed less than or equal	$Z = 1$ or $N \ V$

FIGURE 2.15

Condition codes in ARM.

Example 2.3 shows how to implement an if statement.

Example 2.3

Implementing an if *statement in ARM*

We will use the following if statement as an example:

```
if (a < b) {
    x = 5;
    y = c + d;
    }
else x = c - d;
```

The implementation uses two blocks of code, one for the true case and another for the false case. A branch may either fall through to the true case or branch to the false case:

```
; compute and test the condition
        ADR r4,a        ; get address for a
        LDR r0,[r4]     ; get value of a
        ADR r4,b        ; get address for b
        LDR r1,[r4]     ; get value of b
        CMP r0, r1      ; compare a < b
        BGE fblock      ; if a >= b, take branch
; the true block follows
        MOV r0,#5       ; generate value for x
        ADR r4,x        ; get address for x
        STR r0,[r4]     ; store value of x
        ADR r4,c        ; get address for c
        LDR r0,[r4]     ; get value of c
        ADR r4,d        ; get address for d
        LDR r1,[r4]     ; get value of d
        ADD r0,r0,r1    ; compute c + d
        ADR r4,y        ; get address for y
        STR r0,[r4]     ; store value of y
        B after         ; branch around the false block
; the false block follows
fblock  ADR r4,c        ; get address for c
        LDR r0,[r4]     ; get value of c
        ADR r4,d        ; get address for d
        LDR r1,[r4]     ; get value of d
        SUB r0,r0,r1    ; compute c - d
        ADR r4,x        ; get address for x
        STR r0,[r4]     ; store value of x
after ... ; code after the if statement
```

Example 2.4 illustrates an interesting way to implement multiway conditions.

Example 2.4

Implementing the C switch statement in ARM

The switch statement in C takes the following form:

```
switch (test) {
case 0: ... break;
case 1: ... break;
...
}
```

The above statement could be coded like an if statement by first testing *test* = A, then *test* = B, and so forth. However, it can be more efficiently implemented by using base-plus-offset addressing and building what is known as a **branch table**:

```
          ADR r2,test ; get address for test
          LDR r0,[r2] ; load value for test
          ADR r1,switchtab ; load address for switch table
          LDR r15,[r1,r0,LSL #2]
switchtab DCD case0
          DCD case1
          ...
case0     ... ; code for case 0
          ...
case1     ... ; code for case 1
          ...
```

This implementation uses the value of test as an offset into a table, where the table holds the addresses for the blocks of code that implement the various cases. The heart of this code is the LDR instruction, which packs a lot of functionality into a single instruction:

- It shifts the value of r0 left two bits to turn the offset into a word address.

- It uses base-plus-offset addressing to add the left-shifted value of test (held in r0) to the address of the base of the table held in r1.

- It sets the PC (r15) to the new address computed by the instruction.

Each case is implemented by a block of code that is located elsewhere in memory. The branch table begins at the location named switchtab. The DCD statement is a way of loading a 32-bit address into memory at that point, so the branch table holds the addresses of the starting points of the blocks that correspond to the cases.

The loop is a very common C statement, particularly in signal processing code. Loops can be naturally implemented using conditional branches. Because loops

often operate on values stored in arrays, loops are also a good illustration of another use of the base-plus-offset addressing mode. A simple but common use of a loop is in the FIR filter, which is explained in Application Example 2.1; the loop-based implementation of the FIR filter is described in Example 2.5.

Application Example 2.1

FIR filters

A *finite impulse response (FIR) filter* is a commonly used method for processing signals; we make use of it in Section 5.11. The FIR filter is a simple sum of products:

$$\sum_{1 \le i \le n} c_r x_i \tag{2.1}$$

In use as a filter, the x_is are assumed to be samples of data taken periodically, while the c_is are coefficients. This computation is usually drawn like this:

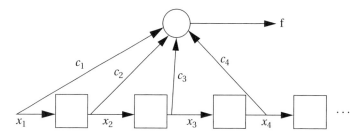

This representation assumes that the samples are coming in periodically and that the FIR filter output is computed once every time a new sample comes in. The boxes represent delay elements that store the recent samples to provide the x_is. The delayed samples are individually multiplied by the c_is and then summed to provide the filter output.

Example 2.5

An FIR filter for the ARM

The C code for the FIR filter of Application Example 2.1 follows:

```
for (i = 0, f = 0; i < N; i++)
        f = f + c[i] * x[i];
```

We can address the arrays c and x using base-plus-offset addressing: We will load one register with the address of the zeroth element of each array and use the register holding *i* as the offset.

The C language [Ker88] defines a for loop as equivalent to a while loop with proper initialization and termination. Using that rule, the for loop can be rewritten as

```
i = 0;
f = 0;
```

```
    while (i < N) {
        f = f + c[i]*x[i];
        i++;
    }
```

Here is the code for the loop:

```
        ; loop initiation code
        MOV r0,#0        ; use r0 for i, set to 0
        MOV r8,#0        ; use a separate index for arrays
        ADR r2,N         ; get address for N
        LDR r1,[r2]      ; get value of N for loop termination test
        MOV r2,#0        ; use r2 for f, set to 0
        ADR r3,c         ; load r3 with address of base of c array
        ADR r5,x         ; load r5 with address of base of x array
        ; loop body
loop LDR r4,[r3,r8]      ; get value of c[i]
        LDR r6,[r5,r8]   ; get value of x[i]
        MUL r4,r4,r6     ; compute c[i]*x[i]
        ADD r2,r2,r4     ; add into running sum f
        ; update loop counter and array index
        ADD r8,r8,#4     ; add one word offset to array index
        ADD r0,r0,#1     ; add 1 to i
        ; test for exit
        CMP r0,r1
        BLT loop         ; if i < N, continue loop
loopend...
```

We have to be careful about numerical accuracy in this type of code, whether it is written in C or assembly language. The result of a 32-bit \times 32-bit multiplication is a 64-bit result. The ARM MUL instruction leaves the lower 32 bits of the result in the destination register. So long as the result fits within 32 bits, this is the desired action. If the input values are such that values can sometimes exceed 32 bits, then we must redesign the code to compute higher-resolution values.

The other important class of C statement to consider is the *function*. A C function returns a value (unless its return type is void); *subroutine* or *procedure* are the common names for such a construct when it does not return a value. Consider this simple use of a function in C:

```
    x = a + b;
    foo(x);
    y = c - d;
```

A function returns to the code immediately after the function call, in this case the assignment to y. A simple branch is insufficient because we would not know where

to return. To properly return, we must save the PC value when the procedure/ function is called and, when the procedure is finished, set the PC to the address of the instruction *just after* the call to the procedure. (You don't want to endlessly execute the procedure, after all.) The branch-and-link instruction is used in the ARM for procedure calls. For instance,

```
BL foo
```

will perform a branch and link to the code starting at location foo (using PC-relative addressing, of course). The branch and link is much like a branch, except that before branching it stores the current PC value in r14. Thus, to return from a procedure, you simply move the value of r14 to r15:

```
MOV r15,r14
```

You should not, of course, overwrite the PC value stored in r14 during the procedure.

But this mechanism only lets us call procedures one level deep. If, for example, we call a C function within another C function, the second function call will overwrite r14, destroying the return address for the first function call. The standard procedure for allowing nested procedure calls (including recursive procedure calls) is to build a stack, as illustrated in Figure 2.16. The C code shows a series of functions that call other functions: f1() calls f2(), which in turn calls f3(). The right side of

```
void f1(int a) {
    f2(a);
}

void f2(int r) {
    f3(r,5);
}

void f3(int x, int y) {
    g = x + y;
}

main() {
    f1(xyz);
}
```

C code

Function call stack

FIGURE 2.16

Nested function calls and stacks.

the figure shows the state of the *procedure call stack* during the execution of f3(). The stack contains one *activation record* for each active procedure. When f3() finishes, it can pop the top of the stack to get its return address, leaving the return address for f2() waiting at the top of the stack for its return.

Most procedures need to pass parameters into the procedure and return values out of the procedure as well as remember their return address.

We can also use the procedure call stack to pass parameters. The conventions used to pass values into and out of procedures is known as *procedure linkage*. To pass parameters into a procedure, the values can be pushed onto the stack just before the procedure call. Once the procedure returns, those values must be popped off the stack by the caller, since they may hide a return address or other useful information on the stack. A procedure may also need to save register values for registers it modifies. The registers can be pushed onto the stack upon entry to the procedure and popped off the stack, restoring the previous values, before returning.

Example 2.6 illustrates the programming of a simple C function.

Example 2.6

Procedure calls in ARM

We use as an example one of the functions from Figure 2.16:

```
void f1(int a) {
    f2(a);
}
```

The ARM C compiler's convention is to use register r13 to point to the top of the stack. We assume that the argument a has been passed into f1() on the stack and that we must push the argument for f2 (which happens to be the same value) onto the stack before calling f2().

Here is some handwritten code for f1(), which includes a call to f2():

```
f1 LDR r0,[r13]    ; load value of a argument into r0 from stack
    ; call f2()
    STR r14,[r13]!    ; store f1's return address on the stack
    STR r0,[r13!]     ; store argument to f2 onto stack
    BL f2             ; branch and link to f2
    ; return from f1()
    SUB r13,#4        ; pop f2's argument off the stack
    LDR r13!,r15      ; restore registers and return
```

We use base-plus-offset addressing to load the value passed into f1() into a register for use by r1. To call f2(), we first push f1()'s return address, stored in r14 by the branch-and-link instruction executed to get into f1(), onto the stack. We then push f2()'s parameter onto the stack. In both cases, we use autoincrement addressing to both store onto the stack and adjust the stack pointer. To return, we must first adjust the stack to get rid of f2()'s parameter that

hides f1()'s return address; we then use autoincrement addressing to pop f1()'s return address off the stack and into the PC (r15).

We will discuss procedure linkage mechanisms for the ARM in more detail in Section 5.4.2.

2.3 TI C55x DSP

The Texas Instruments C55x DSP is a family of digital signal processors designed for relatively high performance signal processing. The family extends on previous generations of TI DSPs; the architecture is also defined to allow several different implementations that comply with the instruction set.

The C55x, like many DSPs, is an ***accumulator architecture***, meaning that many arithmetic operations are of the form *accumulator = operand + accumulator*. Because one of the operands is the accumulator, it need not be specified in the instruction. Accumulator-oriented instructions are also well-suited to the types of operations performed in digital signal processing, such as $a_1x_1 + a_2x_2 + \ldots$. Of course, the C55x has more than one register and not all instructions adhere to the accumulator-oriented format. But we will see that arithmetic and logical operations take a very different form in the C55x than they do in the ARM.

C55x assembly language programs follow the typical format:

```
        MPY *AR0, *CDP+, AC0
label:  MOV #1, T0
```

Assembler mnemonics are case-insensitive. Instruction mnemonics are formed by combining a root with prefixes and/or suffixes. For example, the A prefix denotes an operation performed in addressing mode while the 40 suffix denotes an arithmetic operation performed in 40-bit resolution. We will discuss the prefixes and suffixes in more detail when we describe the instructions.

The C55x also allows operations to be specified in an algebraic form:

```
AC1 = AR0 * coef(*CDP)
```

2.3.1 Processor and Memory Organization

We will use the term register to mean any type of register in the programmer model and the term **accumulator** to mean a register used primarily in the accumulator style.

The C55x supports several data types:

- A ***word*** is 16 bits long.

- A ***longword*** is 32 bits long.

- Instructions are byte-addressable.

- Some instructions operate on addressed bits in registers.

The C55x has a number of registers. Few to none of these registers are general-purpose registers like those of the ARM. Registers are generally used for specialized purposes. Because the C55x registers are less regular, we will discuss them by how they may be used rather than simply listing them.

Most registers are ***memory-mapped***—that is, the register has an address in the memory space. A memory-mapped register can be referred to in assembly language in two different ways: either by referring to its mnemonic name or through its address.

The program counter is PC. The program counter extension register XPC extends the range of the program counter. The return address register RETA is used for subroutines.

The C55x has four 40-bit accumulators AC0, AC1, AC2, and AC3. The low-order bits 0–15 are referred to as AC0L, AC1L, AC2L, and AC3L; the high-order bits 16–31 are referred to as AC0H, AC1H, AC2H, and AC3H; and the guard bits 32–39 are referred to as AC0G, AC1G, AC2G, and AC3G. (Guard bits are used in numerical algorithms like signal processing to provide a larger dynamic range for intermediate calculations.)

The architecture provides six status registers. Three of the status registers, ST0 and ST1 and the processor mode status register PMST, are inherited from the C54x architecture. The C55x adds four registers ST0_55, ST1_55, ST2_55, and ST3_55. These registers provide arithmetic and bit manipulation flags, a data page pointer and auxiliary register pointer, and processor mode bits, among other features.

The stack pointer SP keeps track of the system stack. A separate system stack is maintained through the SSP register. The SPH register is an extended data page pointer for both SP and SSP.

Eight auxiliary registers AR0–AR7 are used by several types of instructions, notably for circular buffer operations. The coefficient data pointer CDP is used to read coefficients for polynomial evaluation instructions; CDPH is the main data page pointer for the CDP.

The circular buffer size register BK47 is used for circular buffer operations for the auxiliary registers AR4-7. Four registers define the start of circular buffers: BSA01 for auxiliary registers AR0 and AR1; BSA23 for AR2 and AR3; BSA45 for AR4 and AR5; BSA67 for AR6 and AR7. The circular buffer size register BK03 is used to address circular buffers that are commonly used in signal processing. BKC is the circular buffer size register for CDP. BSAC is the circular buffer coefficient start address register.

Repeats of single instructions are controlled by the single repeat register CSR. This counter is the primary interface to the program. It is loaded with the required number of iterations. When the repeat starts, the value in CSR is copied into the repeat counter RPTC, which maintains the counts for the current repeat and is decremented during each iteration.

Several registers are used for block repeats—instructions that are executed several times in a row. The block repeat counter BRC0 counts block repeat iterations.

The block repeat start and end registers RSA0L and REA0L keep track of the start and end points of the block.

The block repeat register 1 BRC1 and block repeat save register 1 BRS1 are used to repeat blocks of instructions. There are two repeat start address registers RSA0 and RSA1. Each is divided into low and high parts: RSA0L and RSA0H, for example.

Four temporary registers T0, T1, T2, and T3 are used for various calculations.

Two transition register TRN0 and TRN1 are used for compare-and-extract-extremum instructions. These instructions are used to implement the Viterbi algorithm.

Several registers are used for addressing modes. The memory data page start address registers DP and DPH are used as the base address for data accesses. Similarly, the peripheral data page start address register PDP is used as a base for I/O addresses.

Several registers control interrupts. The interrupt mask registers 0 and 1, named IER0 and IER1, determine what interrupts will be recognized. The interrupt flag registers 0 and 1, named IFR0 and IFR1, keep track of currently pending interrupts. Two other registers, DBIER0 and DBIER1, are used for debugging. Two registers, the interrupt vector register DSP (IVPD) and interrupt vector register host (IVPH) are used as the base address for the interrupt vector table.

The C55x registers are summarized in Figure 2.17.

The C55x supports a 24-bit address space, providing 16 MB of memory as shown in Figure 2.18. Data, program, and I/O accesses are all mapped to the same physical memory. But these three spaces are addressed in different ways. The program space is byte-addressable, so an instruction reference is 24-bit long. Data space is word-addressable, so a data address is 23 bits. (Its least-significant bit is set to 0.) The data space is also divided into 128 pages of 64K words each. The I/O space is 64K words wide, so an I/O address is 16 bits. The situation is summarized in Figure 2.19.

Not all implementations of the C55x may provide all 16 MB of memory on chip. The C5510, for example, provides 352 KB of on-chip memory. The remainder of the memory space is provided by separate memory chips connected to the DSP.

The first 96 words of data page 0 are reserved for the memory-mapped registers. Since the program space is byte-addressable, unlike the word-addressable data space, the first 192 words of the program space are reserved for those same registers.

2.3.2 Addressing Modes

The C55x has three addressing modes:

- Absolute addressing supplies an address in the instruction.

- Direct addressing supplies an offset.

- Indirect addressing uses a register as a pointer.

Absolute addresses may be any of three different types:

- A k16 absolute address is a 16-bit value that is combined with the DPH register to form a 23-bit address.

register mnemonic	description
AC0-AC3	accumulators
AR0-AR7, XAR0-XAR7	auxiliary registers and extensions of auxiliary registers
BK03, BK47, BKC	circular buffer size registers
BRC0-BRC1	block repeat counters
BRS1	BRC1 save register
CDP, CDPH, CDPX	coefficient data register: low (CDP), high (CDPH), full (CDPX)
CFCT	control flow context register
CSR	computed single repeat register
DBIER0-DBIER1	debug interrupt enable registers
DP, DPH, DPX	data page register: low (DP), high (DPH), full (DPX)
IER0-IER1	interrupt enable registers
IFR0-IFR1	interrupt flag registers
IVPD, IVPH	interrupt vector registers
PC, XPC	program counter and program counter extension
PDP	peripheral data page register
RETA	return address register
RPTC	single repeat counter
RSA0-RSA1	block repeat start address registers

FIGURE 2.17

Registers in the TI C55x.

- A k23 absolute address is a 23-bit unsigned number that provides a full data address.
- An I/O absolute address is of the form port (#1234), where the argument to port() is a 16-bit unsigned value that provides the address in the I/O space.

Direct addresses may be any of four different types:

- DP addressing is used to access data pages. The address is calculated as

$$A_{DP} = \text{DPH}[22:15]|(\text{DP} + D_{\text{offset}}). \tag{2.2}$$

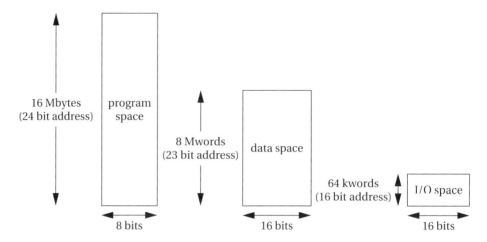

FIGURE 2.18

Address spaces in the TMS320C55x.

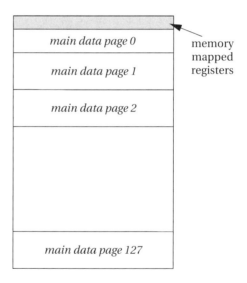

FIGURE 2.19

The C55x memory map.

D_{offset} is calculated by the assembler; its value depends on whether you are accessing a data page value or a memory-mapped register.

- SP addressing is used to access stack values in the data memory. The address is calculated as

$$A_{SP} = \text{SPH}[22:15]|(\text{SP} + S_{\text{offset}}). \tag{2.3}$$

S_{offset} is an offset supplied by the programmer.

■ Register-bit direct addressing accesses bits in registers. The argument @bitoffset is an offset from the least-significant bit of the register. Only a few instructions (register test, set, clear, complement) support this mode.

■ PDP addressing is used to access I/O pages. The 16-bit address is calculated as

$$A_{PDP} = PDP[15:6]\|PDP_{\text{offset}}. \tag{2.4}$$

■ The PDP_{offset} identifies the word within the I/O page. This addressing mode is specified with the port() qualifier.

Indirect addresses may be any of four different types:

■ AR indirect addressing uses an auxiliary register to point to data. This addressing mode is further subdivided into accesses into data, register bits, and I/O. To access a data page, the AR supplies the bottom 16 bits of the address and the top 7 bits are supplied by the top bits of the XAR register. For register bits, the AR supplies a bit number. (As with register-bit direct addressing, this only works on the register bit instructions.) When accessing the I/O space, the AR supplies a 16-bit I/O address. This mode may update the value of the AR register. Updates are specified by modifiers to the register identifier, such as adding + after the register name. Furthermore, the types of modifications allowed depend upon the ARMS bit of status register ST2_55:0 for DSP mode, 1 for control mode. A large number of such updates are possible: examples include *ARn+, which adds 1 to the register for a 16-bit operation and 2 to the register for a 32-bit operation; *(ARn + AR0) writes the value of ARn + AR0 into ARn.

■ Dual AR indirect addressing allows two simultaneous data accesses, either for an instruction that requires two accesses or for executing two instructions in parallel. Depending on the modifiers to the register ID, the register value may be updated.

■ CDP indirect addressing uses the CDP register to access coefficients that may be in data space, register bits, or I/O space. In the case of data space accesses, the top 7 bits of the address come from CDPH and the bottom 16 come from the CDP. For register bits, the CDP provides a bit number. For I/O space accesses specified with port(), the CDP gives a 16 bit I/O address. Depending on the modifiers to the register ID, the CDP register value may be updated.

■ Coefficient indirect addressing is similar to CDP indirect mode, but is used primarily for instructions that require three memory operands per cycle.

Any of the indirect addressing modes may use circular addressing, which is handy for many DSP operations. Circular addressing is specified with the ARnLC bit in status

register ST2_55. For example, if bit AR0LC = 1, then the main data page is supplied by AR0H, the buffer start register is BSA01, and the buffer size register is BK03.

The C55x supports two stacks: one for data and one for the system. Each stack is addressed by a 16-bit address. These two stacks can be relocated to different spots in the memory map by specifying a page using the high register: SP and SPH form XSP, the extended data stack; SSP and SPH form XSSP, the extended system stack. Note that both SP and SSP share the same page register SPH. XSP and XSSP hold 23-bit addresses that correspond to data locations.

The C55x supports three different stack configurations. These configurations depend on how the data and system stacks relate and how subroutine returns are implemented.

- In a dual 16-bit stack with fast return configuration, the data and system stacks are independent. A push or pop on the data stack does not affect the system stack. The RETA and CFCT registers are used to implement fast subroutine returns.

- In a dual 16-bit stack with slow return configuration, the data and system stacks are independent. However, RETA and CFCT are not used for slow subroutine returns; instead, the return address and loop context are stored on the stack.

- In a 32-bit stack with slow return configuration, SP and SSP are both modified by the same amount on any stack operation.

2.3.3 Data Operations

The MOV instruction moves data between registers and memory:

```
MOV src,dst
```

A number of variations of MOV are possible. The instruction can be used to move from memory into a register, from a register to memory, between registers, or from one memory location to another.

The ADD instruction adds a source and destination together and stores the result in the destination:

```
ADD src,dst
```

This instruction produces dst = dst + src. The destination may be an accumulator or another type. Variants allow constants to be added to the destination. Other variants allow the source to be a memory location. The addition may also be performed on two accumulators, one of which has been shifted by a constant number of bits. Other variations are also defined.

A dual addition performs two adds in parallel:

```
ADD dual(Lmem),ACx,ACy
```

This instruction performs HI(ACy) = HI(Lmem) + HI(ACx) and LO(ACy) = LO(Lmem) + LO(ACx). The operation is performed in 40-bit mode, but the lower 16 and upper 24 bits of the result are separated.

The MPY instruction performs an integer multiplication:

```
MPY src,dst
```

Multiplications are performed on 16-bit values. Multiplication may be performed on accumulators, temporary registers, constants, or memory locations. The memory locations may be addressed either directly or using the coefficient addressing mode.

A multiply and accumulate is performed by the MAC instruction. It takes the same basic types of operands as does MPY. In the form

```
MAC ACx,Tx,ACy
```

the instruction performs ACy = ACy + (ACx × Tx).

The compare instruction compares two values and sets a test control flag:

```
CMP Smem == val, TC1
```

The memory location is compared to a constant value. TC1 is set if the two are equal and cleared if they are not equal.

The compare instruction can also be used to compare registers:

```
CMP src RELOP dst, TC1
```

The two registers can be compared using a variety of relational operators RELOP. If the U suffix is used on the instruction, the comparison is performed unsigned.

2.3.4 Flow of Control

The B instruction is an unconditional branch. The branch target may be defined by the low 24 bits of an accumulator

```
B ACx
```

or by an address label

```
B label
```

The BCC instruction is a conditional branch:

```
BCC label, cond
```

The condition code determines the condition to be tested. Condition codes specify registers and the tests to be performed on them:

- Test the value of an accumulator: $<0, <=0, >0, >=0, = 0, !=0$.

- Test the value of the accumulator overflow status bit.

- Test the value of an auxiliary register: $<0, <=0, >0, >=0, = 0, !=0$.

- Test the carry status bit.

- Test the value of a temporary register: $<0, <=0, >0, >=0, = 0, !=0$.

- Test the control flags against 0 (condition prefixed by !) or against 1 (not prefixed by !) for combinations of AND, OR, and NOT.

The C55x allows an instruction or a block of instructions to be repeated. Repeats provide efficient implementation of loops. Repeats may also be nested to provide two levels of repeats.

A single-instruction repeat is controlled by two registers. The single-repeat counter, RPTC, counts the number of additional executions of the instruction to be executed; if RPTC = N, then the instruction is executed a total of N + 1 times. A repeat with a computed number of iterations may be performed using the computed single-repeat register CSR. The desired number of operations is computed and stored in CSR; the value of CSR is then copied into RPTC at the beginning of the repeat.

Block repeats perform a repeat on a block of contiguous instructions. A level 0 block repeat is controlled by three registers: the block repeat counter 0, BRC0, holds the number of times after the initial execution to repeat the instruction; the block repeat start address register 0, RSA0, holds the address of the first instruction in the repeat block; the repeat end address register 0, REA0, holds the address of the last instruction in the repeat block. (Note that, as with a single instruction repeat, if BRCn's value is N, then the instruction or block is executed N + 1 times.)

A level 1 block repeat uses BRC1, RSA1, and REA1. It also uses BRS1, the block repeat save register 1. Each time that the loop repeats, BRC1 is initialized with the value from BRS1. Before the block repeat starts, a load to BRC1 automatically copies the value to BRS1 to be sure that the right value is used for the inner loop executions.

An unconditional subroutine call is performed by the CALL instruction:

```
CALL target
```

The target of the call may be a direct address or an address stored in an accumulator. Subroutines make use of the stack. A subroutine call stores two important registers: the return address and the loop context register. Both these values are pushed onto the stack.

A conditional subroutine call is coded as:

```
CALLCC adrs,cond
```

The address is a direct address; an accumulator value may not be used as the subroutine target. The conditional is the same as with other conditional instructions. As with the unconditional CALL, CALLCC stores the return address and loop context register on the stack.

The C55x provides two types of subroutine returns: *fast-return* and *slow-return*. These vary on where they store the return address and loop context. In a slow return, the return address and loop context are stored on the stack. In a fast return, these two values are stored in registers: the return address register and the control flow context register.

Interrupts use the basic subroutine call mechanism. They are processed in four phases:

1. The interrupt request is received.

2. The interrupt request is acknowledged.

3. Prepare for the interrupt service routine by finishing execution of the current instruction, storing registers, and retrieving the interrupt vector.

4. Processing the interrupt service routine, which concludes with a return-from-interrupt instruction.

The C55x supports 32 interrupt vectors.

Interrupts may be prioritized into 27 levels. The highest-priority interrupt is a hardware and software reset.

Most of the interrupts may be masked using the interrupt flag registers IFR1 and IFR2. Interrupt vectors 2–23, the bus error interrupt, the data log interrupt, and the real-time operating system interrupt can all be masked.

2.3.5 C Coding Guidelines

Some coding guidelines for the C55x [Tex01] not only provide more efficient code but in some cases should be paid attention to in order to ensure that the generated code is correct.

As with all digital signal processing code, the C55x benefits from careful attention to the required sizes of variables. The C55x compiler uses some non-standard lengths of data types: char, short, and int are all 16 bits; long is 32 bits; and long long is 40 bits. The C55x uses IEEE formats for float (32 bits) and double (64 bits). C code should not assume that int and long are the same types, that char is 8 bits long or that long is 64 bits. The int type should be used for fixed-point arithmetic, especially multiplications, and for loop counters.

The C55x compiler makes some important assumptions about operands of multiplications. This code generates a 32-bit result from the multiplication of two 16-bit operands:

```
long result = (long)(int)src1 * (long)(int)src2;
```

Although the operands were coerced to long, the compiler notes that each is 16 bits, so it uses a single-instruction multiplication.

The order of instructions in the compiled code depends in part on the C55x pipeline characteristics. The C compiler schedules code to minimize code conflicts

and to take advantage of parallelism wherever possible. However, if the compiler cannot determine that a set of instructions are independent, it must assume that they are dependent and generate more restrictive, slower code. The restrict keyword can be used to tell the compiler that a given pointer is the only one in the scope that can point to a particular object. The -pm option allows the compiler to perform more global analysis and find more independent sets of instructions.

SUMMARY

When viewed from high above, all CPUs are similar—they read and write memory, perform data operations, and make decisions. However, there are many ways to design an instruction set, as illustrated by the differences between the ARM and the C55x. When designing complex systems, we generally view the programs in high-level language form, which hides many of the details of the instruction set. However, differences in instruction sets can be reflected in nonfunctional characteristics, such as program size and speed.

What We Learned

- Both the von Neumann and Harvard architectures are in common use today.

- The programming model is a description of the architecture relevant to instruction operation.

- ARM is a load-store architecture. It provides a few relatively complex instructions, such as saving and restoring multiple registers.

- The C55x provides a number of architectural features to support the arithmetic loops that are common on digital signal processing code.

FURTHER READING

Books by Jaggar [Jag95] and Furber [Fur96] describe the ARM architecture. The ARM Web site, www.arm.com, contains a large number of documents describing various versions of ARM.

QUESTIONS

Q2-1 What is the difference between a big-endian and little-endian data representation?

Q2-2 What is the difference between the Harvard and von Neumann architectures?

Q2-3 Answer the following questions about the ARM programming model:

 a. How many general-purpose registers are there?

 b. What is the purpose of the CPSR?

 c. What is the purpose of the Z bit?

 d. Where is the program counter kept?

Q2-4 How would the ARM status word be set after these operations?

 a. $2 - 3$

 b. $-2^{32} + 1 - 1$

 c. $-4 + 5$

Q2-5 Write ARM assembly code to implement the following C assignments:

 a. $x = a + b;$

 b. $y = (c - d) + (e - f);$

 c. $z = a^*(b + c) - d^*e;$

Q2-6 What is the meaning of these ARM condition codes?

 a. EQ

 b. NE

 c. MI

 d. VS

 e. GE

 f. LT

Q2-7 Write ARM assembly code to first read and then write a device memory mapped to location 0x2100.

Q2-8 Write in ARM assembly language an interrupt handler that reads a single character from the device at location 0x2200.

Q2-9 Write ARM assembly code to implement the following C conditional:

```
if (x - y < 3) {
    a = b - c;
    x = 0;
    }
else {
    y = 0;
    d = e + f + g;
    }
```

Q2-10 Write ARM assembly language code for the following loops:

 a.
```
for (i = 0; i < 20; i++)
      z[i] = a[i]*b[i];
```

 b.
```
for (i = 0; i < 10; i++)
      for (j = 0; j < 10; j++)
            z[i] = a[i,j] * b[i]
```

Q2-11 Explain the operation of the BL instruction, including the state of ARM registers before and after its operation.

Q2-12 How do you return from an ARM procedure?

Q2-13 In the following code, show the contents of the ARM function call stack just after each C function has been entered and just after the function exits. Assume that the function call stack is empty when main() begins.

```
int foo(int x1, int x2) {
      return x1 + x2;
}

int baz(int x1) {
    return x1 + 1;
}

void scum(int r) {
    for (i = 0; i = 2; i++)
          foo(r + i,5);
}

main() {
    scum(3);
    baz(2);
}
```

Q2-14 What data types does the C55x support?

Q2-15 How many accumulators does the C55x have?

Q2-16 What C55x register holds arithmetic and bit manipulation flags?

Q2-17 What is a block repeat in the C55x?

Q2-18 How are the C55x data and program memory arranged in the physical memory?

Q2-19 Where are C55x memory-mapped registers located in the address space?

Q2-20 What is the AR register used for in the C55x?

Q2-21 What is the difference between DP and PDP addressing modes in the C55x?

Q2-22 How many stacks are supported by the C55x architecture and how are their locations in memory determined?

Q2-23 What register controls single-instruction repeats in the C55x?

Q2-24 What is the difference between slow and fast returns in the C55x?

LAB EXERCISES

L2-1 Write a program that uses a circular buffer to perform FIR filtering.

L2-2 Write a simple loop that lets you exercise the cache. By changing the number of statements in the loop body, you can vary the cache hit rate of the loop as it executes. You should be able to observe changes in the speed of execution by observing the microprocessor bus.

CPUs

- Input and output mechanisms.

- Supervisor mode, exceptions, and traps.

- Memory management and address translation.

- Caches.

- Performance and power consumption of CPUs.

3

INTRODUCTION

This chapter describes aspects of CPUs that do not directly relate to their instruction sets. We consider a number of mechanisms that are important to interfacing to other system elements, such as interrupts and memory management. We also take a first look at aspects of the CPU other than functionality—performance and power consumption are both very important attributes of programs that are only indirectly related to the instructions they use.

In Section 3.1, we study input and output mechanisms such as interrupts. Section 3.2 introduces several mechanisms that are similar to interrupts but are designed to handle internal events. Section 3.3 introduces co-processors that provide optional support for parts of the instruction set. Section 3.4 describes memory systems—both memory management and caches. The next sections look at nonfunctional attributes of execution: Section 3.5 looks at performance, while Section 3.6 considers power consumption. Finally, in Section 3.7 we use a data compressor as an example of a simple yet interesting program.

3.1 PROGRAMMING INPUT AND OUTPUT

The basic techniques for I/O programming can be understood relatively independent of the instruction set. In this section, we cover the basics of I/O programming and place them in the contexts of both the ARM and C55x. We begin by discussing the basic characteristics of I/O devices so that we can understand the requirements they place on programs that communicate with them.

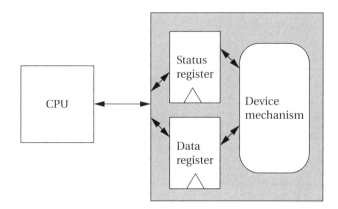

FIGURE 3.1

Structure of a typical I/O device.

3.1.1 Input and Output Devices

Input and output devices usually have some analog or nonelectronic component—for instance, a disk drive has a rotating disk and analog read/write electronics. But the digital logic in the device that is most closely connected to the CPU very strongly resembles the logic you would expect in any computer system.

Figure 3.1 shows the structure of a typical I/O device and its relationship to the CPU. The interface between the CPU and the device's internals (e.g., the rotating disk and read/write electronics in a disk drive) is a set of registers. The CPU talks to the device by reading and writing the registers. Devices typically have several registers:

- *Data registers* hold values that are treated as data by the device, such as the data read or written by a disk.

- *Status registers* provide information about the device's operation, such as whether the current transaction has completed.

Some registers may be read-only, such as a status register that indicates when the device is done, while others may be readable or writable. Application Example 3.1 describes a classic I/O device.

Application Example 3.1

The 8251 UART

The 8251 **UART** (Universal Asynchronous Receiver/Transmitter) [Int82] is the original device used for serial communications, such as the serial port connections on PCs. The 8251 was introduced as a stand-alone integrated circuit for early microprocessors. Today, its functions are typically subsumed by a larger chip, but these more advanced devices still use the basic programming interface defined by the 8251.

The UART is programmable for a variety of transmission and reception parameters. However, the basic format of transmission is simple. Data are transmitted as streams of characters, each of which has the following form:

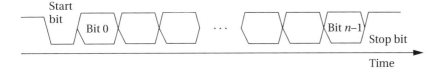

Every character starts with a start bit (a 0) and a stop bit (a 1). The start bit allows the receiver to recognize the start of a new character; the stop bit ensures that there will be a transition at the start of the stop bit. The data bits are sent as high and low voltages at a uniform rate. That rate is known as the **baud rate**; the period of one bit is the inverse of the baud rate.

Before transmitting or receiving data, the CPU must set the UART's mode registers to correspond to the data line's characteristics. The parameters for the serial port are familiar from the parameters for a serial communications program (such as Kermit):

- the baud rate;

- the number of bits per character (5 through 8);

- whether parity is to be included and whether it is even or odd; and

- the length of a stop bit (1, 1.5, or 2 bits).

The UART includes one 8-bit register that buffers characters between the UART and the CPU bus. The Transmitter Ready output indicates that the transmitter is ready to accept a data character; the Transmitter Empty signal goes high when the UART has no characters to send. On the receiver side, the Receiver Ready pin goes high when the UART has a character ready to be read by the CPU.

3.1.2 Input and Output Primitives

Microprocessors can provide programming support for input and output in two ways: *I/O instructions* and *memory-mapped I/O*. Some architectures, such as the Intel x86, provide special instructions (in and out in the case of the Intel x86) for input and output. These instructions provide a separate address space for I/O devices.

But the most common way to implement I/O is by memory mapping—even CPUs that provide I/O instructions can also implement memory-mapped I/O. As the name implies, memory-mapped I/O provides addresses for the registers in each I/O device. Programs use the CPU's normal read and write instructions to communicate with the devices. Example 3.1 illustrates memory-mapped I/O on the ARM.

Example 3.1

Memory-mapped I/O on ARM

We can use the EQU pseudo-op to define a symbolic name for the memory location of our I/O device:

```
DEV1 EQU 0x1000
```

Given that name, we can use the following standard code to read and write the device register:

```
LDR r1,#DEV1 ; set up device address
LDR r0,[r1]  ; read DEV1
LDR r0,#8    ; set up value to write
STR r0,[r1]  ; write 8 to device
```

How can we directly write I/O devices in a high-level language like C? When we define and use a variable in C, the compiler hides the variable's address from us. But we can use pointers to manipulate addresses of I/O devices. The traditional names for functions that read and write arbitrary memory locations are *peek* and *poke*. The peek function can be written in C as:

```
int peek(char *location) {
       return *location; /* de-reference location pointer */
}
```

The argument to peek is a pointer that is de-referenced by the C * operator to read the location. Thus, to read a device register we can write:

```
#define DEV1 0x1000
...
dev_status = peek(DEV1); /* read device register */
```

The poke function can be implemented as:

```
void poke(char *location, char newval) {
      (*location) = newval; /* write to location */
}
```

To write to the status register, we can use the following code:

```
poke(DEV1,8); /* write 8 to device register */
```

These functions can, of course, be used to read and write arbitrary memory locations, not just devices.

3.1.3 Busy-Wait I/O

The most basic way to use devices in a program is ***busy-wait I/O***. Devices are typically slower than the CPU and may require many cycles to complete an operation. If the CPU is performing multiple operations on a single device, such as writing several characters to an output device, then it must wait for one operation to complete before starting the next one. (If we try to start writing the second character before the device has finished with the first one, for example, the device will probably never print the first character.) Asking an I/O device whether it is finished by reading its status register is often called ***polling***.

Example 3.2 illustrates busy-wait I/O.

Example 3.2

Busy-wait I/O programming

In this example we want to write a sequence of characters to an output device. The device has two registers: one for the character to be written and a status register. The status register's value is 1 when the device is busy writing and 0 when the write transaction has completed.

We will use the peek and poke functions to write the busy-wait routine in C. First, we define symbolic names for the register addresses:

```
#define OUT_CHAR 0x1000 /* output device character register */
#define OUT_STATUS 0x1001 /* output device status register */
```

The sequence of characters is stored in a standard C string, which is terminated by a null (0) character. We can use peek and poke to send the characters and wait for each transaction to complete:

```
char *mystring = "Hello, world." /* string to write */
char *current_char; /* pointer to current position in
                       string  */
current_char = mystring; /* point to head of string */
while (*current_char != `\ 0') { /* until null character */
      poke(OUT_CHAR,*current_char); /* send character to
                                       device */
      while (peek(OUT_STATUS) != 0); /* keep checking
                                        status */
      current_char++; /* update character pointer */
}
```

The outer while loop sends the characters one at a time. The inner while loop checks the device status—it implements the busy-wait function by repeatedly checking the device status until the status changes to 0.

Example 3.3 illustrates a combination of input and output.

Example 3.3

Copying characters from input to output using busy-wait I/O

We want to repeatedly read a character from the input device and write it to the output device. First, we need to define the addresses for the device registers:

```
#define IN_DATA 0x1000
#define IN_STATUS 0x1001
#define OUT_DATA 0x1100
#define OUT_STATUS 0x1101
```

The input device sets its status register to 1 when a new character has been read; we must set the status register back to 0 after the character has been read so that the device is ready to read another character. When writing, we must set the output status register to 1 to start writing and wait for it to return to 0. We can use peek and poke to repeatedly perform the read/write operation:

```
while (TRUE) { /* perform operation forever */
      /* read a character into achar */
      while (peek(IN_STATUS) == 0); /* wait until ready */
      achar = (char)peek(IN_DATA); /* read the character */
      /* write achar */
      poke(OUT_DATA,achar);
      poke(OUT_STATUS,1); /* turn on device */
      while (peek(OUT_STATUS) != 0); /* wait until done */
}
```

3.1.4 Interrupts

Basics

Busy-wait I/O is extremely inefficient—the CPU does nothing but test the device status while the I/O transaction is in progress. In many cases, the CPU could do useful work in parallel with the I/O transaction, such as:

- computation, as in determining the next output to send to the device or processing the last input received, and

- control of other I/O devices.

To allow parallelism, we need to introduce new mechanisms into the CPU.

The ***interrupt*** mechanism allows devices to signal the CPU and to force execution of a particular piece of code. When an interrupt occurs, the program counter's value is changed to point to an ***interrupt handler*** routine (also commonly known

as a **device driver**) that takes care of the device: writing the next data, reading data that have just become ready, and so on. The interrupt mechanism of course saves the value of the PC at the interruption so that the CPU can return to the program that was interrupted. Interrupts therefore allow the flow of control in the CPU to change easily between different **contexts,** such as a foreground computation and multiple I/O devices.

As shown in Figure 3.2, the interface between the CPU and I/O device includes the following signals for interrupting:

- the I/O device asserts the **interrupt request** signal when it wants service from the CPU; and

- the CPU asserts the **interrupt acknowledge** signal when it is ready to handle the I/O device's request.

The I/O device's logic decides when to interrupt; for example, it may generate an interrupt when its status register goes into the ready state. The CPU may not be able to immediately service an interrupt request because it may be doing something else that must be finished first—for example, a program that talks to both a high-speed disk drive and a low-speed keyboard should be designed to finish a disk transaction before handling a keyboard interrupt. Only when the CPU decides to acknowledge the interrupt does the CPU change the program counter to point to the device's handler. The interrupt handler operates much like a subroutine, except that it is not called by the executing program. The program that runs when no interrupt is being handled is often called the **foreground program**; when the interrupt handler finishes, it returns to the foreground program, wherever processing was interrupted.

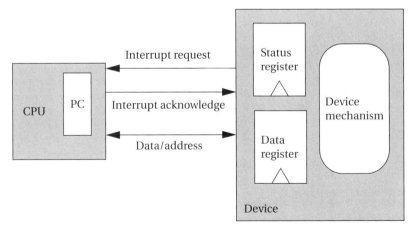

FIGURE 3.2

The interrupt mechanism.

Before considering the details of how interrupts are implemented, let's look at the interrupt style of processing and compare it to busy-wait I/O. Example 3.4 uses interrupts as a basic replacement for busy-wait I/O; Example 3.5 takes a more sophisticated approach that allows more processing to happen concurrently.

Example 3.4

Copying characters from input to output with basic interrupts

As with Example 3.3, we repeatedly read a character from an input device and write it to an output device. We assume that we can write C functions that act as interrupt handlers. Those handlers will work with the devices in much the same way as in busy-wait I/O by reading and writing status and data registers. The main difference is in handling the output—the interrupt signals that the character is done, so the handler does not have to do anything.

We will use a global variable achar for the input handler to pass the character to the foreground program. Because the foreground program doesn't know when an interrupt occurs, we also use a global Boolean variable, gotchar, to signal when a new character has been received. The code for the input and output handlers follows:

```
void input_handler() { /* get a character and put in
                               global */
        achar = peek(IN_DATA); /* get character */
        gotchar = TRUE; /* signal to main program */
        poke(IN_STATUS,0); /* reset status to initiate next
                               transfer */
}
void output_handler() { /* react to character being sent */
        /* don't have to do anything */
}
```

The main program is reminiscent of the busy-wait program. It looks at gotchar to check when a new character has been read and then immediately sends it out to the output device.

```
main() {
        while (TRUE) { /* read then write forever */
                if (gotchar) { /* write a character */
                        poke(OUT_DATA,achar); /* put character
                                                  in device */
                        poke(OUT_STATUS,1); /* set status to
                                                  initiate write */
                        gotchar = FALSE; /* reset flag */
                }
        }
}
```

The use of interrupts has made the main program somewhat simpler. But this program design still does not let the foreground program do useful work. Example 3.5 uses a more sophisticated program design to let the foreground program work completely independently of input and output.

Example 3.5

Copying characters from input to output with interrupts and buffers

Because we do not need to wait for each character, we can make this I/O program more sophisticated than the one in Example 3.4. Rather than reading a single character and then writing it, the program performs reads and writes independently. The read and write routines communicate through the following global variables:

- A character string io_buf will hold a queue of characters that have been read but not yet written.

- A pair of integers buf_start and buf_end will point to the first and last characters read.

- An integer error will be set to 0 whenever io_buf overflows.

The global variables allow the input and output devices to run at different rates. The queue io_buf acts as a wraparound buffer—we add characters to the tail when an input is received and take characters from the tail when we are ready for output. The head and tail wrap around the end of the buffer array to make most efficient use of the array. Here is the situation at the start of the program's execution, where the tail points to the first available character and the head points to the ready character. As seen below, because the head and tail are equal, we know that the queue is empty.

When the first character is read, the tail is incremented after the character is added to the queue, leaving the buffer and pointers looking like the following:

When the buffer is full, we leave one character in the buffer unused. As the next figure shows, if we added another character and updated the tail buffer (wrapping it around to the head of the buffer), we would be unable to distinguish a full buffer from an empty one.

Here is what happens when the output goes past the end of io_buf.

The following code provides the declarations for the above global variables and some service routines for adding and removing characters from the queue. Because interrupt handlers are regular code, we can use subroutines to structure code just as with any program.

```
#define BUF_SIZE 8
char io_buf[BUF_SIZE]; /* character buffer */
int buf_head = 0, buf_tail = 0; /* current position in
                                   buffer */
int error = 0; /* set to 1 if buffer ever overflows */

void empty_buffer() { /* returns TRUE if buffer is empty */
      buf_head == buf_tail;
}

void full_buffer() { /* returns TRUE if buffer is full */
      (buf_tail+1) % BUF_SIZE == buf_head ;
}

int nchars() { /* returns the number of characters in the
                 buffer */
         if (buf_head >= buf_tail) return buf_tail - buf_head;
         else return BUF_SIZE + buf_tail - buf_head;
}

void add_char(char achar) { /* add a character to the buffer
                               head */
```

```
        io_buf[buf_tail++] = achar;
        /* check pointer */
        if (buf_tail == BUF_SIZE)
                buf_tail = 0;
}

char remove_char() { /* take a character from the buffer
                        head */
        char achar;
        achar = io_buf[buf_head++];
        /* check pointer */
        if (buf_head == BUF_SIZE)
                buf_head = 0;
}
```

Assume that we have two interrupt handling routines defined in C, input_handler for the input device and output_handler for the output device. These routines work with the device in much the same way as did the busy-wait routines. The only complication is in starting the output device: If io_buf has characters waiting, the output driver can start a new output transaction by itself. But if there are no characters waiting, an outside agent must start a new output action whenever the new character arrives. Rather than force the foreground program to look at the character buffer, we will have the input handler check to see whether there is only one character in the buffer and start a new transaction.

Here is the code for the input handler:

```
#define IN_DATA 0x1000
#define IN_STATUS 0x1001
void input_handler() {
        char achar;
        if (full_buffer()) /* error */
                error = 1;
        else { /* read the character and update pointer */
                achar = peek(IN_DATA); /* read character */
                add_char(achar); /* add to queue */
        }
        poke(IN_STATUS,0); /* set status register back to 0 */
        /* if buffer was empty, start a new output
           transaction */
        if (nchars() == 1) { /* buffer had been empty until
                                this interrupt */
                poke(OUT_DATA,remove_char()); /* send
                                                character */
                poke(OUT_STATUS,1); /* turn device on */
        }
}
```

```
#define OUT_DATA 0x1100
#define OUT_STATUS 0x1101
void output_handler() {
      if (!empty_buffer()) { /* start a new character */
            poke(OUT_DATA,remove_char()); /* send character */
            poke(OUT_STATUS,1); /* turn device on */
      }
}
```

The foreground program does not need to do anything—everything is taken care of by the interrupt handlers. The foreground program is free to do useful work as it is occasionally interrupted by input and output operations. The following sample execution of the program in the form of a UML sequence diagram shows how input and output are interleaved with the foreground program. (We have kept the last input character in the queue until output is complete to make it clearer when input occurs.) The simulation shows that the foreground program is not executing continuously, but it continues to run in its regular state independent of the number of characters waiting in the queue.

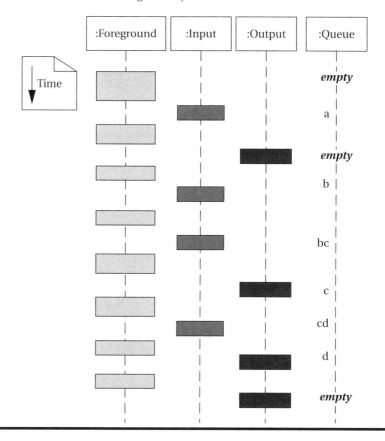

Interrupts allow a lot of concurrency, which can make very efficient use of the CPU. But when the interrupt handlers are buggy, the errors can be very hard to find. The fact that an interrupt can occur at any time means that the same bug can manifest itself in different ways when the interrupt handler interrupts different segments of the foreground program. Example 3.6 illustrates the problems inherent in debugging interrupt handlers.

Example 3.6

Debugging interrupt code

Assume that the foreground code is performing a matrix multiplication operation $y = Ax + b$:

```
for (i = 0; i < M; i++) {
        y[i] = b[i];
        for (j = 0; j < N; j++)
                y[i] = y[i] + A[i,j]*x[j];
        }
```

We use the interrupt handlers of Example 3.5 to perform I/O while the matrix computation is performed, but with one small change: read_handler has a bug that causes it to change the value of j. While this may seem far-fetched, remember that when the interrupt handler is written in assembly language such bugs are easy to introduce. Any CPU register that is written by the interrupt handler must be saved before it is modified and restored before the handler exits. Any type of bug—such as forgetting to save the register or to properly restore it—can cause that register to mysteriously change value in the foreground program.

What happens to the foreground program when j changes value during an interrupt depends on when the interrupt handler executes. Because the value of j is reset at each iteration of the outer loop, the bug will affect only one entry of the result y. But clearly the entry that changes will depend on when the interrupt occurs. Furthermore, the change observed in y depends on not only what new value is assigned to j (which may depend on the data handled by the interrupt code), but also when in the inner loop the interrupt occurs. An interrupt at the beginning of the inner loop will give a different result than one that occurs near the end. The number of possible new values for the result vector is much too large to consider manually—the bug cannot be found by enumerating the possible wrong values and correlating them with a given root cause. Even recognizing the error can be difficult—for example, an interrupt that occurs at the very end of the inner loop will not cause any change in the foreground program's result. Finding such bugs generally requires a great deal of tedious experimentation and frustration.

The CPU implements interrupts by checking the interrupt request line at the beginning of execution of every instruction. If an interrupt request has been asserted, the CPU does not fetch the instruction pointed to by the PC. Instead the CPU sets the PC to a predefined location, which is the beginning of the interrupt

handling routine. The starting address of the interrupt handler is usually given as a pointer—rather than defining a fixed location for the handler, the CPU defines a location in memory that holds the address of the handler, which can then reside anywhere in memory.

Because the CPU checks for interrupts at every instruction, it can respond quickly to service requests from devices. However, the interrupt handler must return to the foreground program without disturbing the foreground program's operation. Since subroutines perform a similar function, it is natural to build the CPU's interrupt mechanism to resemble its subroutine function. Most CPUs use the same basic mechanism for remembering the foreground program's PC as is used for subroutines. The subroutine call mechanism in modern microprocessors is typically a stack, so the interrupt mechanism puts the return address on a stack; some CPUs use the same stack as for subroutines while others define a special stack. The use of a procedure-like interface also makes it easier to provide a high-level language interface for interrupt handlers. The details of the C interface to interrupt handling routines vary both with the CPU and the underlying support software.

Priorities and Vectors

Providing a practical interrupt system requires having more than a simple interrupt request line. Most systems have more than one I/O device, so there must be some mechanism for allowing multiple devices to interrupt. We also want to have flexibility in the locations of the interrupt handling routines, the addresses for devices, and so on. There are two ways in which interrupts can be generalized to handle multiple devices and to provide more flexible definitions for the associated hardware and software:

- *interrupt priorities* allow the CPU to recognize some interrupts as more important than others, and

- *interrupt vectors* allow the interrupting device to specify its handler.

Prioritized interrupts not only allow multiple devices to be connected to the interrupt line but also allow the CPU to ignore less important interrupt requests while it handles more important requests. As shown in Figure 3.3, the CPU provides several different interrupt request signals, shown here as $L1$, $L2$, up to Ln. Typically, the lower-numbered interrupt lines are given higher priority, so in this case, if devices 1, 2, and n all requested interrupts simultaneously, 1's request would be acknowledged because it is connected to the highest-priority interrupt line. Rather than provide a separate interrupt acknowledge line for each device, most CPUs use a set of signals that provide the priority number of the winning interrupt in binary form (so that interrupt level 7 requires 3 bits rather than 7). A device knows that its interrupt request was accepted by seeing its own priority number on the interrupt acknowledge lines.

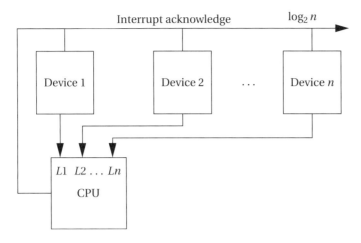

FIGURE 3.3

Prioritized device interrupts.

How do we change the priority of a device? Simply by connecting it to a different interrupt request line. This requires hardware modification, so if priorities need to be changeable, removable cards, programmable switches, or some other mechanism should be provided to make the change easy.

The priority mechanism must ensure that a lower-priority interrupt does not occur when a higher-priority interrupt is being handled. The decision process is known as ***masking***. When an interrupt is acknowledged, the CPU stores in an internal register the priority level of that interrupt. When a subsequent interrupt is received, its priority is checked against the priority register; the new request is acknowledged only if it has higher priority than the currently pending interrupt. When the interrupt handler exits, the priority register must be reset. The need to reset the priority register is one reason why most architectures introduce a specialized instruction to return from interrupts rather than using the standard subroutine return instruction.

The highest-priority interrupt is normally called the ***nonmaskable interrupt (NMI)***. The NMI cannot be turned off and is usually reserved for interrupts caused by power failures—a simple circuit can be used to detect a dangerously low power supply, and the NMI interrupt handler can be used to save critical state in nonvolatile memory, turn off I/O devices to eliminate spurious device operation during power-down, and so on.

Most CPUs provide a relatively small number of interrupt priority levels, such as eight. While more priority levels can be added with external logic, they may not be necessary in all cases. When several devices naturally assume the same priority (such as when you have several identical keypads attached to a single CPU), you can combine polling with prioritized interrupts to efficiently handle the devices.

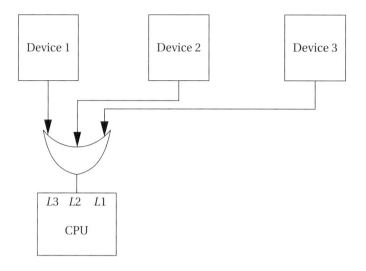

FIGURE 3.4

Using polling to share an interrupt over several devices.

As shown in Figure 3.4, you can use a small amount of logic external to the CPU to generate an interrupt whenever any of the devices you want to group together request service. The CPU will call the interrupt handler associated with this priority; that handler does not know which of the devices actually requested the interrupt. The handler uses software polling to check the status of each device: In this example, it would read the status registers of 1, 2, and 3 to see which of them is ready and requesting service.

Example 3.7 illustrates how priorities affect the order in which I/O requests are handled.

Example 3.7

I/O with prioritized interrupts

Assume that we have devices A, B, and C. A has priority 1 (highest priority), B priority 2, and C priority 3. The following UML sequence diagram shows which interrupt handler is executing as a function of time for a sequence of interrupt requests.

In each case, an interrupt handler keeps running until either it is finished or a higher-priority interrupt arrives. The C interrupt, although it arrives early, does not finish for a long time because interrupts from both A and B intervene—system design must take into account the worst-case combinations of interrupts that can occur to ensure that no device goes without service for too long. When both A and B interrupt simultaneously, A's interrupt gets priority; when A's handler is finished, the priority mechanism automatically answers B's pending interrupt.

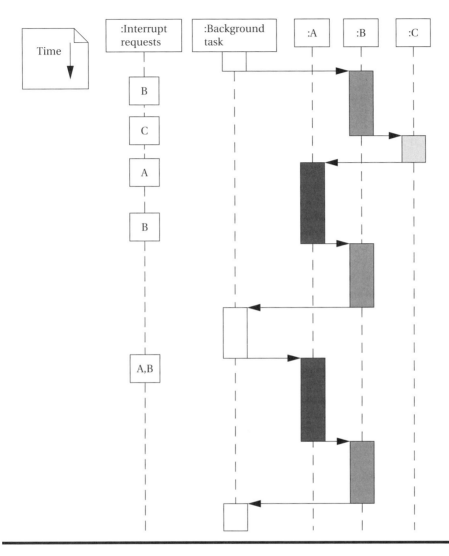

Vectors provide flexibility in a different dimension, namely, the ability to define the interrupt handler that should service a request from a device. Figure 3.5 shows the hardware structure required to support interrupt vectors. In addition to the interrupt request and acknowledge lines, additional interrupt vector lines run from the devices to the CPU. After a device's request is acknowledged, it sends its interrupt vector over those lines to the CPU. The CPU then uses the vector number as an index in a table stored in memory as shown in Figure 3.5. The location referenced in the interrupt vector table by the vector number gives the address of the handler.

There are two important things to notice about the interrupt vector mechanism. First, the device, not the CPU, stores its vector number. In this way, a device

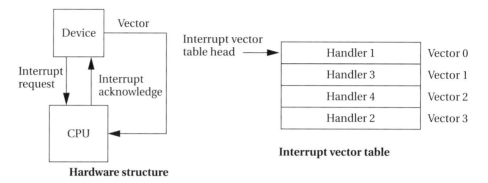

FIGURE 3.5

Interrupt vectors.

can be given a new handler simply by changing the vector number it sends, without modifying the system software. For example, vector numbers can be changed by programmable switches. The second thing to notice is that there is no fixed relationship between vector numbers and interrupt handlers. The interrupt vector table allows arbitrary relationships between devices and handlers. The vector mechanism provides great flexibility in the coupling of hardware devices and the software routines that service them.

Most modern CPUs implement both prioritized and vectored interrupts. Priorities determine which device is serviced first, and vectors determine what routine is used to service the interrupt. The combination of the two provides a rich interface between hardware and software.

Interrupt overhead Now that we have a basic understanding of the interrupt mechanism, we can consider the complete interrupt handling process. Once a device requests an interrupt, some steps are performed by the CPU, some by the device, and others by software. Here are the major steps in the process:

1. ***CPU*** The CPU checks for pending interrupts at the beginning of an instruction. It answers the highest-priority interrupt, which has a higher priority than that given in the interrupt priority register.

2. ***Device*** The device receives the acknowledgment and sends the CPU its interrupt vector.

3. ***CPU*** The CPU looks up the device handler address in the interrupt vector table using the vector as an index. A subroutine-like mechanism is used to save the current value of the PC and possibly other internal CPU state, such as general-purpose registers.

4. ***Software*** The device driver may save additional CPU state. It then performs the required operations on the device. It then restores any saved state and executes the interrupt return instruction.

5. ***CPU*** The interrupt return instruction restores the PC and other automatically saved states to return execution to the code that was interrupted.

Interrupts do not come without a performance penalty. In addition to the execution time required for the code that talks directly to the devices, there is execution time overhead associated with the interrupt mechanisms.

■ The interrupt itself has overhead similar to a subroutine call. Because an interrupt causes a change in the program counter, it incurs a branch penalty. In addition, if the interrupt automatically stores CPU registers, that action requires extra cycles, even if the state is not modified by the interrupt handler.

■ In addition to the branch delay penalty, the interrupt requires extra cycles to acknowledge the interrupt and obtain the vector from the device.

■ The interrupt handler will, in general, save and restore CPU registers that were not automatically saved by the interrupt.

■ The interrupt return instruction incurs a branch penalty as well as the time required to restore the automatically saved state.

The time required for the hardware to respond to the interrupt, obtain the vector, and so on cannot be changed by the programmer. In particular, CPUs vary quite a bit in the amount of internal state automatically saved by an interrupt. The programmer does have control over what state is modified by the interrupt handler and therefore it must be saved and restored. Careful programming can sometimes result in a small number of registers used by an interrupt handler, thereby saving time in maintaining the CPU state. However, such tricks usually require coding the interrupt handler in assembly language rather than a high-level language.

Interrupts in ARM ARM7 supports two types of interrupts: fast interrupt requests (FIQs) and interrupt requests (IRQs). An FIQ takes priority over an IRQ. The interrupt table is always kept in the bottom memory addresses, starting at location 0. The entries in the table typically contain subroutine calls to the appropriate handler.

The ARM7 performs the following steps when responding to an interrupt [ARM99B]:

■ saves the appropriate value of the PC to be used to return,

■ copies the CPSR into a saved program status register (SPSR),

■ forces bits in the CPSR to note the interrupt, and

■ forces the PC to the appropriate interrupt vector.

When leaving the interrupt handler, the handler should:

■ restore the proper PC value,

■ restore the CPSR from the SPSR, and

■ clear interrupt disable flags.

The worst-case latency to respond to an interrupt includes the following components:

- two cycles to synchronize the external request,

- up to 20 cycles to complete the current instruction,

- three cycles for data abort, and

- two cycles to enter the interrupt handling state.

This adds up to 27 clock cycles. The best-case latency is four clock cycles.

Interrupts in C55x Interrupts in the C55x [Tcx04] never take less than seven clock cycles. In many situations, they take 13 clock cycles.

A maskable interrupt is processed in several steps once the interrupt request is sent to the CPU:

- The interrupt flag register (IFR) corresponding to the interrupt is set.

- The interrupt enable register (IER) is checked to ensure that the interrupt is enabled.

- The interrupt mask register (INTM) is checked to be sure that the interrupt is not masked.

- The interrupt flag register (IFR) corresponding to the flag is cleared.

- Appropriate registers are saved as context.

- INTM is set to 1 to disable maskable interrupts.

- DGBM is set to 1 to disable debug events.

- EALLOW is set to 0 to disable access to non-CPU emulation registers.

- A branch is performed to the interrupt service routine (ISR).

The C55x provides two mechanisms—*fast-return* and *slow-return*—to save and restore registers for interrupts and other context switches. Both processes save the return address and loop context registers. The fast-return mode uses RETA to save the return address and CFCT for the loop context bits. The slow-return mode, in contrast, saves the return address and loop context bits on the stack.

3.2 SUPERVISOR MODE, EXCEPTIONS, AND TRAPS

In this section, we consider exceptions and traps. These are mechanisms to handle internal conditions, and they are very similar to interrupts in form. We begin with a discussion of supervisor mode, which some processors use to handle exceptional events and protect executing programs from each other.

3.2.1 Supervisor Mode

As will become clearer in later chapters, complex systems are often implemented as several programs that communicate with each other. These programs may run under the command of an operating system. It may be desirable to provide hardware checks to ensure that the programs do not interfere with each other—for example, by erroneously writing into a segment of memory used by another program. Software debugging is important but can leave some problems in a running system; hardware checks ensure an additional level of safety.

In such cases it is often useful to have a ***supervisor mode*** provided by the CPU. Normal programs run in ***user mode***. The supervisor mode has privileges that user modes do not. For example, we study memory management systems in Section 3.4.2 that allow the addresses of memory locations to be changed dynamically. Control of the memory management unit (MMU) is typically reserved for supervisor mode to avoid the obvious problems that could occur when program bugs cause inadvertent changes in the memory management registers.

Not all CPUs have supervisor modes. Many DSPs, including the C55x, do not provide supervisor modes. The ARM, however, does have such a mode. The ARM instruction that puts the CPU in supervisor mode is called SWI:

```
SWI CODE_1
```

It can, of course, be executed conditionally, as with any ARM instruction. SWI causes the CPU to go into supervisor mode and sets the PC to 0x08. The argument to SWI is a 24-bit immediate value that is passed on to the supervisor mode code; it allows the program to request various services from the supervisor mode.

In supervisor mode, the bottom 5 bits of the CPSR are all set to 1 to indicate that the CPU is in supervisor mode. The old value of the CPSR just before the SWI is stored in a register called the ***saved program status register (SPSR)***. There are in fact several SPSRs for different modes; the supervisor mode SPSR is referred to as SPSR_svc.

To return from supervisor mode, the supervisor restores the PC from register r14 and restores the CPSR from the SPSR_svc.

3.2.2 Exceptions

An ***exception*** is an internally detected error. A simple example is division by zero. One way to handle this problem would be to check every divisor before division to be sure it is not zero, but this would both substantially increase the size of numerical programs and cost a great deal of CPU time evaluating the divisor's value. The CPU can more efficiently check the divisor's value during execution. Since the time at which a zero divisor will be found is not known in advance, this event is similar to an interrupt except that it is generated inside the CPU. The exception mechanism provides a way for the program to react to such unexpected events.

Just as interrupts can be seen as an extension of the subroutine mechanism, exceptions are generally implemented as a variation of an interrupt. Since both deal

with changes in the flow of control of a program, it makes sense to use similar mechanisms. However, exceptions are generated internally.

Exceptions in general require both prioritization and vectoring. Exceptions must be prioritized because a single operation may generate more than one exception—for example, an illegal operand and an illegal memory access. The priority of exceptions is usually fixed by the CPU architecture. Vectoring provides a way for the user to specify the handler for the exception condition. The vector number for an exception is usually predefined by the architecture; it is used to index into a table of exception handlers.

3.2.3 Traps

A *trap*, also known as a ***software interrupt***, is an instruction that explicitly generates an exception condition. The most common use of a trap is to enter supervisor mode. The entry into supervisor mode must be controlled to maintain security—if the interface between user and supervisor mode is improperly designed, a user program may be able to sneak code into the supervisor mode that could be executed to perform harmful operations.

The ARM provides the SWI interrupt for software interrupts. This instruction causes the CPU to enter supervisor mode. An opcode is embedded in the instruction that can be read by the handler.

3.3 CO-PROCESSORS

CPU architects often want to provide flexibility in what features are implemented in the CPU. One way to provide such flexibility at the instruction set level is to allow ***co-processors***, which are attached to the CPU and implement some of the instructions. For example, floating-point arithmetic was introduced into the Intel architecture by providing separate chips that implemented the floating-point instructions.

To support co-processors, certain opcodes must be reserved in the instruction set for co-processor operations. Because it executes instructions, a co-processor must be tightly coupled to the CPU. When the CPU receives a co-processor instruction, the CPU must activate the co-processor and pass it the relevant instruction. Co-processor instructions can load and store co-processor registers or can perform internal operations. The CPU can suspend execution to wait for the co-processor instruction to finish; it can also take a more superscalar approach and continue executing instructions while waiting for the co-processor to finish.

A CPU may, of course, receive co-processor instructions even when there is no coprocessor attached. Most architectures use illegal instruction traps to handle these situations. The trap handler can detect the co-processor instruction and, for example, execute it in software on the main CPU. Emulating co-processor instructions in software is slower but provides compatibility.

The ARM architecture provides support for up to 16 co-processors. Co-processors are able to perform load and store operations on their own registers. They can also move data between the co-processor registers and main ARM registers.

An example ARM co-processor is the floating-point unit. The unit occupies two co-processor units in the ARM architecture, numbered 1 and 2, but it appears as a single unit to the programmer. It provides eight 80-bit floating-point data registers, floating-point status registers, and an optional floating-point status register.

3.4 MEMORY SYSTEM MECHANISMS

Modern microprocessors do more than just read and write a monolithic memory. Architectural features improve both the speed and capacity of memory systems. Microprocessor clock rates are increasing at a faster rate than memory speeds, such that memories are falling further and further behind microprocessors every day. As a result, computer architects resort to *caches* to increase the average performance of the memory system. Although memory capacity is increasing steadily, program sizes are increasing as well, and designers may not be willing to pay for all the memory demanded by an application. *Modern microprocessor units (MMUs)* perform address translations that provide a larger virtual memory space in a small physical memory. In this section, we review both caches and MMUs.

3.4.1 Caches

Caches are widely used to speed up memory system performance. Many microprocessor architectures include caches as part of their definition. The cache speeds up average memory access time when properly used. It increases the variability of memory access times—accesses in the cache will be fast, while access to locations not cached will be slow. This variability in performance makes it especially important to understand how caches work so that we can better understand how to predict cache performance and factor variabilities into system design.

A cache is a small, fast memory that holds copies of some of the contents of main memory. Because the cache is fast, it provides higher-speed access for the CPU; but since it is small, not all requests can be satisfied by the cache, forcing the system to wait for the slower main memory. Caching makes sense when the CPU is using only a relatively small set of memory locations at any one time; the set of active locations is often called the *working set*.

Figure 3.6 shows how the cache support reads in the memory system. A *cache controller* mediates between the CPU and the memory system comprised of the main memory. The cache controller sends a memory request to the cache and main memory. If the requested location is in the cache, the cache controller forwards the location's contents to the CPU and aborts the main memory request; this condition is known as a *cache hit*. If the location is not in the cache, the controller waits for the value from main memory and forwards it to the CPU; this situation is known as a *cache miss*.

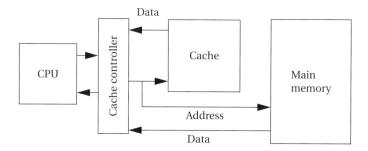

FIGURE 3.6

The cache in the memory system.

We can classify cache misses into several types depending on the situation that generated them:

- a ***compulsory miss*** (also known as a **cold miss**) occurs the first time a location is used,

- a ***capacity miss*** is caused by a too-large working set, and

- a ***conflict miss*** happens when two locations map to the same location in the cache.

Even before we consider ways to implement caches, we can write some basic formulas for memory system performance. Let h be the ***hit rate***, the probability that a given memory location is in the cache. It follows that $1 - h$ is the ***miss rate***, or the probability that the location is not in the cache. Then we can compute the average memory access time as

$$t_{av} = h t_{cache} + (1 - h) t_{main}. \tag{3.1}$$

where t_{cache} is the access time of the cache and t_{main} is the main memory access time. The memory access times are basic parameters available from the memory manufacturer. The hit rate depends on the program being executed and the cache organization, and is typically measured using simulators, as is described in more detail in Section 5.6. The best-case memory access time (ignoring cache controller overhead) is t_{cache}, while the worst-case access time is t_{main}. Given that t_{main} is typically 50–60 ns for DRAM, while t_{cache} is at most a few nanoseconds, the spread between worst-case and best-case memory delays is substantial.

Modern CPUs may use multiple levels of cache as shown in Figure 3.7. The ***first-level cache*** (commonly known as ***L1 cache***) is closest to the CPU, the ***second-level cache (L2 cache)*** feeds the first-level cache, and so on.

The second-level cache is much larger but is also slower. If h_1 is the first-level hit rate and h_2 is the rate at which access hit the second-level cache but not the first-level cache, then the average access time for a two-level cache system is

$$t_{av} = h_1 t_{L1} + h_2 t_{L2} + (1 - h_1 - h_2) t_{main}. \tag{3.2}$$

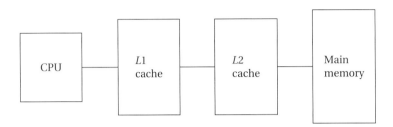

FIGURE 3.7

A two-level cache system.

As the program's working set changes, we expect locations to be removed from the cache to make way for new locations. When set-associative caches are used, we have to think about what happens when we throw out a value from the cache to make room for a new value. We do not have this problem in direct-mapped caches because every location maps onto a unique block, but in a set-associative cache we must decide which set will have its block thrown out to make way for the new block. One possible replacement policy is least recently used (LRU), that is, throw out the block that has been used farthest in the past. We can add relatively small amounts of hardware to the cache to keep track of the time since the last access for each block. Another policy is random replacement, which requires even less hardware to implement.

The simplest way to implement a cache is a *direct-mapped cache*, as shown in Figure 3.8. The cache consists of cache *blocks*, each of which includes a tag to show which memory location is represented by this block, a data field holding the contents of that memory, and a valid tag to show whether the contents of this cache block are valid. An address is divided into three sections. The index is used to select which cache block to check. The tag is compared against the tag value in the block selected by the index. If the address tag matches the tag value in the block, that block includes the desired memory location. If the length of the data field is longer than the minimum addressable unit, then the lowest bits of the address are used as an offset to select the required value from the data field. Given the structure of the cache, there is only one block that must be checked to see whether a location is in the cache—the index uniquely determines that block. If the access is a hit, the data value is read from the cache.

Writes are slightly more complicated than reads because we have to update main memory as well as the cache. There are several methods by which we can do this. The simplest scheme is known as *write-through*—every write changes both the cache and the corresponding main memory location (usually through a write buffer). This scheme ensures that the cache and main memory are consistent, but may generate some additional main memory traffic. We can reduce the number of times we write to main memory by using a *write-back* policy: If we write only when we remove a location from the cache, we eliminate the writes when a location is written several times before it is removed from the cache.

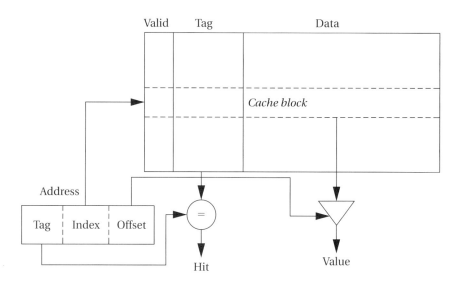

FIGURE 3.8

A direct-mapped cache.

The direct-mapped cache is both fast and relatively low cost, but it does have limits in its caching power due to its simple scheme for mapping the cache onto main memory. Consider a direct-mapped cache with four blocks, in which locations 0, 1, 2, and 3 all map to different blocks. But locations 4, 8, 12, ... all map to the same block as location 0; locations 1, 5, 9, 13, ... all map to a single block; and so on. If two popular locations in a program happen to map onto the same block, we will not gain the full benefits of the cache. As seen in Section 5.6, this can create program performance problems.

The limitations of the direct-mapped cache can be reduced by going to the *set-associative* cache structure shown in Figure 3.9. A set-associative cache is characterized by the number of *banks* or *ways* it uses, giving an *n*-way set-associative cache. A set is formed by all the blocks (one for each bank) that share the same index. Each set is implemented with a direct-mapped cache. A cache request is broadcast to all banks simultaneously. If any of the sets has the location, the cache reports a hit. Although memory locations map onto blocks using the same function, there are *n* separate blocks for each set of locations. Therefore, we can simultaneously cache several locations that happen to map onto the same cache block. The set-associative cache structure incurs a little extra overhead and is slightly slower than a direct-mapped cache, but the higher hit rates that it can provide often compensate.

The set-associative cache generally provides higher hit rates than the direct-mapped cache because conflicts between a small number of locations can be resolved within the cache. The set-associative cache is somewhat slower, so the CPU designer has to be careful that it doesn't slow down the CPU's cycle time too much. A more important problem with set-associative caches for embedded program

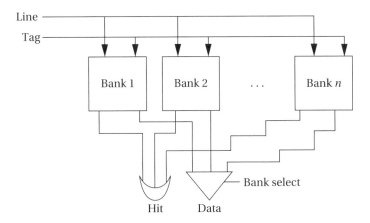

FIGURE 3.9

A set-associative cache.

design is predictability. Because the time penalty for a cache miss is so severe, we often want to make sure that critical segments of our programs have good behavior in the cache. It is relatively easy to determine when two memory locations will conflict in a direct-mapped cache. Conflicts in a set-associative cache are more subtle, and so the behavior of a set-associative cache is more difficult to analyze for both humans and programs. Example 3.8 compares the behavior of direct-mapped and set-associative caches.

Example 3.8

Direct-mapped vs. set-associative caches

For simplicity, let's consider a very simple caching scheme. We use 2 bits of the address as the tag. We compare a direct-mapped cache with four blocks and a two-way set-associative cache with four sets, and we use LRU replacement to make it easy to compare the two caches.

A 3-bit address is used for simplicity. The contents of the memory follow:

Address	Data	Address	Data
000	0101	100	1000
001	1111	101	0001
010	0000	110	1010
011	0110	111	0100

We will give each cache the same pattern of addresses (in binary to simplify picking out the index): 001, 010, 011, 100, 101, and 111.

To understand how the direct-mapped cache works, let's see how its state evolves.

After 001 access:

Block	Tag	Data
00	—	—
01	0	1111
10	—	—
11	—	—

After 010 access:

Block	Tag	Data
00	—	—
01	0	1111
10	0	0000
11	—	—

After 011 access:

Block	Tag	Data
00	—	—
01	0	1111
10	0	0000
11	0	0110

After 100 access (notice that the tag bit for this entry is 1):

Block	Tag	Data
00	1	1000
01	0	1111
10	0	0000
11	0	0110

After 101 access (overwrites the 01 block entry):

Block	Tag	Data
00	1	1000
01	1	0001
10	0	0000
11	0	0110

After 111 access (overwrites the 11 block entry):

Block	Tag	Data
00	1	1000
01	1	0001
10	0	0000
11	1	0100

We can use a similar procedure to determine what ends up in the two-way set-associative cache. The only difference is that we have some freedom when we have to replace a block with new data. To make the results easy to understand, we use a least-recently-used replacement policy. For starters, let's make each way the size of the original direct-mapped cache. The final state of the two-way set-associative cache follows:

Block	Way 0 tag	Way 0 data	Way 1 tag	Way 1 data
00	1	1000	—	—
01	0	1111	1	0001
10	0	0000	—	—
11	0	0110	1	0100

Of course, this is not a fair comparison for performance because the two-way set-associative cache has twice as many entries as the direct-mapped cache. Let's use a two-way, set-associative cache with two sets, giving us four blocks, the same number as in the direct-mapped cache. In this case, the index size is reduced to 1 bit and the tag grows to 2 bits.

Block	Way 0 tag	Way 0 data	Way 1 tag	Way 1 data
0	01	0000	10	1000
1	00	0111	11	0100

In this case, the cache contents are significantly different than for either the direct-mapped cache or the four-block, two-way set-associative cache.

The CPU knows when it is fetching an instruction (the PC is used to calculate the address, either directly or indirectly) or data. We can therefore choose whether to cache instructions, data, or both. If cache space is limited, instructions are the highest priority for caching because they will usually provide the highest hit rates. A cache that holds both instructions and data is called a ***unified cache***.

Various ARM implementations use different cache sizes and organizations [Fur96]. The ARM600 includes a 4-KB, 64-way (wow!) unified instruction/data cache. The StrongARM uses a 16-KB, 32-way instruction cache with a 32-byte block and a 16-KB, 32-way data cache with a 32-byte block; the data cache uses a write-back strategy.

The C5510, one of the models of C55x, uses a 16-K byte instruction cache organized as a two-way set-associative cache with four 32-bit words per line. The instruction cache can be disabled by software if desired. It also includes two RAM sets that are designed to hold large contiguous blocks of code. Each RAM set can hold up to 4-K bytes of code organized as 256 lines of four 32-bit words per line. Each RAM has a tag that specifies what range of addresses are in the RAM; it also includes a tag valid field to show whether the RAM is in use and line valid bits for each line.

3.4.2 Memory Management Units and Address Translation

A MMU translates addresses between the CPU and physical memory. This translation process is often known as ***memory mapping*** since addresses are mapped from a logical space into a physical space. MMUs in embedded systems appear primarily in the host processor. It is helpful to understand the basics of MMUs for embedded systems complex enough to require them.

Many DSPs, including the C55x, do not use MMUs. Since DSPs are used for compute-intensive tasks, they often do not require the hardware assist for logical address spaces.

Early computers used MMUs to compensate for limited address space in their instruction sets. When memory became cheap enough that physical memory could be larger than the address space defined by the instructions, MMUs allowed software to manage multiple programs in a single physical memory, each with its own address space.

Because modern CPUs typically do not have this limitation, MMUs are used to provide ***virtual addressing***. As shown in Figure 3.10, the MMU accepts logical addresses from the CPU. Logical addresses refer to the program's abstract address space but do not correspond to actual RAM locations. The MMU translates them from tables to physical addresses that do correspond to RAM. By changing the MMU's tables, you can change the physical location at which the program resides without modifying the program's code or data. (We must, of course, move the program in main memory to correspond to the memory mapping change.)

Furthermore, if we add a secondary storage unit such as flash or a disk, we can eliminate parts of the program from main memory. In a virtual memory system, the MMU keeps track of which logical addresses are actually resident in main memory; those that do not reside in main memory are kept on the secondary storage device.

FIGURE 3.10

A virtually addressed memory system.

When the CPU requests an address that is not in main memory, the MMU generates an exception called a ***page fault***. The handler for this exception executes code that reads the requested location from the secondary storage device into main memory. The program that generated the page fault is restarted by the handler only after

- the required memory has been read back into main memory, and
- the MMU's tables have been updated to reflect the changes.

Of course, loading a location into main memory will usually require throwing something out of main memory. The displaced memory is copied into secondary storage before the requested location is read in. As with caches, LRU is a good replacement policy.

There are two styles of address translation: ***segmented*** and ***paged***. Each has advantages and the two can be combined to form a segmented, paged addressing scheme. As illustrated in Figure 3.11, segmenting is designed to support a large, arbitrarily sized region of memory, while pages describe small, equally sized regions. A segment is usually described by its start address and size, allowing different segments to be of different sizes. Pages are of uniform size, which simplifies the hardware required for address translation. A segmented, paged scheme is created by dividing each segment into pages and using two steps for address translation. Paging introduces the possibility of ***fragmentation*** as program pages are scattered around physical memory.

In a simple segmenting scheme, shown in Figure 3.12, the MMU would maintain a segment register that describes the currently active segment. This register would point to the base of the current segment. The address extracted from an instruction (or from any other source for addresses, such as a register) would be used as the offset for the address. The physical address is formed by adding the segment base to the offset. Most segmentation schemes also check the physical address against the upper limit of the segment by extending the segment register to include the segment size and comparing the offset to the allowed size.

The translation of paged addresses requires more MMU state but a simpler calculation. As shown in Figure 3.13, the logical address is divided into two sections, including a page number and an offset. The page number is used as an index into a page table, which stores the physical address for the start of each page. However,

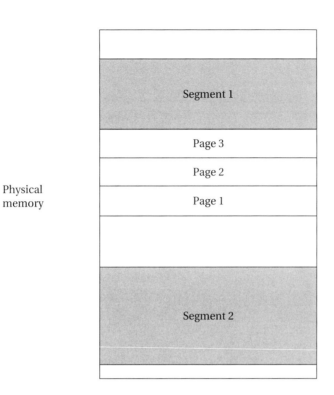

FIGURE 3.11

Segments and pages.

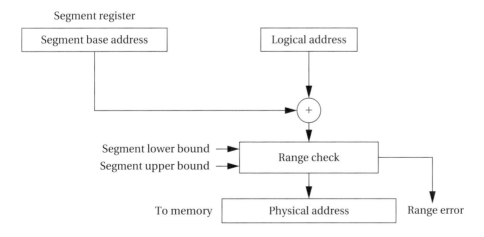

FIGURE 3.12

Address translation for a segment.

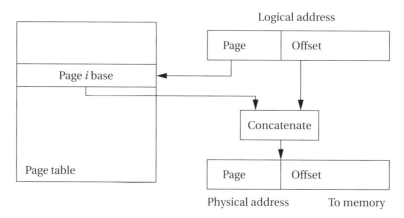

FIGURE 3.13

Address translation for a page.

since all pages have the same size and it is easy to ensure that page boundaries fall on the proper boundaries, the MMU simply needs to concatenate the top bits of the page starting address with the bottom bits from the page offset to form the physical address. Pages are small, typically between 512 bytes and 4 KB. As a result, the page table is large for an architecture with a large address space. The page table is normally kept in main memory, which means that an address translation requires memory access.

The page table may be organized in several ways, as shown in Figure 3.14. The simplest scheme is a flat table. The table is indexed by the page number and each entry holds the page descriptor. A more sophisticated method is a tree. The root entry of the tree holds pointers to pointer tables at the next level of the tree; each pointer table is indexed by a part of the page number. We eventually (after three levels, in this case) arrive at a descriptor table that includes the page descriptor we are interested in. A tree-structured page table incurs some overhead for the pointers, but it allows us to build a partially populated tree. If some part of the address space is not used, we do not need to build the part of the tree that covers it.

The efficiency of paged address translation may be increased by caching page translation information. A cache for address translation is known as a ***translation lookaside buffer (TLB)***. The MMU reads the TLB to check whether a page number is currently in the TLB cache and, if so, uses that value rather than reading from memory.

Virtual memory is typically implemented in a paging or segmented, paged scheme so that only page-sized regions of memory need to be transferred on a page fault. Some extensions to both segmenting and paging are useful for virtual memory:

- At minimum, a present bit is necessary to show whether the logical segment or page is currently in physical memory.

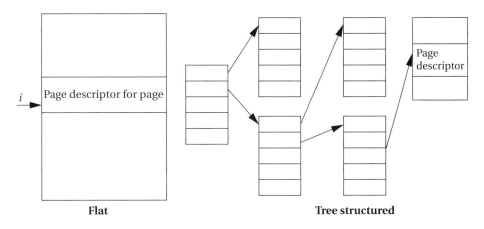

FIGURE 3.14

Alternative schemes for organizing page tables.

- A dirty bit shows whether the page/segment has been written to. This bit is maintained by the MMU, since it knows about every write performed by the CPU.

- Permission bits are often used. Some pages/segments may be readable but not writable. If the CPU supports modes, pages/segments may be accessible by the supervisor but not in user mode.

A data or instruction cache may operate either on logical or physical addresses, depending on where it is positioned relative to the MMU.

A MMU is an optional part of the ARM architecture. The ARM MMU supports both virtual address translation and memory protection; the architecture requires that the MMU be implemented when cache or write buffers are implemented. The ARM MMU supports the following types of memory regions for address translation:

- a *section* is a 1-MB block of memory,

- a *large page* is 64 KB, and

- a *small page* is 4 KB.

An address is marked as section mapped or page mapped. A two-level scheme is used to translate addresses. The first-level table, which is pointed to by the Translation Table Base register, holds descriptors for section translation and pointers to the second-level tables. The second-level tables describe the translation of both large and small pages. The basic two-level process for a large or small page is illustrated in Figure 3.15. The details differ between large and small pages, such as the size of the second-level table index. The first- and second-level pages also contain access control bits for virtual memory and protection.

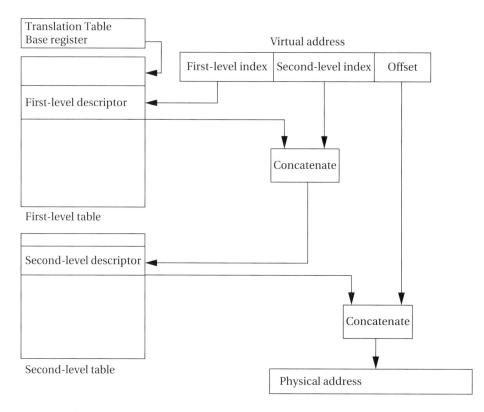

FIGURE 3.15

ARM two-stage address translation.

3.5 CPU PERFORMANCE

Now that we have an understanding of the various types of instructions that CPUs can execute, we can move on to a topic particularly important in embedded computing: How fast can the CPU execute instructions? In this section, we consider three factors that can substantially influence program performance: pipelining and caching.

3.5.1 Pipelining

Modern CPUs are designed as ***pipelined*** machines in which several instructions are executed in parallel. Pipelining greatly increases the efficiency of the CPU. But like any pipeline, a CPU pipeline works best when its contents flow smoothly. Some sequences of instructions can disrupt the flow of information in the pipeline and, temporarily at least, slow down the operation of the CPU.

The ARM7 has a three-stage pipeline:

■ **Fetch** the instruction is fetched from memory.

■ **Decode** the instruction's opcode and operands are decoded to determine what function to perform.

■ **Execute** the decoded instruction is executed.

Each of these operations requires one clock cycle for typical instructions. Thus, a normal instruction requires three clock cycles to completely execute, known as the **latency** of instruction execution. But since the pipeline has three stages, an instruction is completed in every clock cycle. In other words, the pipeline has a **throughput** of one instruction per cycle. Figure 3.16 illustrates the position of instructions in the pipeline during execution using the notation introduced by Hennessy and Patterson [Hen06]. A vertical slice through the timeline shows all instructions in the pipeline at that time. By following an instruction horizontally, we can see the progress of its execution.

The C55x includes a seven-stage pipeline [Tex00B]:

1. **Fetch**.

2. **Decode**.

3. **Address** computes data and branch addresses.

4. **Access 1** reads data.

5. **Access 2** finishes data read.

6. **Read stage** puts operands onto internal busses.

7. **Execute** performs operations.

RISC machines are designed to keep the pipeline busy. CISC machines may display a wide variation in instruction timing. Pipelined RISC machines typically have more regular timing characteristics—most instructions that do not have pipeline hazards display the same latency.

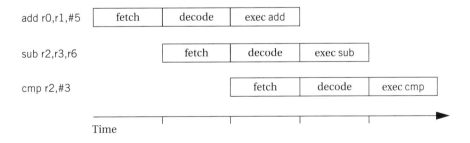

FIGURE 3.16

Pipelined execution of ARM instructions.

The one-cycle-per-instruction completion rate does not hold in every case, however. The simplest case for extended execution is when an instruction is too complex to complete the execution phase in a single cycle. A multiple load instruction is an example of an instruction that requires several cycles in the execution phase. Figure 3.17 illustrates a ***data stall*** in the execution of a sequence of instructions starting with a load multiple (LDMIA) instruction. Since there are two registers to load, the instruction must stay in the execution phase for two cycles. In a multiphase execution, the decode stage is also occupied, since it must continue to remember the decoded instruction. As a result, the SUB instruction is fetched at the normal time but not decoded until the LDMIA is finishing. This delays the fetching of the third instruction, the CMP.

Branches also introduce ***control stall*** delays into the pipeline, commonly referred to as the ***branch penalty***, as shown in Figure 3.18. The decision whether to take the conditional branch BNE is not made until the third clock cycle of that instruction's execution, which computes the branch target address. If the branch is taken, the succeeding instruction at PC+4 has been fetched and started to be decoded. When the branch is taken, the branch target address is used to fetch the branch target instruction. Since we have to wait for the execution cycle to complete before knowing the target, we must throw away two cycles of work on instructions

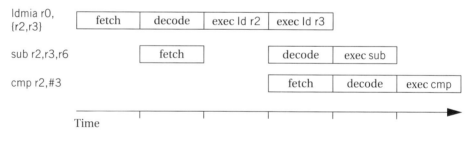

FIGURE 3.17

Pipelined execution of multicycle ARM instruction.

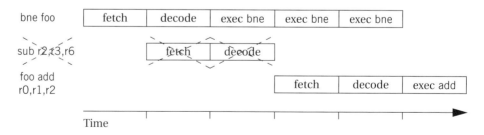

FIGURE 3.18

Pipelined execution of a branch in ARM.

in the path not taken. The CPU uses the two cycles between starting to fetch the branch target and starting to execute that instruction to finish housekeeping tasks related to the execution of the branch.

One way around this problem is to introduce the ***delayed branch***. In this style of branch instruction, some number of instructions directly after the branch are always executed, whether or not the branch is taken. This allows the CPU to keep the pipeline full during execution of the branch. However, some of those instructions after the delayed branch may be no-ops. Any instruction in the delayed branch window must be valid for both execution paths, whether or not the branch is taken. If there are not enough instructions to fill the delayed branch window, it must be filled with no-ops.

Let's use this knowledge of instruction execution time to evaluate the execution time of some C code, as shown in Example 3.9.

Example 3.9

Execution time of a for loop on the ARM

We will use the C code for the FIR filter of Application Example 2.1:

```
for (i = 0, f = 0; i < N; i++)
    f = f + c[i] * x[i];
```

We repeat the ARM code for this loop:

```
; loop initiation code
MOV r0,#0    ; use r0 for i, set to 0
MOV r8,#0    ; use a separate index for arrays
ADR r2,N     ; get address for N
LDR r1,[r2]  ; get value of N for loop termination test
MOV r2,#0    ; use r2 for f, set to 0
ADR r3,c     ; load r3 with address of base of c array
ADR r5,x     ; load r5 with address of base of x array
; loop body
loop  LDR r4,[r3,r8]    ; get value of c[i]
      LDR r6,[r5,r8]    ; get value of x[i]
      MUL r4,r4,r6      ; compute c[i]*x[i]
      ADD r2,r2,r4      ; add into running sum f
      ; update loop counter and array index
      ADD r8,r8,#4      ; add one word offset to array index
      ADD r0,r0,#1      ; add 1 to i
      ; test for exit
      CMP r0,r1
      BLT loop          ; if i < N, continue loop
loopend...
```

Inspection of the code shows that the only instruction that may take more than one cycle is the conditional branch in the loop test. We can count the number of instructions and associated number of clock cycles in each block as follows:

Block	Variable	# Instructions	# Cycles
Initiation	t_{init}	7	7
Body	t_{body}	4	4
Update	t_{update}	2	2
Test	t_{test}	2	2 best case, 4 worst case

The unconditional branch at the end of the update block always incurs a branch penalty of two cycles. The BLT instruction in the test block incurs a pipeline delay of two cycles when the branch is taken. That happens for all but the last iteration, when the instruction has an execution time of $t_{test,worst}$; the last iteration executes in time $t_{test,best}$. We can write a formula for the total execution time of the loop in cycles as

$$t_{loop} = t_{init} + N(t_{body} + t_{update}) + (N - 1)t_{test,worst} + t_{test,best}. \tag{3.3}$$

3.5.2 Caching

We have already discussed caches functionally. Although caches are invisible in the programming model, they have a profound effect on performance. We introduce caches because they substantially reduce memory access time when the requested location is in the cache. However, the desired location is not always in the cache since it is considerably smaller than main memory. As a result, caches cause the time required to access memory to vary considerably. The extra time required to access a memory location not in the cache is often called the **cache miss penalty**. The amount of variation depends on several factors in the system architecture, but a cache miss is often several clock cycles slower than a cache hit.

The time required to access a memory location depends on whether the requested location is in the cache. However, as we have seen, a location may not be in the cache for several reasons.

- At a compulsory miss, the location has not been referenced before.

- At a conflict miss, two particular memory locations are fighting for the same cache line.

- At a capacity miss, the program's working set is simply too large for the cache.

The contents of the cache can change considerably over the course of execution of a program. When we have several programs running concurrently on the CPU,

we can have very dramatic changes in the cache contents. We need to examine the behavior of the programs running on the system to be able to accurately estimate performance when caches are involved. We consider this problem in more detail in Section 5.6.

3.6 CPU POWER CONSUMPTION

Power consumption is, in some situations, as important as execution time. In this section we study the characteristics of CPUs that influence power consumption and mechanisms provided by CPUs to control how much power they consume.

First, it is important to distinguish between *energy* and *power*. Power is, of course, energy consumption per unit time. Heat generation depends on power consumption. Battery life, on the other hand, most directly depends on energy consumption. Generally, we will use the term *power* as shorthand for energy and power consumption, distinguishing between them only when necessary.

The high-level power consumption characteristics of CPUs and other system components are derived from the circuits used to build those components. Today, virtually all digital systems are built with **complementary metal oxide semi-conductor (CMOS)** circuitry. The detailed circuit characteristics are best left to a study of VLSI design [Wol08], but the basic sources of CMOS power consumption are easily identified and briefly described below.

- *Voltage drops*: The dynamic power consumption of a CMOS circuit is proportional to the square of the power supply voltage (V^2). Therefore, by reducing the power supply voltage to the lowest level that provides the required performance, we can significantly reduce power consumption. We also may be able to add parallel hardware and even further reduce the power supply voltage while maintaining required performance [Cha92].

- *Toggling*: A CMOS circuit uses most of its power when it is changing its output value. This provides two ways to reduce power consumption. By reducing the speed at which the circuit operates, we can reduce its power consumption (although not the total energy required for the operation, since the result is available later). We can actually reduce energy consumption by eliminating unnecessary changes to the inputs of a CMOS circuit—eliminating unnecessary glitches at the circuit outputs eliminates unnecessary power consumption.

- *Leakage*: Even when a CMOS circuit is not active, some charge leaks out of the circuit's nodes through the substrate. The only way to eliminate leakage current is to remove the power supply. Completely disconnecting the power supply eliminates power consumption, but it usually takes a significant amount of time to reconnect the system to the power supply and reinitialize its internal state so that it once again performs properly.

As a result, we see the following power-saving strategies used in CMOS CPUs.

- CPUs can be used at reduced voltage levels. For example, reducing the power supply from 1 to 0.9 V causes the power consumption to drop by $1^2 0.9^2 = 1.2$ X.

- The CPU can be operated at a lower clock frequency to reduce power (but not energy) consumption.

- The CPU may internally disable certain function units that are not required for the currently executing function. This reduces energy consumption.

- Some CPUs allow parts of the CPU to be totally disconnected from the power supply to eliminate leakage currents.

There are two types of power management features provided by CPUs. A **static power management** mechanism is invoked by the user but does not otherwise depend on CPU activities. An example of a static mechanism is a **power-down mode** intended to save energy. This mode provides a high-level way to reduce unnecessary power consumption. The mode is typically entered with an instruction. If the mode stops the interpretation of instructions, then it clearly cannot be exited by execution of another instruction. Power-down modes typically end upon receipt of an interrupt or other event. A **dynamic power management** mechanism takes actions to control power based upon the dynamic activity in the CPU. For example, the CPU may turn off certain sections of the CPU when the instructions being executed do not need them. Application Example 3.2 describes the static and dynamic energy efficiency features of one of the PowerPC chips.

Application Example 3.2

Energy efficiency features in the PowerPC 603

The PowerPC 603 [Gar94] was designed specifically for low-power operation while retaining high performance. It typically dissipates 2.2 W running at 80 MHz. The architecture provides three low-power modes—doze, nap, and sleep—that provide static power management capabilities for use by the programs and operating system.

The 603 also uses a variety of dynamic power management techniques for power minimization that are performed automatically, without program intervention. The CPU is a two-issue, out-of-order superscalar processor. It uses the dynamic techniques summarized below to reduce power consumption.

- An execution unit that is not being used can be shut down.

- The cache, an 8-KB, two-way set-associative cache, was organized into subarrays so that at most two out of eight subarrays will be accessed on any given clock cycle. A variety of circuit techniques were also used in the cache to reduce power consumption.

Not all units in the CPU are active all the time; idling them when they are not being used can save power. The table below shows the percentage of time various units in the 603 were idle for the SPEC integer and floating-point benchmarks [Gar94].

Unit	Specint92 (% idle)	Specfp92 (% idle)
Data cache	29	28
Instruction cache	29	17
Load-store	35	17
Fixed-point	38	76
Floating-point	99	30
System register	89	97

Idle units are turned off automatically by switching off their clocks. Various stages of the pipeline are turned on and off, depending on which stages are necessary at the current time. Measurements comparing the chip's power consumption with and without dynamic power management show that dynamic techniques provide significant power savings.

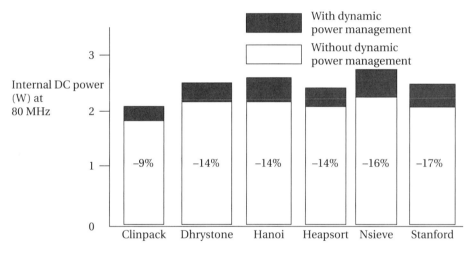

From [Gar94].

A power-down mode provides the opportunity to greatly reduce power consumption because it will typically be entered for a substantial period of time. However, going into and especially out of a power-down mode is not free—it costs both time and energy. The power-down or power-up transition consumes time and energy in order to properly control the CPU's internal logic. Modern pipelined processors require complex control that must be properly initialized to avoid corrupting data in the pipeline. Starting up the processor must also be done carefully to avoid power surges that could cause the chip to malfunction or even damage it.

The modes of a CPU can be modeled by a ***power state machine*** [Ben00]. An example is shown in Figure 3.19. Each state in the machine represents a different mode of the machine, and every state is labeled with its average power consumption. The example machine has two states: run mode with power consumption P_{run} and

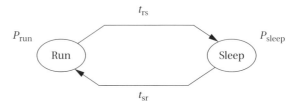

FIGURE 3.19

A power state machine for a processor.

sleep mode with power consumption P_{sleep}. Transitions show how the machine can go from state to state; each transition is labeled with the time required to go from the source to the destination state. In a more complex example, it may not be possible to go from a particular state to another particular state—traversing a sequence of states may be necessary. Application Example 3.3 describes the power-down modes of the Strong ARM SA-1100.

Application Example 3.3

Power-saving modes of the StrongARM SA-1100

The StrongARM SA-1100 [Int99] is designed to provide sophisticated power management capabilities that are controlled by the on-chip power manager. The processor takes two power supplies, as seen in the following figure:

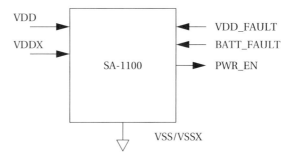

VDD is the main power supply for the core CPU and is nominally 3.3 V. The VDDX supply is used for the pins and other logic such as the power manager; it is normally at 1.5 V. (The two supplies share a common ground.) The system can supply two inputs about the status of the power supply. VDD_FAULT tells the CPU that the main power supply is not being properly regulated, while BATT_FAULT indicates that the battery has been removed or is low. Either of these events can cause the CPU to go into a low-power mode. In low-power operation, the VDD supply can be turned off (the VDDX supply always remains on). When resuming operation, the PWR_EN signal is used by the CPU to tell the external power supply to ramp up the VDD power supply.

A system power manager can both monitor the CPU and other devices and control their operation to gracefully transition between power modes. It provides several registers that allow programs to control power modes, determine why power modes were entered, determine the current state of power management modes, and so on.

The SA-1100 provides the three power modes described below.

- **Run** mode is normal operation and has the highest power consumption.

- **Idle** mode saves power by stopping the CPU clock. The system unit modules—real-time clock, operating system timer, interrupt control, general-purpose I/O, and power manager—all remain operational. Idle mode is entered by executing a three-instruction sequence. The CPU returns to run mode upon receiving an interrupt from one of the internal system units or from a peripheral or by resetting the CPU. This causes the machine to restart the CPU clock and to resume execution where it left off.

- **Sleep** mode shuts off most of the chip's activity. Entering sleep mode causes the system to shut down on-chip activity, reset the CPU, and negate the PWR_EN pin to tell the external electronics that the chip's power supply should be driven to 0 V. A separate I/O power supply remains on and supplies power to the power manager so that the CPU can be awakened from sleep mode; the low-speed clock keeps the power manager running at low speeds sufficient to manage sleep mode. The CPU software should set several registers to prepare for sleep mode. Sleep mode is entered by forcing the sleep bit in the power manager control register; it can also be entered by a power supply fault. The sleep shutdown sequence happens in three steps, each of which requires about 30 μs. The machine wakes up from sleep state on a preprogrammed wake-up event. The wake-up sequence has three steps: the PWR_EN pin is asserted to turn on the external power supply and waits for about 10 ms; the 3.686-MHz oscillator is ramped up to speed; and the internal reset is negated and the CPU boot sequence begins.

Here is the power state machine of the SA-1100 [Ben00]:

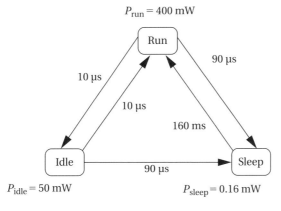

$P_{run} = 400$ mW

Run

90 μs

10 μs

10 μs

160 ms

Idle

Sleep

90 μs

$P_{idle} = 50$ mW

$P_{sleep} = 0.16$ mW

From [Ben00].

The sleep mode saves over three orders of magnitude of power consumption. However, the time required to reenter run mode from sleep is over a tenth of a second.

The SA-1100 has a companion chip, the SA-1111, that provides an integrated set of peripherals. That chip has its own power management modes that complement the SA-1100.

Design Example

3.7 DATA COMPRESSOR

Our design example for this chapter is a data compressor that takes in data with a constant number of bits per data element and puts out a compressed data stream in which the data is encoded in variable-length symbols. Because this chapter concentrates on CPUs, we focus on the data compression routine itself.

3.7.1 Requirements and Algorithm

We use the ***Huffman coding*** technique, which is introduced in Application Example 3.4.

We require some understanding of how our compression code fits into a larger system. Figure 3.20 shows a collaboration diagram for the data compression process. The data compressor takes in a sequence of **input symbols** and then produces a stream of **output symbols**. Assume for simplicity that the input symbols are one byte in length. The output symbols are variable length, so we have to choose a format in which to deliver the output data. Delivering each coded symbol separately is tedious, since we would have to supply the length of each symbol and use external code to pack them into words. On the other hand, bit-by-bit delivery is almost certainly too slow. Therefore, we will rely on the data compressor to pack the coded symbols into an array. There is not a one-to-one relationship between the input and output symbols, and we may have to wait for several input symbols before a packed output word comes out.

Application Example 3.4

Huffman coding for text compression

Text compression algorithms aim at statistical reductions in the volume of data. One commonly used compression algorithm is Huffman coding [Huf52], which makes use of information

FIGURE 3.20

UML collaboration diagram for the data compressor.

on the frequency of characters to assign variable-length codes to characters. If shorter bit sequences are used to identify more frequent characters, then the length of the total sequence will be reduced.

In order to be able to decode the incoming bit string, the code characters must have unique prefixes: No code may be a prefix of a longer code for another character. As a simple example of Huffman coding, assume that these characters have the following probabilities P of appearance in a message:

Character	P	Character	P
A	0.45	D	0.08
B	0.24	E	0.07
C	0.11	F	0.05

We build the code from the bottom up. After sorting the characters by probability, we create a new symbol by adding a bit. We then compute the joint probability of finding either one of those characters and re-sort the table. The result is a tree that we can read top down to find the character codes. The coding tree for our example appears below.

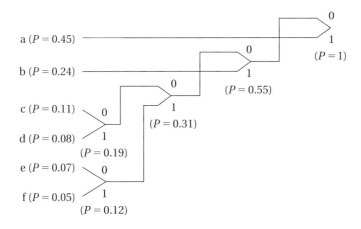

Reading the codes off the tree from the root to the leaves, we obtain the following coding of the characters:

Character	Code	Character	Code
A	1	D	0001
B	01	E	0010
C	0000	F	0011

Once the code has been constructed, which in many applications is done off-line, the codes can be stored in a table for encoding. This makes encoding simple, but clearly the encoded bit rate can vary significantly depending on the input character sequence. On the decoding side, since we do not know a priori the length of a character's bit sequence, the computation time required to decode a character can vary significantly.

The data compressor as discussed above is not a complete system, but we can create at least a partial requirements list for the module as seen below. We used the abbreviation N/A for not applicable to describe some items that do not make sense for a code module.

Name	Data compression module
Purpose	Code module for Huffman data compression
Inputs	Encoding table, uncoded byte-size input symbols
Outputs	Packed compressed output symbols
Functions	Huffman coding
Performance	Requires fast performance
Manufacturing cost	N/A
Power	N/A
Physical size and weight	N/A

3.7.2 Specification

Let's refine the description of Figure 3.20 to come up with a more complete speci-fication for our data compression module. That collaboration diagram concentrates on the steady-state behavior of the system. For a fully functional system, we have to provide the following additional behavior.

- We have to be able to provide the compressor with a new symbol table.

- We should be able to flush the symbol buffer to cause the system to release all pending symbols that have been partially packed. We may want to do this when we change the symbol table or in the middle of an encoding session to keep a transmitter busy.

A class description for this refined understanding of the requirements on the module is shown in Figure 3.21. The class's *buffer* and *current-bit* behaviors keep track of the state of the encoding, and the *table* attribute provides the current symbol table. The class has three methods as follows:

- *Encode* performs the basic encoding function. It takes in a 1-byte input sym-bol and returns two values: a boolean showing whether it is returning a full buffer and, if the boolean is true, the full buffer itself.

```
┌─────────────────────────────────────┐
│ Data-compressor                     │
├─────────────────────────────────────┤
│ buffer: data-buffer                 │
│ table: symbol-table                 │
│ current-bit: integer                │
├─────────────────────────────────────┤
│ encode(): boolean, data-buffer      │
│ flush()                             │
│ new-symbol-table()                  │
└─────────────────────────────────────┘
```

FIGURE 3.21

Definition of the Data-compressor class.

```
┌─────────────────────────────────┐   ┌─────────────────────────────────────┐
│ Data-buffer                     │   │ Symbol-table                        │
├─────────────────────────────────┤   ├─────────────────────────────────────┤
│ databuf[databuflen]: character  │   │ symbols[nsymbols]: data-buffer      │
│ len: integer                    │   │                                     │
├─────────────────────────────────┤   ├─────────────────────────────────────┤
│ insert()                        │   │ value(): symbol                     │
│ length()                        │   │ load()                              │
└─────────────────────────────────┘   └─────────────────────────────────────┘
```

FIGURE 3.22

Additional class definitions for the data compressor.

- **New-symbol-table** installs a new symbol table into the object and throws away the current contents of the internal buffer.

- **Flush** returns the current state of the buffer, including the number of valid bits in the buffer.

We also need to define classes for the data buffer and the symbol table. These classes are shown in Figure 3.22. The *data-buffer* will be used to hold both packed symbols and unpacked ones (such as in the symbol table). It defines the buffer itself and the length of the buffer. We have to define a data type because the longest encoded symbol is longer than an input symbol. The longest Huffman code for an eight-bit input symbol is 256 bits. (Ending up with a symbol this long happens only when the symbol probabilities have the proper values.) The insert function packs a new symbol into the upper bits of the buffer; it also puts the remaining bits in a new buffer if the current buffer is overflowed. The *Symbol-table* class indexes

the encoded version of each symbol. The class defines an access behavior for the table; it also defines a *load* behavior to create a new symbol table. The relationships between these classes are shown in Figure 3.23—a data compressor object includes one buffer and one symbol table.

Figure 3.24 shows a state diagram for the *encode* behavior. It shows that most of the effort goes into filling the buffers with variable-length symbols. Figure 3.25

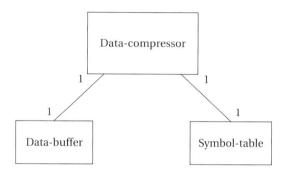

FIGURE 3.23

Relationships between classes in the data compressor.

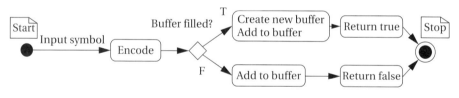

FIGURE 3.24

State diagram for encode behavior.

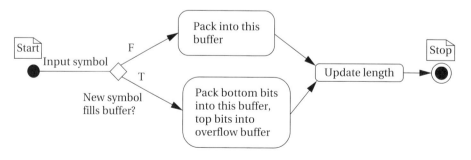

FIGURE 3.25

State diagram for insert behavior.

shows a state diagram for *insert*. It shows that we must consider two cases—the new symbol does not fill the current buffer or it does.

3.7.3 Program Design

Since we are only building an encoder, the program is fairly simple. We will use this as an opportunity to compare object-oriented and non-OO implementations by coding the design in both C++ and C.

OO design in C++

First is the object-oriented design using C++, since this implementation most closely mirrors the specification. The first step is to design the data buffer. The data buffer needs to be as long as the longest symbol. We also need to implement a function that lets us merge in another data_buffer, shifting the incoming buffer by the proper amount.

```
const int databuflen = 8;   /* as long in bytes as
                                 longest symbol */
const int bitsperbyte = 8; /* definition of byte */
const int bytemask = 0xff; /* use to mask to 8 bits for
                                 safety */
const char lowbitsmask [bitsperbyte] = { 0, 1, 3, 7, 15, 31,
                                           63, 127};
   /* used to keep low bits in a byte */
typedef char boolean; /* for clarity */
#define TRUE 1
#define FALSE 0

class data_buffer {
        char databuf[databuflen];
        int len;
        int length_in_chars() { return len/bitsperbyte; }
    /* length in bytes rounded down-used in implementation */
public:
        void insert(data_buffer, data_buffer&);
        int length() { return len; } /* returns number of bits
                                        in symbol */
        int length_in_bytes() { return (int)ceil(len/8.0); }
        void initialize(); /* initializes the data
                              structure */
        void data_buffer::fill(data_buffer, int);
           /* puts upper bits of symbol into buffer */
        data_buffer& operator = (data_buffer&);
           /* assignment operator */
```

```
            data_buffer() { initialize(); } /* C++ constructor */
            ~data_buffer() { } /* C++ destructor */
    };

    data_buffer empty_buffer; /* use this to initialize other
                            data_buffers */

    void data_buffer::insert(data_buffer newval, data_buffer&
                            newbuf) {
    /* This function puts the lower bits of a symbol (newval)
        into an existing buffer without overflowing the buffer.
        Puts spillover, if any, into newbuf. */

    int i, j, bitstoshift, maxbyte;
    /* precalculate number of positions to shift up */
    bitstoshift = length() - length_in_bytes()*bitsperbyte;
    /* compute how many bytes to transfer-can't run past end of
        this buffer */
    maxbyte = newval.length() + length() >
                databuflen*bitsperbyte ?
            databuflen : newval.length_in_chars();
    for (i = 0; i < maxbyte; i++) {
            /* add lower bits of this newval byte */
            databuf[i + length_in_chars()] | =
                    (newval.databuf[i] << bitstoshift) &
                                            byte-mask;
            /* add upper bits of this newval byte */
            databuf[i + length_in_chars() + 1] | =
                    (newval.databuf[i] >> (bitsperbyte -
                                            bitstoshift)) &
                    lowbitsmask[bitsperbyte - bitstoshift];
    }
    /* fill up new buffer if necessary */
    if (newval.length() + length() > databuflen*bitsperbyte) {
            /* precalculate number of positions to shift down */
            bitstoshift = length() % bitsperbyte;
            for (i = maxbyte, j = 0; i++, j++;
                i <= newval.length_in_chars()) {
                    newbuf.databuf[j] =
                            (newval.databuf[i] >> bitstoshift) &
                                            bytemask;
                    newbuf.databuf[j] | =
                            newval.databuf[i + 1] &
                            lowbitsmask[bitstoshift];
                }
```

```
        }
        /* update length */
        len = len + newval.length() > databuflen*bitsperbyte ?
                    databuflen*bitsperbyte : len +
                                            newval.length();
}

data_buffer& data_buffer::operator=(data_buffer& e) {
        /* assignment operator for data buffer */
        int i;
        /* copy the buffer itself */
        for (i = 0; i < databuflen; i++)
                databuf[i] = e.databuf[i];
        /* set length */
        len = e.len;
        /* return */
        return e;
}
void data_buffer::fill(data_buffer newval, int shiftamt) {
        /* This function puts the upper bits of a symbol
           (newval) into the buffer. */

        int i, bitstoshift, maxbyte;
        /* precalculate number of positions to shift up */
        bitstoshift = length() - length_in_bytes()*bitsperbyte;
        /* compute how many bytes to transfer-can't run past
           end of this buffer */
        maxbyte = newval.length_in_chars() > databuflen ?
            databuflen : newval.length_in_chars();
        for (i = 0; i < maxbyte; i++) {
                /* add lower bits of this newval byte */
                databuf[i + length_in_chars()] =
                newval.databuf[i] << bitstoshift;
                /* add upper bits of this newval byte */
                databuf[i + length_in_chars() + 1] =
                    newval.databuf[i] >> (bitsperbyte -
                                            bitstoshift);

        }
}

void data_buffer::initialize() {
        /* Initialization code for data_buffer. */
        int i;
```

```
                    /* initialize buffer to all zero bits */
                    for (i = 0; i < databuflen; i++)
                            databuf[i] = 0;
                    /* initialize length to zero */
                    len = 0;
        }
```

The code for data_buffer is relatively complex, and not all of its complexity was reflected in the state diagram of Figure 3.25. That does not mean the specification was bad, but only that it was written at a higher level of abstraction.

The symbol table code can be implemented relatively easily as shown below.

```
        const int nsymbols = 256;
        class symbol_table {
                data_buffer symbols[nsymbols];
        public:
                data_buffer value(int i) { return symbols[i]; }
                void load(symbol_table&);
                symbol_table() { } /* C++ constructor */
                ~symbol_table() { } /* C++ destructor */
        };

        void symbol_table::load(symbol_table& newsyms) {
                int i;
                for (i = 0; i < nsymbols; i++) {
                        symbols[i] = newsyms.symbols[i];
                }
        }
```

Now let's create the class definition for data_compressor:

```
        typedef char boolean; /* for clarity */
        class data_compressor {
                data_buffer buffer;
                int current_bit;
                symbol_table table;

        public:
                boolean encode(char, data_buffer&);
                void new_symbol_table(symbol_table newtable)
                        { table = newtable; current_bit = 0;
                        buffer = empty_buffer; }
                int flush(data_buffer& buf)
                        { int temp = current_bit; buf = buffer;
                        buffer = empty_buffer; current_bit = 0;
                                return temp; }
                data_compressor() { } /* C++ constructor */
```

```
        ~data_compressor() { } /* C++ destructor */
  };
```

Now let's implement the encode() method. The main challenge here is managing the buffer.

```
boolean data_compressor::encode(char isymbol, data_buffer&
                                fullbuf) {
      data_buffer temp;
      int overlen;

      /* look up the new symbol */            .
      temp = table.value(isymbol); /* the symbol itself */
      /* will this symbol overflow the buffer? */
      overlen = temp.length() + current_bit -
        buffer.length(); /* amount of overflow */
      if ( overlen > 0 ) { /* we did in fact overflow */
              data_buffer nextbuf;
              buffer.insert(temp,nextbuf);
              /* return the full buffer and keep the next
                 partial buffer */
              fullbuf = buffer;
              buffer = nextbuf;
              return TRUE;
      } else { /* no overflow */
              data_buffer no_overflow;
              buffer.insert(temp,no_overflow);
              /* won't use this argument */
              if (current_bit == buffer.length()) {
              /* return current buffer */
                      fullbuf = buffer;
                      buffer.initialize(); /* initialize the
                                              buffer */
                      return TRUE;
                      }
              else return FALSE; /* buffer isn't full yet */
      }
  }
```

OO design in C

How would we have to modify the implementation for C? We have two choices in implementation, based on whether we want to support multiple simultaneous data compressors. If we want to strictly adhere to the specification, we must be able to run several simultaneous compressors, since in the object-oriented specification we can create as many new *data-compressor* objects as we want.

We may not have the luxury of coding the algorithm in C++. While C is almost universally supported on embedded processors, support for languages that support object orientation such as C++ or Java is not so universal. How would we have to structure C code to provide multiple instantiations of the data compressor? The fundamental point is that we cannot rely on any global variables—all of the object state must be replicable. We can do this relatively easily, making the code only a little more cumbersome. We create a structure that holds the data part of the object as follows:

```
struct data_compressor_struct {
        data_buffer buffer;
        int current_bit;
        sym_table table;
}
```

typedef struct data_compressor_struct data_compressor,

```
*data_compressor_ptr; /* data type declaration for
                        convenience */
```

We would, of course, have to do something similar for the other classes. Depending on how strict we want to be, we may want to define data access functions to get to fields in the various structures we create. C would permit us to get to those struct fields without using the access functions, but using the access functions would give us a little extra freedom to modify the structure definitions later.

We then implement the class methods as C functions, passing in a pointer to the data_compressor object we want to operate on. Appearing below is the beginning of the modified encode method showing how we make explicit all references to the data in the object.

```
typedef char boolean; /* for clarity */
#define TRUE 1
#define FALSE 0

boolean data_compressor_encode(data_compressor_ptr mycmprs,
char isymbol, data_buffer *fullbuf) {
        data_buffer temp;
        int len, overlen;

        /* look up the new symbol */
        temp = mycmprs->table[isymbol].value; /* the symbol
                                                 itself */
        len = mycmprs->table[isymbol].length; /* its value */
        . . .
```

(For C++ afficionados, the above amounts to making explicit the C++ *this* pointer.)

If, on the other hand, we did not care about the ability to run multiple compressions simultaneously, we can make the functions a little more readable by using global variables for the class variables:

```
static data_buffer buffer;
static int current_bit;
static sym_table table;
```

We have used the C static declaration to ensure that these globals are not defined outside the file in which they are defined; this gives us a little added modularity. We would, of course, have to update the specification so that it makes clear that only one compressor object can be running at a time. The functions that implement the methods can then operate directly on the globals as seen below.

```
boolean data_compressor_encode(char isymbol, data_buffer*
                               fullbuf) {
        data_buffer temp;
        int len, overlen;

        /* look up the new symbol */
        temp = table[isymbol].value; /* the symbol itself */
        len = table[isymbol].length; /* its value */
        ...
```

Notice that this code does not need the structure pointer argument, making it resemble the C++ code a little more closely. However, horrible bugs will ensue if we try to run two different compressions at the same time through this code.

What can we say about the efficiency of this code? Efficiency has many aspects covered in more detail in Chapter 5. For the moment, let's consider instruction selection, that is, how well the compiler does in choosing the right instructions to implement the operations. Bit manipulations such as we do here often raise concerns about efficiency. But if we have a good compiler and we select the right data types, instruction selection is usually not a problem. If we use data types that do not require data type transformations, a good compiler can select the right instructions to efficiently implement the required operations.

3.7.4 Testing

How do we test this program module to be sure it works? We consider testing much more thoroughly in Section 5.10. In the meantime, we can use common sense to come up with some testing techniques.

One way to test the code is to run it and look at the output without considering how the code is written. In this case, we can load up a symbol table, run some symbols through it, and see whether we get the correct result. We can get the symbol table from outside sources (such as the tables of Application Example 3.4)

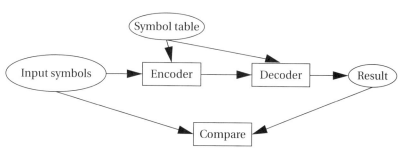

FIGURE 3.26

A test of the encoder.

or by writing a small program to generate it ourselves. We should test several different symbol tables. We can get an idea of how thoroughly we are covering the possibilities by looking at the encoding trees—if we choose several very different looking encoding trees, we are likely to cover more of the functionality of the module. We also want to test enough symbols for each symbol table. One way to help automate testing is to write a Huffman decoder. As illustrated in Figure 3.26, we can run a set of symbols through the encoder, and then through the decoder, and simply make sure that the input and output are the same. If they are not, we have to check both the encoder and decoder to locate the problem, but since most practical systems will require both in any case, this is a minor concern.

Another way to test the code is to examine the code itself and try to identify potential problem areas. When we read the code, we should look for places where data operations take place to see that they are performed properly. We also want to look at the conditionals to identify different cases that need to be exercised. Some ideas of things to look out for are listed below.

- Is it possible to run past the end of the symbol table?

- What happens when the next symbol does not fill up the buffer?

- What happens when the next symbol exactly fills up the buffer?

- What happens when the next symbol overflows the buffer?

- Do very long encoded symbols work properly? How about very short ones?

- Does flush() work properly?

Testing the internals of code often requires building ***scaffolding code***. For example, we may want to test the insert method separately, which would require building a program that calls the method with the proper values. If our programming language comes with an interpreter, building such scaffolding is easier because we do not have to create a complete executable, but we often want to automate such tests even with interpreters because we will usually execute them several times.

SUMMARY

Numerous mechanisms must be used to implement complete computer systems. For example, interrupts have little direct visibility in the instruction set, but they are very important to input and output operations. Similarly, memory management is invisible to most of the program but is very important to creating a working system.

Although we are not directly concerned with the details of computer architecture, characteristics of the underlying CPU hardware have a major impact on programs. When designing embedded systems, we are typically concerned about characteristics such as execution speed or power consumption. Having some understanding of the factors that determine performance and power will help you later as you develop techniques for optimizing programs to meet these criteria.

What We Learned

- Two major styles of I/O are polled and interrupt driven.

- Interrupts may be vectorized and prioritized.

- Supervisor mode helps protect the computer from program errors and provides a mechanism for controlling multiple programs.

- An exception is an internal error; a trap or software interrupt is explicitly generated by an instruction. Both are handled similarly to interrupts.

- A cache provides fast storage for a small number of main memory locations. Caches may be direct mapped or set associative.

- A memory management unit translates addresses from logical to physical addresses.

- Co-processors provide a way to optionally implement certain instructions in hardware.

- Program performance can be influenced by pipelining, superscalar execution, and the cache. Of these, the cache introduces the most variability into instruction execution time.

- CPUs may provide static (independent of program behavior) or dynamic (influenced by currently executing instructions) methods for managing power consumption.

FURTHER READING

As with instruction sets, the ARM and C55x manuals provide good descriptions of exceptions, memory management, and caches for those processors. Patterson and Hennessy [Pat07] provide a thorough description of computer architecture, including pipelining, caches, and memory management.

QUESTIONS

Q3-1 Why do most computer systems use memory-mapped I/O?

Q3-2 Write ARM code that tests a register at location ds1 and continues execution only when the register is nonzero.

Q3-3 Write ARM code that waits for the low-order bit of device register ds1 to become 1 and then reads a value from register dd1.

Q3-4 Implement peek() and poke() in assembly language for ARM.

Q3-5 Draw a UML sequence diagram for a busy-wait read of a device. The diagram should include the program running on the CPU and the device.

Q3-6 Draw a UML sequence diagram for a busy-wait write of a device. The diagram should include the program running on the CPU and the device.

Q3-7 Draw a UML sequence diagram for copying characters from an input to an output device using busy-wait I/O. The diagram should include the two devices and the two busy-wait I/O handlers.

Q3-8 When would you prefer to use busy-wait I/O over interrupt-driven I/O?

Q3-9 Draw a UML sequence diagram for an interrupt-driven read of a device. The diagram should include the background program, the handler, and the device.

Q3-10 Draw a UML sequence diagram for an interrupt-driven write of a device. The diagram should include the background program, the handler, and the device.

Q3-11 Draw a UML sequence diagram for a vectored interrupt-driven read of a device. The diagram should include the background program, the interrupt vector table, the handler, and the device.

Q3-12 Draw a UML sequence diagram for copying characters from an input to an output device using interrupt-driven I/O. The diagram should include the two devices and the two I/O handlers.

Q3-13 Draw a UML sequence diagram of a higher-priority interrupt that happens during a lower-priority interrupt handler. The diagram should include the device, the two handlers, and the background program.

Q3-14 Draw a UML sequence diagram of a lower-priority interrupt that happens during a higher-priority interrupt handler. The diagram should include the device, the two handlers, and the background program.

Q3-15 Draw a UML sequence diagram of a nonmaskable interrupt that happens during a low-priority interrupt handler. The diagram should include the device, the two handlers, and the background program.

Q3-16 Three devices are attached to a microprocessor: Device 1 has highest priority and device 3 has lowest priority. Each device's interrupt handler takes 5 time units to execute. Show what interrupt handler (if any) is executing at each time given the sequence of device interrupts displayed below.

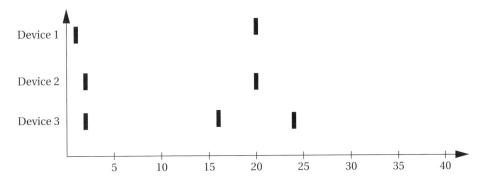

Q3-17 Draw a UML sequence diagram that shows how an ARM processor goes into supervisor mode. The diagram should include the supervisor mode program and the user mode program.

Q3-18 Draw a UML sequence diagram that shows how an ARM processor handles a floating-point exception. The diagram should include the user program, the exception handler, and the exception handler table.

Q3-19 Provide examples of how each of the following can occur in a typical program:

 a. Compulsory miss.
 b. Capacity miss.
 c. Conflict miss.

Q3-20 What is the average memory access time of a machine whose hit rate is 93%, with a cache access time of 5 ns and a main memory access time of 80 ns?

Q3-21 If we want an average memory access time of 6.5 ns, our cache access time is 5 ns, and our main memory access time is 80 ns, what cache hit rate must we achieve?

Q3-22 Assume that a system has a two-level cache: The level 1 cache has a hit rate of 90% and the level 2 cache has a hit rate of 97%. The level 1 cache access time is 4 ns, the level 2 access time is 15 ns, and the level 3 access time is 80 ns. What is the average memory access time?

Q3-23 In the two-way, set-associative cache with four banks of Example 3.8, show the state of the cache after each memory access, as was done for the direct-mapped cache. Use an LRU replacement policy.

Q3-24 The following code is executed by an ARM processor with each instruction executed exactly once:

```
        MOV r0,#0           ; use r0 for i, set to 0
        LDR r1,#10          ; get value of N for loop
                              termination test
        MOV r2,#0           ; use r2 for f, set to 0
        ADR r3,c            ; load r3 with address of
                              base of c array
        ADR r5,x            ; load r5 with address of
                              base of x array
        ; loop test
loop    CMP r0,r1
        BGE loopend         ; if i >= N, exit loop
        ; loop body
        LDR r4,[r3,r0]      ; get value of c[i]
        LDR r6,[r5,r0]      ; get value of x[i]
        MUL r4,r4,r6        ; compute c[i]*x[i]
        ADD r2,r2,r4        ; add into running sum f
        ; update loop counter
        ADD r0,r0,#1        ; add 1 to i
        B loop              ; unconditional branch to top
                              of loop
```

Show the contents of the instruction cache for these configurations, assuming each line holds one ARM instruction:

a. Direct-mapped, four lines.

b. Direct-mapped, eight lines.

c. Two-way set-associative, four lines per set.

Q3-25 Show a UML state diagram for a paged address translation using a flat page table.

Q3-26 Show a UML state diagram for a paged address translation using a three-level, tree-structured page table.

Q3-27 What are the stages in an ARM pipeline?

Q3-28 What are the stages in the C55x pipeline?

Q3-29 What is the difference between latency and throughput?

Q3-30 Draw two pipeline diagrams showing what happens when an ARM BZ instruction is taken and not taken, respectively.

Q3-31 Name three mechanisms by which a CMOS microprocessor consumes power.

Q3-32 Provide a user-level example of

 a. Static power management.

 b. Dynamic power management.

Q3-33 Why can't you use the same mechanism to return from a sleep power-saving state as you do from an idle power-saving state?

LAB EXERCISES

L3-1 Write a simple loop that lets you exercise the cache. By changing the number of statements in the loop body, you can vary the cache hit rate of the loop as it executes. If your microprocessor fetches instructions from off-chip memory, you should be able to observe changes in the speed of execution by observing the microprocessor bus.

L3-2 Try to measure the time required to respond to an interrupt.

Bus-Based Computer Systems

4

- CPU buses, I/O devices, and interfacing.
- The CPU system as a framework for understanding design methodology.
- System-level performance and power consumption.
- Development environments and debugging.
- An alarm clock design.

INTRODUCTION

In this chapter, we concentrate on *bus-based computer systems* created using microprocessors, I/O devices, and memory components. The microprocessor is an important element of the embedded computing system, but it cannot do its job without memories and I/O devices. We need to understand how to interconnect microprocessors and devices using the CPU bus. Luckily, there are many similarities between the platforms required for different applications, so we can extract some generally useful principles by examining a few basic concepts.

In the next section, we study the CPU bus, which forms the backbone of the hardware system. Because memories are very important components of embedded platforms, Section 4.2 studies types of memory devices. Section 4.3 introduces a variety of types of I/O devices. Section 4.4 introduces basic techniques for interfacing memories and I/O devices to the CPU bus. Section 4.5 focuses on the structure of the complete platform, while Section 4.6 considers development and debugging. Section 4.7 looks at system-level performance analysis for bus-based systems. Section 4.8 wraps up with an alarm clock as a design example.

4.1 THE CPU BUS

A computer system encompasses much more than the CPU; it also includes memory and I/O devices. The *bus* is the mechanism by which the CPU communicates with memory and devices. A bus is, at a minimum, a collection of wires, but the bus also

defines a protocol by which the CPU, memory, and devices communicate. One of the major roles of the bus is to provide an interface to memory. (Of course, I/O devices also connect to the bus.) Based on understanding of the bus, we study the characteristics of memory components in this section.

4.1.1 Bus Protocols

The basic building block of most bus protocols is the ***four-cycle handshake***, illustrated in Figure 4.1. The handshake ensures that when two devices want to communicate, one is ready to transmit and the other is ready to receive. The handshake uses a pair of wires dedicated to the handshake: ***enq*** (meaning enquiry) and ***ack*** (meaning acknowledge). Extra wires are used for the data transmitted during the handshake. The four cycles are described below.

1. *Device 1* raises its output to signal an enquiry, which tells *device 2* that it should get ready to listen for data.

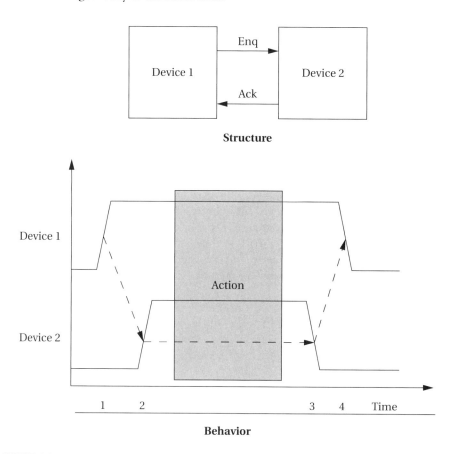

FIGURE 4.1

The four-cycle handshake.

2. When *device 2* is ready to receive, it raises its output to signal an acknowl-
 edgment. At this point, *devices 1* and *2* can transmit or receive.

3. Once the data transfer is complete, *device 2* lowers its output, signaling that
 it has received the data.

4. After seeing that *ack* has been released, *device 1* lowers its output.

At the end of the handshake, both handshaking signals are low, just as they were
at the start of the handshake. The system has thus returned to its original state in
readiness for another handshake-enabled data transfer.

Microprocessor buses build on the handshake for communication between the
CPU and other system components. The term *bus* is used in two ways. The most
basic use is as a set of related wires, such as address wires. However, the term may
also mean a protocol for communicating between components. To avoid confusion,
we will use the term **bundle** to refer to a set of related signals. The fundamental
bus operations are reading and writing. Figure 4.2 shows the structure of a typical
bus that supports reads and writes. The major components follow:

- *Clock* provides synchronization to the bus components,

- *R/W* is true when the bus is reading and false when the bus is writing,

- *Address* is an *a*-bit bundle of signals that transmits the address for an access,

- *Data* is an *n*-bit bundle of signals that can carry data to or from the CPU, and

- *Data ready* signals when the values on the data bundle are valid.

All transfers on this basic bus are controlled by the CPU—the CPU can read or
write a device or memory, but devices or memory cannot initiate a transfer. This is
reflected by the fact that *R/W* and address are unidirectional signals, since only the
CPU can determine the address and direction of the transfer.

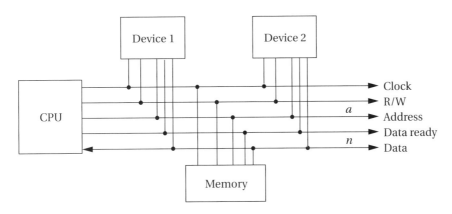

FIGURE 4.2

A typical microprocessor bus.

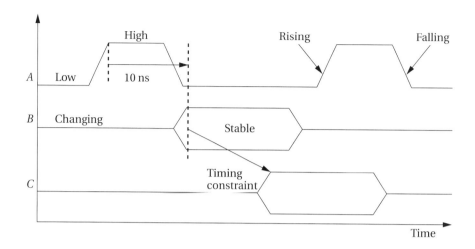

FIGURE 4.3

Timing diagram notation.

The behavior of a bus is most often specified as a ***timing diagram***. A timing diagram shows how the signals on a bus vary over time, but since values like the address and data can take on many values, some standard notation is used to describe signals, as shown in Figure 4.3. A's value is known at all times, so it is shown as a standard waveform that changes between zero and one. B and C alternate between ***changing*** and ***stable*** states. A stable signal has, as the name implies, a stable value that could be measured by an oscilloscope, but the exact value of that signal does not matter for purposes of the timing diagram. For example, an address bus may be shown as stable when the address is present, but the bus's timing requirements are independent of the exact address on the bus. A signal can go between a known 0/1 state and a stable/changing state. A changing signal does not have a stable value. Changing signals should not be used for computation. To be sure that signals go to their proper values at the proper times, timing diagrams sometimes show ***timing constraints***. We draw timing constraints in two different ways, depending on whether we are concerned with the amount of time between events or only the order of events. The timing constraint from A to B, for example, shows that A must go high before B becomes stable. The constraint from A to B also has a time value of 10 ns, indicating that A goes high at least 10 ns before B goes stable.

Figure 4.4 shows a timing diagram for the example bus. The diagram shows a read and a write. Timing constraints are shown only for the read operation, but similar constraints apply to the write operation. The bus is normally in the read mode since that does not change the state of any of the devices or memories. The CPU can then ignore the bus data lines until it wants to use the results of a read. Notice also that the direction of data transfer on bidirectional lines is not specified in the timing diagram. During a read, the external device or memory is sending a value on the data lines, while during a write the CPU is controlling the data lines.

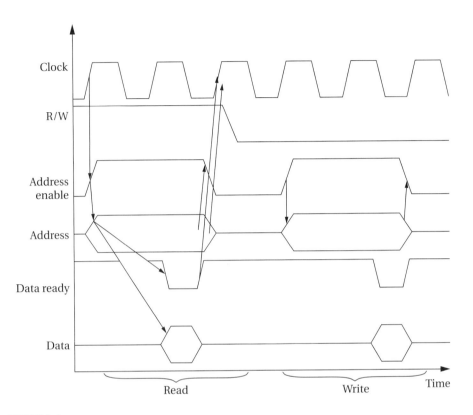

FIGURE 4.4

Timing diagram for the example bus.

With practice, we can see the sequence of operations for a read on the timing diagram as follows:

- A read or write is initiated by setting address enable high after the clock starts to rise. We set $R/W = 1$ to indicate a read, and the address lines are set to the desired address.

- One clock cycle later, the memory or device is expected to assert the data value at that address on the data lines. Simultaneously, the external device specifies that the data are valid by pulling down the *data ready* line. This line is *active low*, meaning that a logically true value is indicated by a low voltage, in order to provide increased immunity to electrical noise.

- The CPU is free to remove the address at the end of the clock cycle and must do so before the beginning of the next cycle. The external device has a similar requirement for removing the data value from the data lines.

The write operation has a similar timing structure. The read/write sequence does illustrate that timing constraints are required on the transition of the R/W signal

between read and write states. The signal must, of course, remain stable within a read or write. As a result there is a restricted time window in which the CPU can change between read and write modes.

The handshake that tells the CPU and devices when data are to be transferred is formed by data ready for the acknowledge side, but is implicit for the enquiry side. Since the bus is normally in read mode, enq does not need to be asserted, but the acknowledge must be provided by *data ready*.

The *data ready* signal allows the bus to be connected to devices that are slower than the bus. As shown in Figure 4.5, the external device need not immediately assert *data ready*. The cycles between the minimum time at which data can be

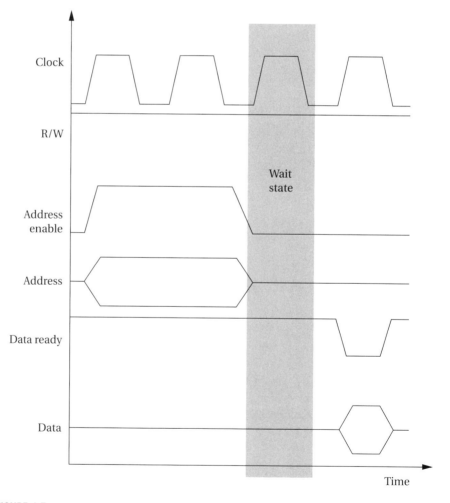

FIGURE 4.5

A wait state on a read operation.

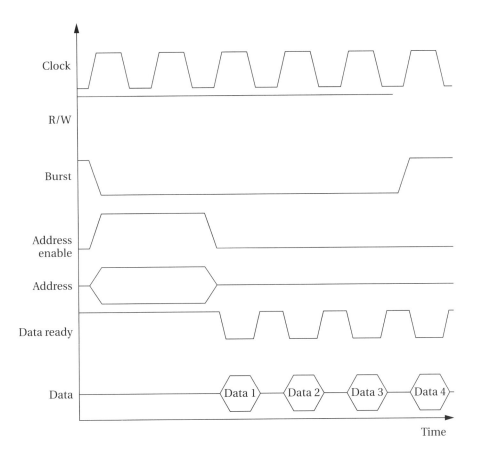

FIGURE 4.6

A burst read transaction.

asserted and when it is actually asserted are known as **wait states**. Wait states are commonly used to connect slow, inexpensive memories to buses.

We can also use the bus handshaking signals to perform **burst transfers**, as illustrated in Figure 4.6. In this burst read transaction, the CPU sends one address but receives a sequence of data values. We add an extra line to the bus, called *burst9* here, which signals when a transaction is actually a burst. Releasing the *burst9* signal tells the device that enough data has been transmitted. To stop receiving data after the end of *data 4*, the CPU releases the *burst9* signal at the end of *data 3* since the device requires some time to recognize the end of the burst. Those values come from successive memory locations starting at the given address.

Some buses provide **disconnected transfers**. In these buses, the request and response are separate. A first operation requests the transfer. The bus can then be used for other operations. The transfer is completed later, when the data are ready.

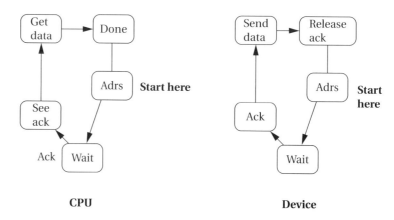

FIGURE 4.7

State diagrams for the bus read transaction.

The state machine view of the bus transaction is also helpful and a useful complement to the timing diagram. Figure 4.7 shows the CPU and device state machines for the read operation. As with a timing diagram, we do not show all the possible values of address and data lines but instead concentrate on the transitions of control signals. When the CPU decides to perform a read transaction, it moves to a new state, sending bus signals that cause the device to behave appropriately. The device's state transition graph captures its side of the protocol.

Some buses have data bundles that are smaller than the natural word size of the CPU. Using fewer data lines reduces the cost of the chip. Such buses are easiest to design when the CPU is natively addressable. A more complicated protocol hides the smaller data sizes from the instruction execution unit in the CPU. Byte addresses are sequentially sent over the bus, receiving one byte at a time; the bytes are assembled inside the CPU's bus logic before being presented to the CPU proper.

Some buses use multiplexed address and data. As shown in Figure 4.8, additional control lines are provided to tell whether the value on the address/data lines is an address or data. Typically, the address comes first on the combined address/data lines, followed by the data. The address can be held in a register until the data arrive so that both can be presented to the device (such as a RAM) at the same time.

4.1.2 DMA

Standard bus transactions require the CPU to be in the middle of every read and write transaction. However, there are certain types of data transfers in which the CPU does not need to be involved. For example, a high-speed I/O device may want to transfer a block of data into memory. While it is possible to write a program that alternately reads the device and writes to memory, it would be faster to eliminate the CPU's involvement and let the device and memory communicate directly. This

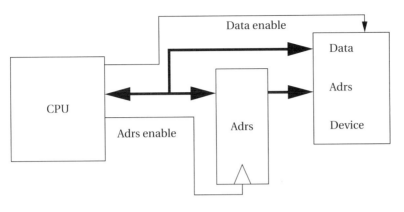

FIGURE 4.8

Bus signals for multiplexing address and data.

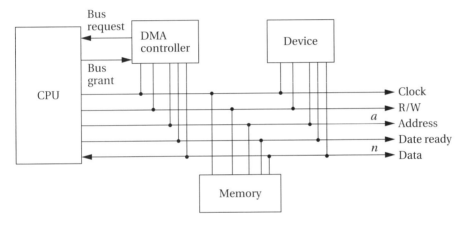

FIGURE 4.9

A bus with a DMA controller.

capability requires that some unit other than the CPU be able to control operations on the bus.

Direct memory access (DMA) is a bus operation that allows reads and writes not controlled by the CPU. A DMA transfer is controlled by a *DMA controller*, which requests control of the bus from the CPU. After gaining control, the DMA controller performs read and write operations directly between devices and memory.

Figure 4.9 shows the configuration of a bus with a DMA controller. The DMA requires the CPU to provide two additional bus signals:

- The *bus request* is an input to the CPU through which DMA controllers ask for ownership of the bus.

- The *bus grant* signals that the bus has been granted to the DMA controller.

A device that can initiate its own bus transfer is known as a ***bus master***. Devices that do not have the capability to be ***bus masters*** do not need to connect to a bus request and bus grant. The DMA controller uses these two signals to gain control of the bus using a classic four-cycle handshake. The bus request is asserted by the DMA controller when it wants to control the bus, and the bus grant is asserted by the CPU when the bus is ready.

The CPU will finish all pending bus transactions before granting control of the bus to the DMA controller. When it does grant control, it stops driving the other bus signals: R/W, address, and so on. Upon becoming bus master, the DMA controller has control of all bus signals (except, of course, for bus request and bus grant).

Once the DMA controller is bus master, it can perform reads and writes using the same bus protocol as with any CPU-driven bus transaction. Memory and devices do not know whether a read or write is performed by the CPU or by a DMA controller. After the transaction is finished, the DMA controller returns the bus to the CPU by deasserting the bus request, causing the CPU to deassert the bus grant.

The CPU controls the DMA operation through registers in the DMA controller. A typical DMA controller includes the following three registers:

- A starting address register specifies where the transfer is to begin.

- A length register specifies the number of words to be transferred.

- A status register allows the DMA controller to be operated by the CPU.

The CPU initiates a DMA transfer by setting the starting address and length registers appropriately and then writing the status register to set its start transfer bit. After the DMA operation is complete, the DMA controller interrupts the CPU to tell it that the transfer is done.

What is the CPU doing during a DMA transfer? It cannot use the bus. As illustrated in Figure 4.10, if the CPU has enough instructions and data in the cache and registers, it may be able to continue doing useful work for quite some time and may not notice the DMA transfer. But once the CPU needs the bus, it stalls until the DMA controller returns bus mastership to the CPU.

To prevent the CPU from idling for too long, most DMA controllers implement modes that occupy the bus for only a few cycles at a time. For example, the transfer may be made 4, 8, or 16 words at a time. As illustrated in Figure 4.11, after each block, the DMA controller returns control of the bus to the CPU and goes to sleep for a preset period, after which it requests the bus again for the next block transfer.

4.1.3 System Bus Configurations

A microprocessor system often has more than one bus. As shown in Figure 4.12, high-speed devices may be connected to a high-performance bus, while lower-speed

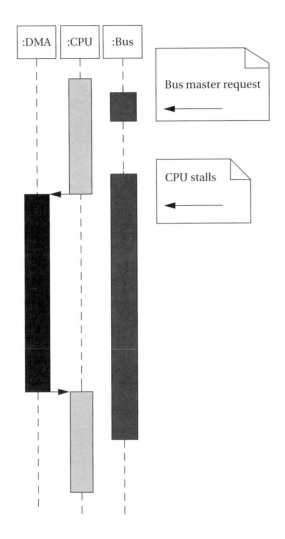

FIGURE 4.10

UML sequence diagram of system activity around a DMA transfer.

devices are connected to a different bus. A small block of logic known as a **bridge** allows the buses to connect to each other. There are several good reasons to use multiple buses and bridges:

- Higher-speed buses may provide wider data connections.
- A high-speed bus usually requires more expensive circuits and connectors. The cost of low-speed devices can be held down by using a lower-speed, lower-cost bus.

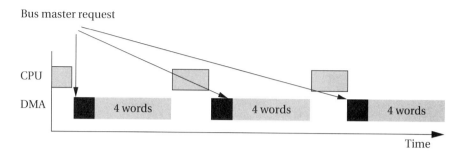

FIGURE 4.11

Cyclic scheduling of a DMA request.

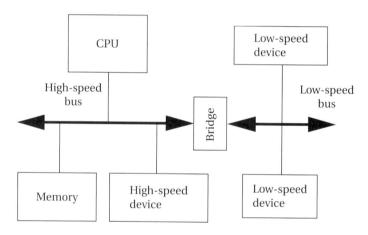

FIGURE 4.12

A multiple bus system.

■ The bridge may allow the buses to operate independently, thereby providing some parallelism in I/O operations.

In Section 4.5.3, we see that PCs often use this methodology.

Let's consider the operation of a bus bridge between what we will call a fast bus and a slow bus as illustrated in Figure 4.13. The bridge is a slave on the fast bus and the master of the slow bus. The bridge takes commands from the fast bus on which it is a slave and issues those commands on the slow bus. It also returns the results from the slow bus to the fast bus—for example, it returns the results of a read on the slow bus to the fast bus.

The upper sequence of states handles a write from the fast bus to the slow bus. These states must read the data from the fast bus and set up the handshake for the slow bus. Operations on the fast and slow sides of the bus bridge should

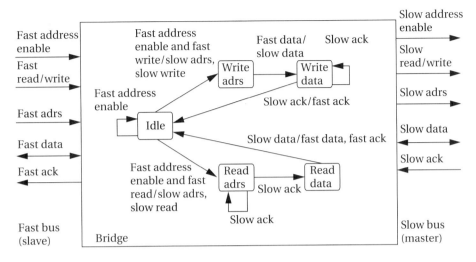

FIGURE 4.13

UML state diagram of bus bridge operation.

be overlapped as much as possible to reduce the latency of bus-to-bus transfers. Similarly, the bottom sequence of states reads from the slow bus and writes the data to the fast bus.

The bridge serves as a protocol translator between the two bridges as well. If the bridges are very close in protocol operation and speed, a simple state machine may be enough. If there are larger differences in the protocol and timing between the two buses, the bridge may need to use registers to hold some data values temporarily.

4.1.4 AMBA Bus

Since the ARM CPU is manufactured by many different vendors, the bus provided off-chip can vary from chip to chip. ARM has created a separate bus specification for single-chip systems. The AMBA bus [ARM99A] supports CPUs, memories, and peripherals integrated in a system-on-silicon. As shown in Figure 4.14, the AMBA specification includes two buses. The AMBA high-performance bus (AHB) is optimized for high-speed transfers and is directly connected to the CPU. It supports several high-performance features: pipelining, burst transfers, split transactions, and multiple bus masters.

A bridge can be used to connect the AHB to an AMBA peripherals bus (APB). This bus is designed to be simple and easy to implement; it also consumes relatively little power. The AHB assumes that all peripherals act as slaves, simplifying the logic required in both the peripherals and the bus controller. It also does not perform pipelined operations, which simplifies the bus logic.

FIGURE 4.14

Elements of the ARM AMBA bus system.

4.2 MEMORY DEVICES

In this section, we introduce the basic types of memory components that are commonly used in embedded systems. Now that we understand the operation of the bus, we are able to understand the pinouts of these memories and how values are read and written. We also need to understand the varieties of memory cells that are used to build memories. There are several varieties of both read-only and read/write memories, each with its own advantages. After discussing some basic characteristics of memories, we describe RAMs and then ROMs.

4.2.1 Memory Device Organization

The most basic way to characterize a memory is by its capacity, such as 256 MB. However, manufacturers usually make several versions of a memory of a given size, each with a different data width. For example, a 256-MB memory may be available in two versions:

- As a 64 M \times 4-bit array, a single memory access obtains an 8-bit data item, with a maximum of 2^{26} different addresses.

- As a 32 M \times 8-bit array, a single memory access obtains a 1-bit data item, with a maximum of 2^{23} different addresses.

The height/width ratio of a memory is known as its **aspect ratio**. The best aspect ratio depends on the amount of memory required.

Internally, the data are stored in a two-dimensional array of memory cells as shown in Figure 4.15. Because the array is stored in two dimensions, the n-bit address received by the chip is split into a row and a column address (with $n = r + c$).

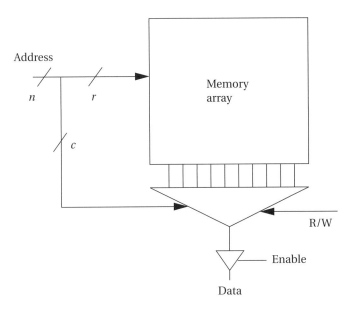

FIGURE 4.15

Internal organization of a memory device.

The row and column select a particular memory cell. If the memory's external width is 1 bit, the column address selects a single bit; for wider data widths, the column address can be used to select a subset of the columns. Most memories include an *enable* signal that controls the tri-stating of data onto the memory's pins. We will see in Section 4.4.1 how the enable pin can be used to easily build large memories from multiple banks of memory chips. A read/write signal (R/W in the figure) on read/write memories controls the direction of data transfer; memory chips do not typically have separate read and write data pins.

4.2.2 Random-Access Memories

Random-access memories can be both read and written. They are called random access because, unlike magnetic disks, addresses can be read in any order. Most bulk memory in modern systems is *dynamic RAM (DRAM)*. DRAM is very dense; it does, however, require that its values be **refreshed** periodically since the values inside the memory cells decay over time.

The dominant form of dynamic RAM today is the *synchronous DRAMs (SDRAMs)*, which uses clocks to improve DRAM performance. SDRAMs use Row Address Select (RAS) and Column Address Select (CAS) signals to break the address into two parts, which select the proper row and column in the RAM array. Signal transitions are relative to the SDRAM clock, which allows the internal SDRAM operations to be pipelined.

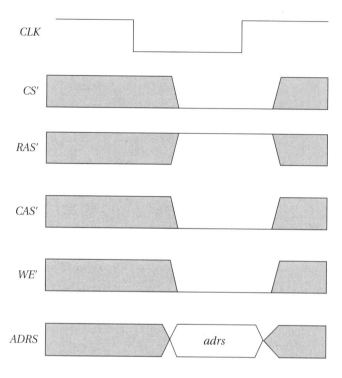

FIGURE 4.16

Timing diagram for a read on a synchronous DRAM.

As shown in Figure 4.16, transitions on the control signals are related to a clock [Mic00]. RAS′ and CAS′ can therefore become valid at the same time. The address lines are not shown in full detail here; some address lines may not be active depending on the mode in use. SDRAMs use a separate refresh signal to control refreshing. DRAM has to be refreshed roughly once per millisecond. Rather than refresh the entire memory at once, DRAMs refresh part of the memory at a time. When a section of memory is being refreshed, it cannot be accessed until the refresh is complete. The memory refresh occurs over fairly few seconds so that each section is refreshed every few microseconds.

SDRAMs include registers that control the mode in which the SDRAM operates. SDRAMs support burst modes that allow several sequential addresses to be accessed by sending only one address. SDRAMs generally also support an interleaved mode that exchanges pairs of bytes.

Even faster synchronous DRAMs, known as **_double-data rate (DDR)_** SDRAMs or **DDR2** and **DDR3** SDRAMs, are now in use. The details of DDR operation are beyond the scope of this book, but the basic capabilities of DDR memories are similar to those of single-rate SDRAMs; DDRs simply use sophisticated circuit techniques to perform more operations per clock cycle.

SIMMs and DIMMs

Memory for PCs is generally purchased as *single in-line memory modules (SIMMs)* or *double in-line memory modules (DIMMs)*. A SIMM or DIMM is a small circuit board that fits into a standard memory socket. A DIMM has two sets of leads compared to the SIMM's one. Memory chips are soldered to the circuit board to supply the desired memory.

4.2.3 Read-Only Memories

Read-only memories (ROMs) are preprogrammed with fixed data. They are very useful in embedded systems since a great deal of the code, and perhaps some data, does not change over time. Read-only memories are also less sensitive to radiation-induced errors.

There are several varieties of ROM available. The first-level distinction to be made is between *factory-programmed ROM* (sometimes called *mask-programmed ROM*) and *field-programmable ROM*. Factory-programmed ROMs are ordered from the factory with particular programming. ROMs can typically be ordered in lots of a few thousand, but clearly factory programming is useful only when the ROMs are to be installed in some quantity.

Field-programmable ROMs, on the other hand, can be programmed in the lab. *Flash memory* is the dominant form of field-programmable ROM and is electrically erasable. Flash memory uses standard system voltage for erasing and programming, allowing it to be reprogrammed inside a typical system. This allows applications such as automatic distribution of upgrades—the flash memory can be reprogrammed while downloading the new memory contents from a telephone line. Early flash memories had to be erased in their entirety; modern devices allow memory to be erased in blocks. Most flash memories today allow certain blocks to be protected. A common application is to keep the boot-up code in a protected block but allow updates to other memory blocks on the device. As a result, this form of flash is commonly known as *boot-block flash*.

4.3 I/O DEVICES

In this section we survey some input and output devices commonly used in embedded computing systems. Some of these devices are often found as on-chip devices in micro-controllers; others are generally implemented separately but are still commonly used. Looking at a few important devices now will help us understand both the requirements of device interfacing in this chapter and the uses of devices in programming in this and later chapters.

4.3.1 Timers and Counters

Timers and *counters* are distinguished from one another largely by their use, not their logic. Both are built from adder logic with registers to hold the current

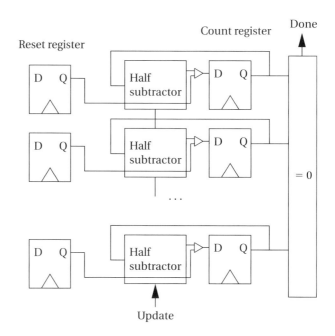

FIGURE 4.17

Internals of a counter/timer.

value, with an increment input that adds one to the current register value. However, a timer has its count connected to a periodic clock signal to measure time intervals, while a counter has its count input connected to an aperiodic signal in order to count the number of occurrences of some external event. Because the same logic can be used for either purpose, the device is often called a ***counter/timer***.

Figure 4.17 shows enough of the internals of a counter/timer to illustrate its operation. An n-bit counter/timer uses an n-bit register to store the current state of the count and an array of *half subtractors* to decrement the count when the count signal is asserted. Combinational logic checks when the count equals zero; the *done* output signals the zero count. It is often useful to be able to control the time-out, rather than require exactly 2^n events to occur. For this purpose, a reset register provides the value with which the count register is to be loaded. The counter/timer provides logic to load the reset register. Most counters provide both cyclic and acyclic modes of operation. In the cyclic mode, once the counter reaches the *done* state, it is automatically reloaded and the counting process continues. In acyclic mode, the counter/timer waits for an explicit signal from the microprocessor to resume counting.

A ***watchdog timer*** is an I/O device that is used for internal operation of a system. As shown in Figure 4.18, the watchdog timer is connected into the CPU bus and also to the CPU's reset line. The CPU's software is designed to periodically reset

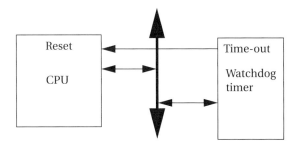

FIGURE 4.18

A watchdog timer.

the watchdog timer, before the timer ever reaches its time-out limit. If the watchdog timer ever does reach that limit, its time-out action is to reset the processor. In that case, the presumption is that either a software flaw or hardware problem has caused the CPU to misbehave. Rather than diagnose the problem, the system is reset to get it operational as quickly as possible.

4.3.2 A/D and D/A Converters

Analog/digital (A/D) and *digital/analog (D/A)* converters (typically known as **ADCs** and **DACs**, respectively) are often used to interface nondigital devices to embedded systems. The design of A/D and D/A converters themselves is beyond the scope of this book; we concentrate instead on the interface to the microprocessor bus. Because A/D conversion requires more complex circuitry, it requires a somewhat more complex interface.

Analog/digital conversion requires sampling the analog input before converting it to digital form. A control signal causes the A/D converter to take a sample and digitize it.

There are several different types of A/D converter circuits, some of which take a constant amount of time, while the conversion time of others depends on the sampled value. Variable-time converters provide a done signal so that the microprocessor knows when the value is ready.

A typical A/D interface has, in addition to its analog inputs, two major digital inputs. A data port allows A/D registers to be read and written, and a clock input tells when to start the next conversion.

D/A conversion is relatively simple, so the D/A converter interface generally includes only the data value. The input value is continuously converted to analog form.

4.3.3 Keyboards

A keyboard is basically an array of switches, but it may include some internal logic to help simplify the interface to the microprocessor. In this chapter, we build our understanding from a single switch to a microprocessor-controlled keyboard.

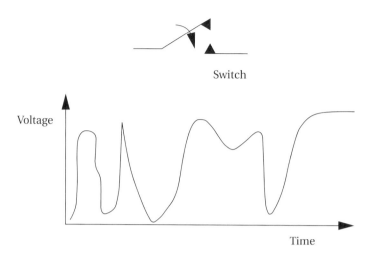

FIGURE 4.19

Switch bouncing.

A switch uses a mechanical contact to make or break an electrical circuit. The major problem with mechanical switches is that they **bounce** as shown in Figure 4.19. When the switch is depressed by pressing on the button attached to the switch's arm, the force of the depression causes the contacts to bounce several times until they settle down. If this is not corrected, it will appear that the switch has been pressed several times, giving false inputs. A hardware **debouncing** circuit can be built using a one-shot timer. Software can also be used to debounce switch inputs. A raw keyboard can be assembled from several switches. Each switch in a raw keyboard has its own pair of terminals, making raw keyboards impractical when a large number of keys is required.

More expensive keyboards, such as those used in PCs, actually contain a microprocessor to preprocess button inputs. PC keyboards typically use a 4-bit microprocessor to provide the interface between the keys and the computer. The microprocessor can provide debouncing, but it also provides other functions as well. An **encoded keyboard** uses some code to represent which switch is currently being depressed. At the heart of the encoded keyboard is the scanned array of switches shown in Figure 4.20. Unlike a raw keyboard, the scanned keyboard array reads only one row of switches at a time. The demultiplexer at the left side of the array selects the row to be read. When the scan input is 1, that value is transmitted to one terminal of each key in the row. If the switch is depressed, the 1 is sensed at that switch's column. Since only one switch in the column is activated, that value uniquely identifies a key. The row address and column output can be used for encoding, or circuitry can be used to give a different encoding.

A consequence of encoding the keyboard is that combinations of keys may not be represented. For example, on a PC keyboard, the encoding must be chosen so

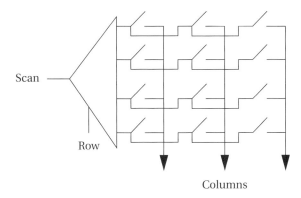

FIGURE 4.20

A scanned key array.

that combinations such as control-Q can be recognized and sent to the PC. Another consequence is that ***rollover*** may not be allowed. For example, if you press "a," and then press "b" before releasing "a," in most applications you want the keyboard to send an "a" followed by a "b." Rollover is very common in typing at even modest rates. A naive implementation of the encoder circuitry will simply throw away any character depressed after the first one until all the keys are released. The keyboard microcontroller can be programmed to provide **n-*key rollover***, so that rollover keys are sensed, put on a stack, and transmitted in sequence as keys are released.

4.3.4 LEDs

Light-emitting diodes (LEDs) are often used as simple displays by themselves, and arrays of LEDs may form the basis of more complex displays. Figure 4.21 shows how to connect an LED to a digital output. A resistor is connected between the output pin and the LED to absorb the voltage difference between the digital output voltage and the 0.7 V drop across the LED. When the digital output goes to 0, the LED voltage is in the device's off region and the LED is not on.

4.3.5 Displays

A display device may be either directly driven or driven from a frame buffer. Typically, displays with a small number of elements are driven directly by logic, while large displays use a RAM frame buffer.

The *n*-digit array, shown in Figure 4.22, is a simple example of a display that is usually directly driven. A single-digit display typically consists of seven segments; each segment may be either an LED or a ***liquid crystal display (LCD)*** element. This display relies on the digits being visible for some time after the drive to the digit is removed, which is true for both LEDs and LCDs. The digit input is used to choose which digit is currently being updated, and the selected digit activates its

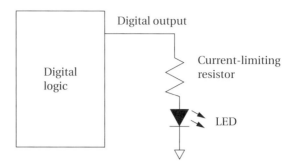

FIGURE 4.21

An LED connected to a digital output.

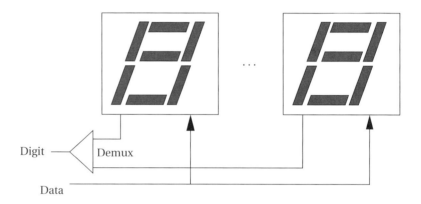

FIGURE 4.22

An *n*-digit display.

display elements based on the current data value. The display's driver is responsible for repeatedly scanning through the digits and presenting the current value of each to the display.

A *frame buffer* is a RAM that is attached to the system bus. The microprocessor writes values into the frame buffer in whatever order is desired. The pixels in the frame buffer are generally written to the display in *raster order* (by tradition, the screen is in the fourth quadrant) by reading pixels sequentially.

Many large displays are built using LCD. Each pixel in the display is formed by a single liquid crystal. LCD displays present a very different interface to the system because the array of pixel LCDs can be randomly accessed. Early LCD panels were called *passive matrix* because they relied on a two-dimensional grid of wires to address the pixels. Modern LCD panels use an *active matrix* system that puts a transistor at each pixel to control access to the LCD. Active matrix displays provide higher contrast and a higher-quality display.

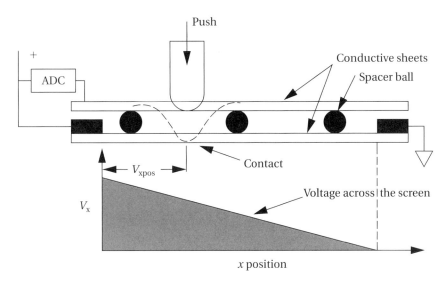

FIGURE 4.23

Cross section of a resistive touchscreen.

4.3.6 Touchscreens

A **touchscreen** is an input device overlaid on an output device. The touchscreen registers the position of a touch to its surface. By overlaying this on a display, the user can react to information shown on the display.

The two most common types of touchscreens are resistive and capacitive. A resistive touchscreen uses a two-dimensional voltmeter to sense position. As shown in Figure 4.23, the touchscreen consists of two conductive sheets separated by spacer balls. The top conductive sheet is flexible so that it can be pressed to touch the bottom sheet. A voltage is applied across the sheet; its resistance causes a voltage gradient to appear across the sheet. The top sheet samples the conductive sheet's applied voltage at the contact point. An analog/digital converter is used to measure the voltage and resulting position. The touchscreen alternates between x and y position sensing by alternately applying horizontal and vertical voltage gradients.

4.4 COMPONENT INTERFACING

Building the logic to interface a device to a bus is not too difficult but does take some attention to detail. We first consider interfacing memory components to the bus, since that is relatively simple, and then use those concepts to interface to other types of devices.

4.4.1 Memory Interfacing

If we can buy a memory of the exact size we need, then the memory structure is simple. If we need more memory than we can buy in a single chip, then we must construct the memory out of several chips. We may also want to build a memory that is wider than we can buy on a single chip; for example, we cannot generally buy a 32-bit-wide memory chip. We can easily construct a memory of a given width (32 bits, 64 bits, etc.) by placing RAMs in parallel.

We also need logic to turn the bus signals into the appropriate memory signals. For example, most busses won't send address signals in row and column form. We also need to generate the appropriate refresh signals.

4.4.2 Device Interfacing

Some I/O devices are designed to interface directly to a particular bus, forming *glueless interfaces*. But *glue logic* is required when a device is connected to a bus for which it is not designed.

An I/O device typically requires a much smaller range of addresses than a memory, so addresses must be decoded much more finely. Some additional logic is required to cause the bus to read and write the device's registers. Example 4.1 shows one style of interface logic.

Example 4.1

A glue logic interface

Below is an interfacing scheme for a simple I/O device.

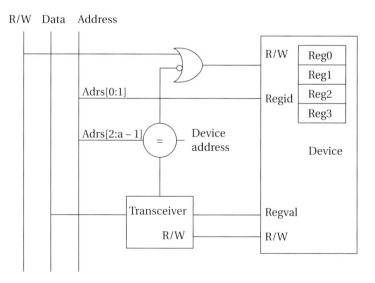

The device has four registers that can be read and written by presenting the register number on the *regid* pins, asserting R/W as required, and reading or writing the value on the *regval* pins. To interface to the bus, the bottom two bits of the address are used to refer to registers within the device, and the remaining bits are used to identify the device itself. The top bits of the address are sent to a comparator for testing against the device address. The device's address can be set with switches to allow the address to be easily changed. When the bus address matches the device's, the result is used to enable a *transceiver* for the data pins. When the *transceiver* is disabled, the *regval* pins are disconnected from the data bus. The comparator's output is also used to modify the *R/W* signal: The device's *R/W* pin is given the value (bus *R/W* + not-equal address), so that when the comparator's result is not 1, the device's *R/W* pin always receives a 1 to avoid inadvertently writing the device registers.

4.5 DESIGNING WITH MICROPROCESSORS

In this section we concentrate on how to create an initial working embedded system and how to ensure that the system works properly. Section 4.5.1 considers possible architectures for embedded computing systems. Section 4.5.2 studies techniques for designing the hardware components of embedded systems. Section 4.5.3 describes the use of the PC as an embedded computing platform.

4.5.1 System Architecture

We know that an architecture is a set of elements and the relationships between them that together form a single unit. The architecture of an embedded computing system is the blueprint for implementing that system—it tells you what components you need and how you put them together.

The architecture of an embedded computing system includes both hardware and software elements. Let's consider each in turn.

The hardware architecture of an embedded computing system is the more obvious manifestation of the architecture since you can touch it and feel it. It includes several elements, some of which may be less obvious than others.

- *CPU* An embedded computing system clearly contains a microprocessor. But which one? There are many different architectures, and even within an architecture we can select between models that vary in clock speed, bus data width, integrated peripherals, and so on. The choice of the CPU is one of the most important, but it cannot be made without considering the software that will execute on the machine.

- *Bus* The choice of a bus is closely tied to that of a CPU, since the bus is an integral part of the microprocessor. But in applications that make intensive use of the bus due to I/O or other data traffic, the bus may be more of a limiting

factor than the CPU. Attention must be paid to the required data bandwidths to be sure that the bus can handle the traffic.

- *Memory* Once again, the question is not whether the system will have memory but the characteristics of that memory. The most obvious characteristic is total size, which depends on both the required data volume and the size of the program instructions. The ratio of ROM to RAM and selection of DRAM versus SRAM can have a significant influence on the cost of the system. The speed of the memory will play a large part in determining system performance.

- *Input and output devices* The user's view of the input and output mechanisms may not correspond to the devices connected to the microprocessor. For example, a set of switches and knobs on a front panel may all be controlled by a single microcontroller, which is in turn connected to the main CPU. For a given function, there may be several different devices of varying sophistication and cost that can do the job. The difficulty of using a particular device, such as the amount of glue logic required to interface it, may also play a role in final device selection.

You may not think of programs as having architectures, but well-designed programs do have structure that represents an architecture. A fundamental task in software architecture design is *partitioning*—breaking the functionality into pieces in a way that makes it easy to implement, test, and modify.

Most embedded systems will do more than one thing—for example, processing streams of data and handling the user interface. Mixing together different types of functionality into a single code module leads to *spaghetti code*, which has poorly structured control flow, excessive use of global data, and generally unreliable programs.

Breaking the system's functionality into pieces that roughly correspond to the major modes of operation and functions of the device is often a good choice. First, different types of functionality often require different programming styles, so that they will naturally fall into different procedures in the code. Second, the functionality boundaries often correspond to performance requirements. Since at least some of the software components will almost certainly have to finish executing within a given deadline, it is important to be able to identify the code that must satisfy the deadline and to measure the performance of that code.

It is also important to remember that some of the functionality may in fact be implemented in the I/O devices. You may have a choice between using a simple, inexpensive device that requires more software support or a more sophisticated and expensive device that can perform more functions automatically. (An example in the digital audio domain is μ-law scaling, which can be done automatically by some analog/digital converters.) Using DMA to move data rather than a programmed loop is another example of using hardware to substitute for software. Most of the functionality will be in the software, but careful consideration of the hardware architecture can help simplify the software and make it easier for the software to meet its performance requirements.

4.5.2 Hardware Design

The design complexity of the hardware platform can vary greatly, from a totally off-the-shelf solution to a highly customized design.

At the board level, the first step is to consider *evaluation boards* supplied by the microprocessor manufacturer or another company working in collaboration with the manufacturer. Evaluation boards are sold for many microprocessor systems; they typically include the CPU, some memory, a serial link for downloading programs, and some minimal number of I/O devices. Figure 4.24 shows an ARM evaluation board manufactured by Sharp. The evaluation board may be a complete solution or provide what you need with only slight modifications. If the evaluation board is supplied by the microprocessor vendor, its design (netlist, board layout, etc.) may be available from the vendor; companies provide such information to make it easy for customers to use their microprocessors. If the evaluation board comes from a third party, it may be possible to contract them to design a new board with your required modifications, or you can start from scratch on a new board design.

The other major task is the choice of memory and peripheral components. In the case of I/O devices, there are two alternatives for each device: selecting a

JTAG port

CPU

Serial port

Interrupt switch

Power supply

Reset switch

FIGURE 4.24

An ARM evaluation board.

component from a catalog or designing one yourself. When shopping for devices from a catalog, it is important to read data sheets carefully—it may not be trivial to figure out whether the device does what you need it to do. You should also consider the amount of glue logic required to connect the device to your bus. Simple peripheral logic can be implemented in ***programmable logic devices (PLDs)***, while more complex units can be built from ***field-programmable gate arrays (FPGAs)***.

4.5.3 The PC as a Platform

Personal computers are often used as platforms for embedded computing. A PC offers several important advantages—it is a predesigned hardware platform with a great many features, a wide variety of I/O devices can be purchased for it, and it provides a rich programming environment. Because a PC-based system does not use custom hardware, it also carries the resulting disadvantages. It is larger, more power-hungry, and more expensive than a custom hardware platform would be. However, for low-volume applications and environments such as factories and offices where size and power are not critical, using a PC to build an embedded system often makes a lot of sense. The term *personal computer* has come to apply to a variety of machines, including IBM-compatibles, Macs, and others. In this section, we describe a generic PC architecture with some discussion of features relevant to different types of PCs. A detailed discussion of any of these platforms is beyond the scope of this book.

As shown in Figure 4.25, a typical PC includes several major hardware components:

■ The CPU provides basic computational facilities.

■ RAM is used for program storage.

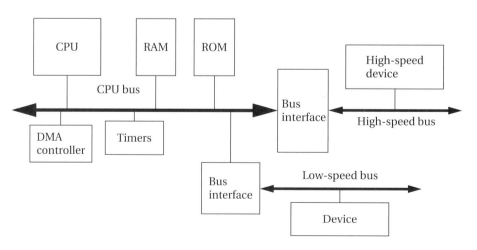

FIGURE 4.25

Hardware architecture of a typical PC.

- ROM holds the boot program.

- A DMA controller provides DMA capabilities.

- Timers are used by the operating system for a variety of purposes.

- A high-speed bus, connected to the CPU bus through a bridge, allows fast devices to communicate efficiently with the rest of the system.

- A low-speed bus provides an inexpensive way to connect simpler devices and may be necessary for backward compatibility as well.

PCI (*Peripheral Component Interconnect*) is the dominant high-performance system bus today. PCI uses high-speed data transmission techniques and efficient protocols to achieve high throughput. The original PCI standard allowed operation up to 33 MHz; at that rate, it could achieve a maximum transfer rate of 264 MB/s using 64-bit transfers. The revised PCI standard allows the bus to run up to 66 MHz, giving a maximum transfer rate of 524 MB/s with 64-bit wide transfers.

PCI uses wide buses with many data and address bits along with multiple control bits. The width of the bus both increases the cost of an interface to the bus and makes the physical connection to the bus more complicated. As a result, PC manufacturers have introduced serial buses to provide high-speed transfers while keeping the cost of connecting to the bus relatively low. *USB* (*Universal Serial Bus*) and **IEEE 1394** are the two major high-speed serial buses. Both of these buses offer high transfer rates using simple connectors. They also allow devices to be chained together so that users don't have to worry about the order of devices on the bus or other details of connection.

A PC also provides a standard software platform that provides interfaces to the underlying hardware as well as more advanced services. At the bottom of the software platform structure in most PCs is a minimal set of software in ROM. This software is designed to load the complete operating system from some other device (disk, network, etc.), and it may also provide low-level hardware interfaces. In the IBM-compatible PC, the low-level software is known as the *basic input/output system* (*BIOS*). The BIOS provides low-level hardware drivers as well as booting facilities. The operating system provides high-level drivers, control of executing processes, user interfaces, and so on. Because the PC software environment is so rich, developing embedded code for a PC target is much easier than when a host must be connected to a CPU in a development target. However, if the software is delivered directly on a standard version of the operating system, the resulting software package will require significant amounts of RAM as well as occupy a large disk image. Developers often create pared down versions of the operating system for delivering embedded code on PC platforms.

Both the IBM-compatible PC and the Mac provide a combination of hardware and software that allows devices to provide their own configuration information. On the IBM-compatible PC, this is known as the Plug-and-Play standard developed by Microsoft. These standards make it possible to plug in a device and have it work directly, without hardware or software intervention from the user.

It is now possible to put all the components (except for memory) for a standard PC on a single chip. A single-chip PC makes the development of certain types of embedded systems much easier, providing the rich software development of a PC with the low cost of a single-chip hardware platform.

The ability to integrate a CPU and devices on a single chip has allowed manufacturers to provide single-chip systems that do not conform to board-level standards. Application Example 4.1 describes one such single-chip system, the Intel StrongARM SA-1100.

Application Example 4.1

System organization of the Intel StrongARM SA-1100 and SA-1111

The StrongARM SA-1100 provides a number of functions besides the ARM CPU:

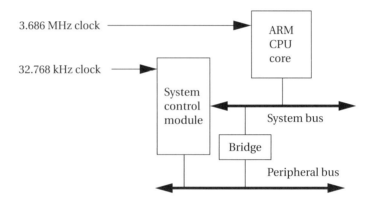

The chip contains two on-chip buses: a high-speed system bus and a lower-speed peripheral bus. The chip also uses two different clocks. A 3.686 MHz clock is used to drive the CPU and high-speed peripherals, and a 32.768 kHz clock is an input to the system control module. The system control module contains the following peripheral devices:

- A real-time clock

- An operating system timer

- 28 general-purpose I/Os (GPIOs)

- An interrupt controller

- A power manager controller

- A reset controller that handles resetting the processor.

The 32.768 kHz clock's frequency is chosen to be useful in timing real-time events. The slower clock is also used by the power manager to provide continued operation of the manager at a lower clock rate and therefore lower power consumption.

The SA-1111 is a companion chip that provides a suite of I/O functions. It connects to the SA-1100 through its system bus and provides several functions: a USB host controller; PS/2 ports for keyboards, mice, and so on; a PCMCIA interface; pulse-width modulation outputs; a serial port for digital audio; and an SSP serial port for telecom interfacing.

4.6 DEVELOPMENT AND DEBUGGING

In this section we take a step back from the platform and consider how it is used during design. We first consider how we can build an effective means for programming and testing an embedded system using hosts. We then see how hosts and other techniques can be used for debugging embedded systems.

4.6.1 Development Environments

A typical embedded computing system has a relatively small amount of everything, including CPU horsepower, memory, I/O devices, and so forth. As a result, it is common to do at least part of the software development on a PC or workstation known as a *host* as illustrated in Figure 4.26. The hardware on which the code will finally run is known as the *target*. The host and target are frequently connected by a USB link, but a higher-speed link such as Ethernet can also be used.

The target must include a small amount of software to talk to the host system. That software will take up some memory, interrupt vectors, and so on, but it should

FIGURE 4.26

Connecting a host and a target system.

generally leave the smallest possible footprint in the target to avoid interfering with the application software. The host should be able to do the following:

- load programs into the target,
- start and stop program execution on the target, and
- examine memory and CPU registers.

A ***cross-compiler*** is a compiler that runs on one type of machine but generates code for another. After compilation, the executable code is downloaded to the embedded system by a serial link or perhaps burned in a PROM and plugged in. We also often make use of host-target debuggers, in which the basic hooks for debugging are provided by the target and a more sophisticated user interface is created by the host.

A PC or workstation offers a programming environment that is in many ways much friendlier than the typical embedded computing platform. But one problem with this approach emerges when debugging code talks to I/O devices. Since the host almost certainly will not have the same devices configured in the same way, the embedded code cannot be run as is on the host. In many cases, a ***testbench program*** can be built to help debug the embedded code. The ***testbench*** generates inputs to simulate the actions of the input devices; it may also take the output values and compare them against expected values, providing valuable early debugging help. The embedded code may need to be slightly modified to work with the testbench, but careful coding (such as using the #ifdef directive in C) can ensure that the changes can be undone easily and without introducing bugs.

4.6.2 Debugging Techniques

A good deal of software debugging can be done by compiling and executing the code on a PC or workstation. But at some point it inevitably becomes necessary to run code on the embedded hardware platform. Embedded systems are usually less friendly programming environments than PCs. Nonetheless, the resourceful designer has several options available for debugging the system.

The serial port found on most evaluation boards is one of the most important debugging tools. In fact, it is often a good idea to design a serial port into an embedded system even if it will not be used in the final product; the serial port can be used not only for development debugging but also for diagnosing problems in the field.

Another very important debugging tool is the ***breakpoint***. The simplest form of a breakpoint is for the user to specify an address at which the program's execution is to break. When the PC reaches that address, control is returned to the monitor program. From the monitor program, the user can examine and/or modify CPU registers, after which execution can be continued. Implementing breakpoints does

not require using exceptions or external devices. Programming Example 4.1 shows how to use instructions to create breakpoints.

Programming Example 4.1

Breakpoints

A breakpoint is a location in memory at which a program stops executing and returns to the debugging tool or monitor program. Implementing breakpoints is very simple—you simply replace the instruction at the breakpoint location with a subroutine call to the monitor. In the following code, to establish a breakpoint at location 0x40c in some ARM code, we've replaced the branch (B) instruction normally held at that location with a subroutine call (BL) to the breakpoint handling routine:

```
0 x 400 MUL r4,r4,r6              0 x 400 MUL r4,r4,r6
0 x 404 ADD r2,r2,r4     ⟶        0 x 404 ADD r2,r2,r4
0 x 408 ADD r0,r0,#1              0 x 408 ADD r0,r0,#1
0 x 40c B loop                   0 x 40c BL bkpoint
```

When the breakpoint handler is called, it saves all the registers and can then display the CPU state to the user and take commands.

To continue execution, the original instruction must be replaced in the program. If the breakpoint can be erased, the original instruction can simply be replaced and control returned to that instruction. This will normally require fixing the subroutine return address, which will point to the instruction after the breakpoint. If the breakpoint is to remain, then the original instruction can be replaced and a new temporary breakpoint placed at the next instruction (taking jumps into account, of course). When the temporary breakpoint is reached, the monitor puts back the original breakpoint, removes the temporary one, and resumes execution.

The Unix *dbx* debugger shows the program being debugged in source code form, but that capability is too complex to fit into some embedded systems. Very simple monitors will require you to specify the breakpoint as an absolute address, which requires you to know how the program was linked. A more sophisticated monitor will read the symbol table and allow you to use labels in the assembly code to specify locations.

Never underestimate the importance of LEDs in debugging. As with serial ports, it is often a good idea to design a few to indicate the system state even if they will not normally be seen in use. LEDs can be used to show error conditions, when the code enters certain routines, or to show idle time activity. LEDs can be entertaining as well—a simple flashing LED can provide a great sense of accomplishment when it first starts to work.

When software tools are insufficient to debug the system, hardware aids can be deployed to give a clearer view of what is happening when the system is running. The ***microprocessor in-circuit emulator (ICE)*** is a specialized hardware tool that can help debug software in a working embedded system. At the heart of an

in-circuit emulator is a special version of the microprocessor that allows its internal registers to be read out when it is stopped. The in-circuit emulator surrounds this specialized microprocessor with additional logic that allows the user to specify breakpoints and examine and modify the CPU state. The CPU provides as much debugging functionality as a debugger within a monitor program, but does not take up any memory. The main drawback to in-circuit emulation is that the machine is specific to a particular microprocessor, even down to the pinout. If you use several microprocessors, maintaining a fleet of in-circuit emulators to match can be very expensive.

The *logic analyzer* [Ald73] is the other major piece of instrumentation in the embedded system designer's arsenal. Think of a logic analyzer as an array of inexpensive oscilloscopes—the analyzer can sample many different signals simultaneously (tens to hundreds) but can display only 0, 1, or changing values for each. All these logic analysis channels can be connected to the system to record the activity on many signals simultaneously. The logic analyzer records the values on the signals into an internal memory and then displays the results on a display once the memory is full or the run is aborted. The logic analyzer can capture thousands or even millions of samples of data on all of these channels, providing a much larger time window into the operation of the machine than is possible with a conventional oscilloscope.

A typical logic analyzer can acquire data in either of two modes that are typically called *state* and *timing modes*. To understand why two modes are useful and the difference between them, it is important to remember that an oscilloscope trades reduced resolution on the signals for the longer time window. The measurement resolution on each signal is reduced in both voltage and time dimensions. The reduced voltage resolution is accomplished by measuring logic values (0, 1, x) rather than analog voltages. The reduction in timing resolution is accomplished by sampling the signal, rather than capturing a continuous waveform as in an analog oscilloscope.

State and timing mode represent different ways of sampling the values. Timing mode uses an internal clock that is fast enough to take several samples per clock period in a typical system. State mode, on the other hand, uses the system's own clock to control sampling, so it samples each signal only once per clock cycle. As a result, timing mode requires more memory to store a given number of system clock cycles. On the other hand, it provides greater resolution in the signal for detecting glitches. Timing mode is typically used for glitch-oriented debugging, while state mode is used for sequentially oriented problems.

The internal architecture of a logic analyzer is shown in Figure 4.27. The system's data signals are sampled at a latch within the logic analyzer; the latch is controlled by either the system clock or the internal logic analyzer sampling clock, depending on whether the analyzer is being used in state or timing mode. Each sample is copied into a vector memory under the control of a state machine. The latch, timing circuitry, sample memory, and controller must be designed to run at high speed

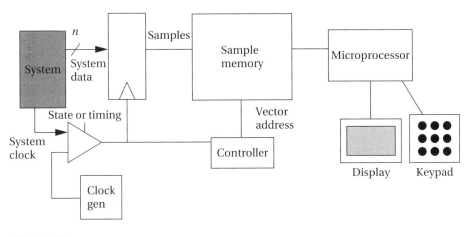

FIGURE 4.27

Architecture of a logic analyzer.

since several samples per system clock cycle may be required in timing mode. After the sampling is complete, an embedded microprocessor takes over to control the display of the data captured in the sample memory.

Logic analyzers typically provide a number of formats for viewing data. One format is a timing diagram format. Many logic analyzers allow not only customized displays, such as giving names to signals, but also more advanced display options. For example, an inverse assembler can be used to turn vector values into microprocessor instructions.

The logic analyzer does not provide access to the internal state of the components, but it does give a very good view of the externally visible signals. That information can be used for both functional and timing debugging.

4.6.3 Debugging Challenges

Logical errors in software can be hard to track down, but errors in real-time code can create problems that are even harder to diagnose. Real-time programs are required to finish their work within a certain amount of time; if they run too long, they can create very unexpected behavior. Example 4.2 demonstrates one of the problems that can arise.

Example 4.2

A timing error in real-time code

Let's consider a simple program that periodically takes an input from an analog/digital converter, does some computations on it, and then outputs the result to a digital/analog converter.

To make it easier to compare input to output and see the results of the bug, we assume that the computation produces an output equal to the input, but that a bug causes the computation to run 50% longer than its given time interval. A sample input to the program over several sample periods follows:

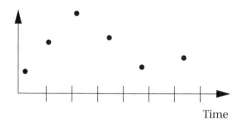

Time

If the program ran fast enough to meet its deadline, the output would simply be a time-shifted copy of the input. But when the program runs over its allotted time, the output will become very different. Exactly what happens depends in part on the behavior of the A/D and D/A converters, so let's make some assumptions. First, the A/D converter holds its current sample in a register until the next sample period, and the D/A converter changes its output whenever it receives a new sample. Next, a reasonable assumption about interrupt systems is that, when an interrupt is not satisfied and the device interrupts again, the device's old value will disappear and be replaced by the new value. The basic situation that develops when the interrupt routine runs too long is something like this:

1. The A/D converter is prompted by the timer to generate a new value, saves it in the register, and requests an interrupt.

2. The interrupt handler runs too long from the last sample.

3. The A/D converter gets another sample at the next period.

4. The interrupt handler finishes its first request and then immediately responds to the second interrupt. It never sees the first sample and only gets the second one.

Thus, assuming that the interrupt handler takes 1.5 times longer than it should, here is how it would process the sample input:

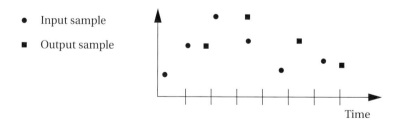

- Input sample
- Output sample

Time

The output waveform is seriously distorted because the interrupt routine grabs the wrong samples and puts the results out at the wrong times.

The exact results of missing real-time deadlines depend on the detailed characteristics of the I/O devices and the nature of the timing violation. This makes debugging real-time problems especially difficult. Unfortunately, the best advice is that if a system exhibits truly unusual behavior, missed deadlines should be suspected. In-circuit emulators, logic analyzers, and even LEDs can be useful tools in checking the execution time of real-time code to determine whether it in fact meets its deadline.

4.7 SYSTEM-LEVEL PERFORMANCE ANALYSIS

Bus-based systems add another layer of complication to performance analysis. The CPU, bus, and memory or I/O device all act as independent elements that can operate in parallel. In this section, we will develop some basic techniques for analyzing the performance of bus-based systems.

4.7.1 System-Level Performance Analysis

System-level performance involves much more than the CPU. We often focus on the CPU because it processes instructions, but any part of the system can affect total system performance. More precisely, the CPU provides an upper bound on performance, but any other part of the system can slow down the CPU. Merely counting instruction execution times is not enough.

Consider the simple system of Figure 4.28. We want to move data from memory to the CPU to process it. To get the data from memory to the CPU we must:

- read from the memory;
- transfer over the bus to the cache; and
- transfer from the cache to the CPU.

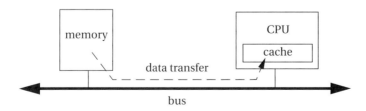

FIGURE 4.28

System level data flows and performance.

The time required to transfer from the cache to the CPU is included in the instruction execution time, but the other two times are not.

The most basic measure of performance we are interested in is **bandwidth**—the rate at which we can move data. Ultimately, if we are interested in real-time performance, we are interested in real-time performance measured in seconds. But often the simplest way to measure performance is in units of clock cycles. However, different parts of the system will run at different clock rates. We have to make sure that we apply the right clock rate to each part of the performance estimate when we convert from clock cycles to seconds.

Bandwidth questions often come up when we are transferring large blocks of data. For simplicity, let's start by considering the bandwidth provided by only one system component, the bus. Consider an image of 320×240 pixels, with each pixel composed of 3 bytes of data. This gives a grand total of $230, 400$ bytes of data. If these images are video frames, we want to check if we can push one frame through the system within the $1/30$ s that we have to process a frame before the next one arrives.

Let us assume that we can transfer one byte of data every microsecond, which implies a bus speed of 1 MHz. In this case, we would require $230, 400 \, \mu s = 0.23$ s to transfer one frame. That is more than the 0.033 s allotted to the data transfer. We would have to increase the transfer rate by $7\times$ to satisfy our performance requirement.

We can increase bandwidth in two ways: We can increase the clock rate of the bus or we can increase the amount of data transferred per clock cycle. For example, if we increased the bus to carry four bytes or 32 bits per transfer, we would reduce the transfer time to 0.058 s. If we could also increase the bus clock rate to 2 MHz, then we would reduce the transfer time to 0.029 s, which is within our time budget for the transfer.

How do we know how long it takes to transfer one unit of data? To determine that, we have to look at the data sheet for the bus. As we saw in Section 4.1.1, a bus transfer generally takes more than one bus cycle. Burst transfers, which move to contiguous locations, may be more efficient per byte. We also need to know the width of the bus—how many bytes per transfer. Finally, we need to know the bus clock period, which in general will be different from the CPU clock period.

Let's call the bus clock period P and the bus width W. We will put W in units of bytes but we could use other measures of width as well. We want to write formulas for the time required to transfer N bytes of data. We will write our basic formulas in units of bus cycles T, then convert those bus cycle counts to real time t using the bus clock period P:

$$t = TP. \tag{4.1}$$

As shown in Figure 4.29, a basic bus transfer transfers a W-wide set of bytes. The data transfer itself takes D clock cycles. (Ideally, $D = 1$, but a memory that introduces wait states is one example of a transfer that could require $D > 1$ cycles.)

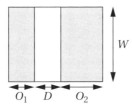

FIGURE 4.29

Times and data volumes in a basic bus transfer.

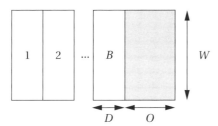

FIGURE 4.30

Times and data volumes in a burst bus transfer.

Addresses, handshaking, and other activities constitute overhead that may occur before (O_1) or after (O_2) the data. For simplicity, we will lump the overhead into $O = O_1 + O_2$. This gives a total transfer time in clock cycles of:

$$T_{\text{basic}}(N) = (D + O)\frac{N}{W}. \tag{4.2}$$

As shown in Figure 4.30, a burst transaction performs B transfers of W bytes each. Each of those transfers will require D clock cycles. The bus also introduces O cycles of overhead per burst. This gives

$$T_{\text{burst}}(N) = (BD + O)\frac{N}{BW}. \tag{4.3}$$

Bandwidth questions also come up in situations that we do not normally think of as communications. Transferring data into and out of components also raises questions of bandwidth. The simplest illustration of this problem is memory.

The width of a memory determines the number of bits we can read from the memory in one cycle. That is a form of data bandwidth. We can change the types of memory components we use to change the memory bandwidth; we may also be able to change the format of our data to accommodate the memory components.

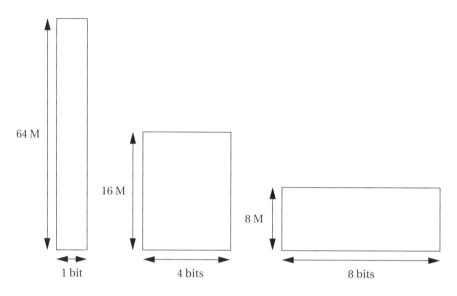

FIGURE 4.31

Memory aspect ratios.

A single memory chip is not solely specified by the number of bits it can hold. As shown in Figure 4.31, memories of the same size can have different **aspect ratios**. For example, a 64-MB memory that is 1-bit-wide will present 64 million addresses of 1-bit data. The same size memory in a 4-bit-wide format will have 16 distinct addresses and an 8-bit-wide memory will have 8 million distinct addresses.

Memory chips do not come in extremely wide aspect ratios. However, we can build wider memories by using several chips. By choosing chips with the right aspect ratio, we can build a memory system with the total amount of storage that we want and that presents the data width that we want.

The memory system width may also be determined by the memory modules we use. Rather than buy memory chips individually, we may buy memory as SIMMs or DIMMs. These memories are wide but generally only come in fairly standard widths.

Which aspect ratio is preferable for the overall memory system depends in part on the format of the data that we want to store in the memory and the speed with which it must be accessed, giving rise to bandwidth analysis.

We also have to consider the time required to read or write a memory. Once again, we refer to the component data sheets to find these values. Access times depend quite a bit on the type of memory chip used as we saw in Section 4.2.2. Page modes operate similarly to burst modes in buses. If the memory is not synchronous, we can still refer the times between events back to the bus clock cycle to determine the number of clock cycles required for an access.

The basic form of the equation for memory transfer time is that of Eq. 4.3, where O is determined by the page mode overhead and D is the time between successive transfers.

However, the situation is slightly more complex if the data types do not fit naturally into the width of the memory. Let's say that we want to store color video pixels in our memory. A standard pixel is 38-bit color values (red, green, blue, for example). A 24-bit-wide memory would allow us to read or write an entire pixel value in one access. An 8-bit-wide memory, in contrast, would require three accesses for the pixel. If we have a 32-bit-wide memory, we have two main choices: We could waste one byte of each transfer or use that byte to store unrelated data, or we could pack the pixels. In the latter case, the first read would get all of the first pixel and one byte of the second pixel; the second transfer would get the last two bytes of the second pixel and the first two bytes of the third pixel; and so forth. The total number of accesses required to read E data elements of w bits each out of a memory of width W is:

$$ A = \left[\left(\frac{E}{w} \right) \bmod W \right] + 1. \tag{4.4} $$

The next example applies our bandwidth models to a simple design problem.

Example 4.3

Performance bottlenecks in a bus-based system

Consider a simple bus-based system:

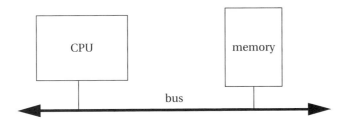

We want to transfer data between the CPU and the memory over the bus. We need to be able to read a 320×240 video frame into the CPU at the rate of 30 frames/s, for a total of 612,000 bytes/s. Which will be the bottleneck and limit system performance: the bus or the memory?

Let's assume that the bus has a 1-MHz clock rate (period of 10^{-6} sec) and is 2 bytes wide, with $D = 1$ and $O = 3$. This gives a total transfer time of

$$ T_{\text{basic}} = (1 + 3) \frac{612{,}000}{2} = 1{,}224{,}000 \text{ cycles,} \tag{4.5} $$

$$t = T_{\text{basic}} \cdot P = 1,224,000 \cdot 1 \times 10^{-6} = 1.224 \text{ sec.} \tag{4.6}$$

Since the total time to transfer one second's worth of frames is more than 1 s, the bus is not fast enough for our application.

The memory provides a burst mode with $B = 4$ but is only 4 bits wide, giving $W = 0.5$. For this memory, $D = 1$ and $O = 4$. The clock period for this memory is 10^{-7} s. Then

$$T_{\text{mem}} = (4 \cdot 1 + 4)\frac{612,000}{4 \cdot 0.5} = 2,448,000 \text{ cycles,} \tag{4.7}$$

$$t = T_{\text{mem}} \cdot P = 2,448,000 \cdot 1 \times 10^{-7} = 0.2448 \text{ sec} \tag{4.8}$$

The memory requires <1 s to transfer the 30 frames that must be transmitted in 1 s, so it is fast enough.

One way to explore design trade-offs is to build a spreadsheet:

Bus		Memory	
Clock period	1.00E − 06	Clock period	1.00E − 08
W	2	W	0.5
D	1	D	1
O	3	O	4
		B	4
N	612000	N	612000
T_{basic}	1224000	T_{mem}	2448000
t	1.22E + 00	t	2.45E − 02

If we insert the formulas for bandwidth into the spreadsheet, we can change values like bus width and clock rate and instantly see their effects on available bandwidth.

4.7.2 Parallelism

Computer systems have multiple components. When the hardware and software are properly designed, those systems can operate independently for at least part of the time. When different components of the system operate in parallel, we can get more work done in a given amount of time.

Direct memory access is a prime example of parallelism. DMA was designed to off-load memory transfers from the CPU. The CPU can do other useful work while the DMA transfer is running.

Figure 4.32 shows the paths of data transfers without and with DMA when transferring from memory to a device. Without DMA, the data must go through the CPU;

transfer without DMA

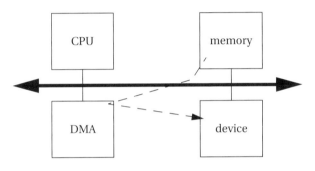

transfer without DMA

FIGURE 4.32

DMA transfers and parallelism.

the CPU cannot do useful work at that time. Our bandwidth analysis illuminates an important point about that transfer time—the CPU is tied up for the amount of time required for the bus transfer. Since buses often operate at slower clock rates than the CPU, that time can be considerable. We can significantly increase system performance by overlapping operations on the different units of the system. The timing diagrams of Figure 4.33 show timing diagrams for two versions of a computation. The top timing diagram shows activity in the system when the CPU first performs some setup operations, then waits for the bus transfer to complete, then resumes its work. In the bottom timing diagram, we have rewritten the program on the CPU so that its main work is broken into two sections. In this case, once the first transfer is done, the CPU can start working on that data. Meanwhile, thanks to DMA, the second transfer happens on the bus at the same time. Once that data arrives and the first calculation is finished, the CPU can

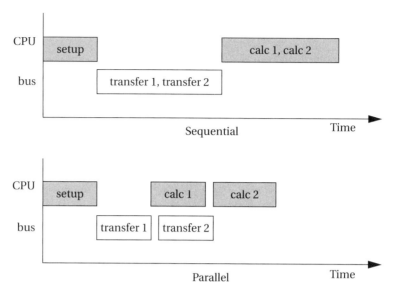

FIGURE 4.33

Sequential and parallel schedules in a bus-based system.

go on to the second part of the computation. The result is that the entire computation finishes considerably earlier than in the sequential case.

Design Example

4.8 ALARM CLOCK

Our first system design example will be an alarm clock. We use a microprocessor to read the clock's buttons and update the time display. Since we now have an understanding of I/O, we work through the steps of the methodology to go from a concept to a completed and tested system.

4.8.1 Requirements

The basic functions of an alarm clock are well understood and easy to enumerate. Figure 4.34 illustrates the front panel design for the alarm clock. The time is shown as four digits in 12-h format; we use a light to distinguish between AM and PM. We use several buttons to set the clock time and alarm time. When we press the *hour* and *minute* buttons, we advance the hour and minute, respectively, by one. When setting the time, we must hold down the *set time* button while we hit the *hour* and *minute* buttons; the *set alarm* button works in a similar fashion. We turn the alarm on and off with the *alarm on* and *alarm off* buttons. When the alarm is activated, the *alarm ready* light is on. A separate speaker provides the audible alarm.

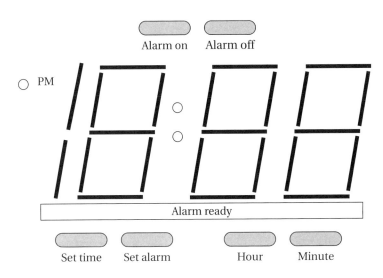

Alarm on Alarm off

PM

Alarm ready

Set time Set alarm Hour Minute

FIGURE 4.34

Front panel of the alarm clock.

We are now ready to create the requirements table.

Name	Alarm clock.
Purpose	A 24-h digital clock with a single alarm.
Inputs	Six push buttons: set time, set alarm, hour, minute, alarm on, alarm off.
Outputs	Four-digit, clock-style output. PM indicator light. Alarm ready light. Buzzer.
Functions	Default mode: The display shows the current time. PM light is on from noon to midnight.
	Hour and minute buttons are used to advance time and alarm, respectively. Pressing one of these buttons increments the hour/minute once.
	Depress set time button: This button is held down while hour/minute buttons are pressed to set time. New time is automatically shown on display.
	Depress set alarm button: While this button is held down, display shifts to current alarm setting; depressing hour/minute buttons sets alarm value in a manner similar to setting time.
	Alarm on: puts clock in alarm-on state, causes clock to turn on buzzer when current time reaches alarm time, turns on alarm ready light.

(Continued)

	Alarm off: turns off buzzer, takes clock out of alarm-on state, turns off alarm ready light.
Performance	Displays hours and minutes but not seconds. Should be accurate within the accuracy of a typical microprocessor clock signal. (Excessive accuracy may unreasonably drive up the cost of generating an accurate clock.)
Manufacturing cost	Consumer product range. Cost will be dominated by the microprocessor system, not the buttons or display.
Power	Powered by AC through a standard power supply.
Physical size and weight	Small enough to fit on a nightstand with expected weight for an alarm clock.

4.8.2 Specification

The basic function of the clock is simple, but we do need to create some classes and associated behaviors to clarify exactly how the user interface works.

Figure 4.35 shows the basic classes for the alarm clock. Borrowing a term from mechanical watches, we call the class that handles the basic clock operation the *Mechanism* class. We have three classes that represent physical elements: *Lights** for all the digits and lights, *Buttons** for all the buttons, and *Speaker** for the sound output. The *Buttons** class can easily be used directly by *Mechanism*. As discussed below, the physical display must be scanned to generate the digits output, so we introduce the *Display* class to abstract the physical lights.

The details of the low-level user interface classes are shown in Figure 4.36. The *Buzzer** class allows the buzzer to be turned off; we will use analog electronics to generate the buzz tone for the speaker. The *Buttons** class provides read-only access to the current state of the buttons. The *Lights** class allows us to drive the lights. However, to save pins on the display, *Lights** provides signals for only one digit, along with a set of signals to indicate which digit is currently being addressed.

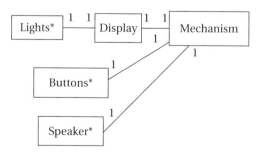

FIGURE 4.35

Class diagram for the alarm clock.

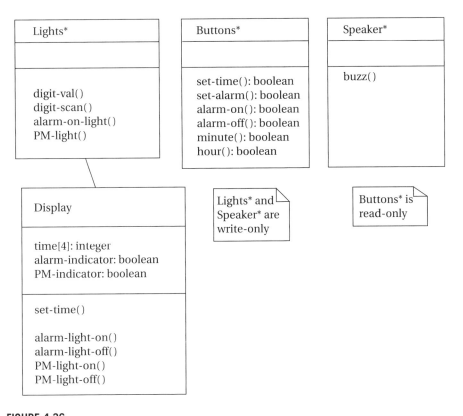

FIGURE 4.36

Details of low-level class for the alarm clock.

We generate the display by scanning the digits periodically. That function is performed by the *Display* class, which makes the display appear as an unscanned, continuous display to the rest of the system.

The *Mechanism* class is described in Figure 4.37. This class keeps track of the current time, the current alarm time, whether the alarm has been turned on, and whether it is currently buzzing. The clock shows the time only to the minute, but it keeps internal time to the second. The time is kept as discrete digits rather than a single integer to simplify transferring the time to the display. The class provides two behaviors, both of which run continuously. First, *scan-keyboard* is responsible for looking at the inputs and updating the alarm and other functions as requested by the user. Second, *update-time* keeps the current time accurate.

Figure 4.38 shows the state diagram for *update-time*. This behavior is straightforward, but it must do several things. It is activated once per second and must update the seconds clock. If it has counted 60 s, it must then update the displayed time; when it does so, it must roll over between digits and keep track of AM-to-PM and PM-to-AM transitions. It sends the updated time to the display object. It also

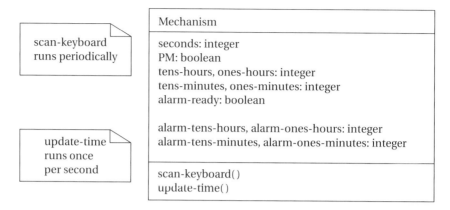

FIGURE 4.37

The Mechanism class.

compares the time with the alarm setting and sets the alarm buzzing under proper conditions.

The state diagram for *scan-keyboard* is shown in Figure 4.39. This function is called periodically, frequently enough so that all the user's button presses are caught by the system. Because the keyboard will be scanned several times per second, we do not want to register the same button press several times. If, for example, we advanced the minutes count on every keyboard scan when the *set-time* and *minutes* buttons were pressed, the time would be advanced much too fast. To make the buttons respond more reasonably, the function computes button activations—it compares the current state of the button to the button's value on the last scan, and it considers the button activated only when it is on for this scan but was off for the last scan. Once computing the activation values for all the buttons, it looks at the activation combinations and takes the appropriate actions. Before exiting, it saves the current button values for computing activations the next time this behavior is executed.

4.8.3 System Architecture

The software and hardware architectures of a system are always hard to completely separate, but let's first consider the software architecture and then its implications on the hardware.

The system has both periodic and aperiodic components—the current time must obviously be updated periodically, and the button commands occur occasionally.

It seems reasonable to have the following two major software components:

- An interrupt-driven routine can update the current time. The current time will be kept in a variable in memory. A timer can be used to interrupt periodically and update the time. As seen in the subsequent discussion of the hardware

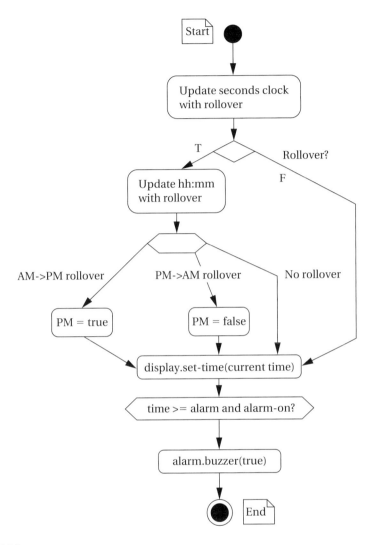

FIGURE 4.38

State diagram for update-time.

architecture, the display must be sent the new value when the minute value changes. This routine can also maintain the PM indicator.

- A foreground program can poll the buttons and execute their commands. Since buttons are changed at a relatively slow rate, it makes no sense to add the hardware required to connect the buttons to interrupts. Instead, the fore-ground program will read the button values and then use simple conditional tests to implement the commands, including setting the current time, setting

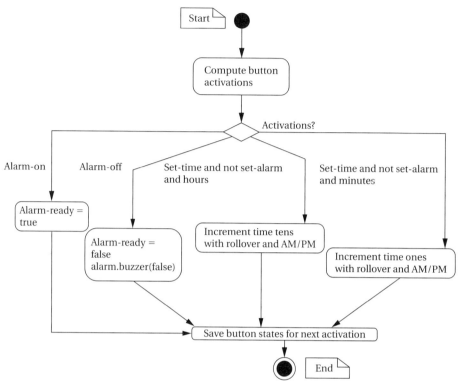

FIGURE 4.39

State diagram for scan-keyboard.

the alarm, and turning off the alarm. Another routine called by the foreground program will turn the buzzer on and off based on the alarm time.

An important question for the interrupt-driven current time handler is how often the timer interrupts occur. A 1-min interval would be very convenient for the software, but a one-minute timer would require a large number of counter bits. It is more realistic to use a one-second timer and to use a program variable to count the seconds in a minute.

The foreground code will be implemented as a while loop:

```
while (TRUE) {
    read_buttons(button_values);/* read inputs */
    process_command(button_values);/* do commands */
    check_alarm();/* decide whether to turn on the alarm */
}
```

The loop first reads the buttons using read_buttons(). In addition to reading the current button values from the input device, this routine must preprocess the

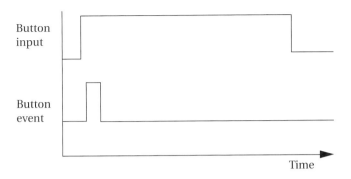

FIGURE 4.40

Preprocessing button inputs.

button values so that the user interface code will respond properly. The buttons will remain depressed for many sample periods since the sample rate is much faster than any person can push and release buttons. We want to make sure that the clock responds to this as a single depression of the button, not one depression per sample interval. As shown in Figure 4.40, this can be done by performing a simple edge detection on the button input—the button event value is 1 for one sample period when the button is depressed and then goes back to 0 and does not return to 1 until the button is depressed and then released. This can be accomplished by a simple two-state machine.

The process_command() function is responsible for responding to button events. The check_alarm() function checks the current time against the alarm time and decides when to turn on the buzzer. This routine is kept separate from the command processing code since the alarm must go on when the proper time is reached, independent of the button inputs.

We have determined from the software architecture that we will need a timer connected to the CPU. We will also need logic to connect the buttons to the CPU bus. In addition to performing edge detection on the button inputs, we must also of course debounce the buttons.

The final step before starting to write code and build hardware is to draw the state transition graph for the clock's commands. That diagram will be used to guide the implementation of the software components.

4.8.4 Component Design and Testing

The two major software components, the interrupt handler and the foreground code, can be implemented relatively straightforwardly. Since most of the functionality of the interrupt handler is in the interruption process itself, that code is best tested on the microprocessor platform. The foreground code can be more easily tested on the PC or workstation used for code development. We can create a testbench

for this code that generates button depressions to exercise the state machine. We will also need to simulate the advancement of the system clock. Trying to directly execute the interrupt handler to control the clock is probably a bad idea—not only would that require some type of emulation of interrupts, but it would require us to count interrupts second by second. A better testing strategy is to add testing code that updates the clock, perhaps once per four iterations of the foreground while loop.

The timer will probably be a stock component, so we would then focus on implementing logic to interface to the buttons, display, and buzzer. The buttons will require debouncing logic. The display will require a register to hold the current display value in order to drive the display elements.

4.8.5 System Integration and Testing

Because this system has a small number of components, system integration is relatively easy. The software must be checked to ensure that debugging code has been turned off. Three types of tests can be performed. First, the clock's accuracy can be checked against a reference clock. Second, the commands can be exercised from the buttons. Finally, the buzzer's functionality should be verified.

SUMMARY

The microprocessor is only one component in an embedded computing system—memory and I/O devices are equally important. The microprocessor bus serves as the glue that binds all these components together. Hardware platforms for embedded systems are often built around common platforms with appropriate amounts of memory and I/O devices added on; low-level monitor software also plays an important role in these systems.

What We Learned

- CPU buses are built on handshaking protocols.

- A variety of memory components are available, which vary widely in speed, capacity, and other capabilities.

- An I/O device uses logic to interface to the bus so that the CPU can read and write the device's registers.

- Embedded systems can be debugged using a variety of hardware and software methods.

- System-level performance depends not just on the CPU, but the memory and bus as well.

FURTHER READING

Shanley and Anderson [Min95] describe the PCI bus in detail. Dahlin [Dah00] describes how to interface to a touchscreen. Collins [Col97] describes the design of microprocessor in-circuit emulators. Earnshaw et al. [Ear97] describe an advanced debugging environment for the ARM architecture.

QUESTIONS

Q4-1 Draw a UML sequence diagram that shows a four-cycle handshake between a bus master and a device.

Q4-2 Draw a timing diagram with the following signals (where $[t_1, t_2]$ is the time interval starting at t_1 and ending at t_2):

 a. Signal A is stable [0, 10], changing [10, 15], stable [15, 30].
 b. Signal B is 1 [0, 5], falling [5, 7], 0 [7, 20], changing [20, 30].
 c. Signal C is changing [0, 10], 0 [10, 15], rising [15, 18], 1 [18, 25], changing [25, 30].

Q4-3 Draw a timing diagram for a write operation with no wait states.

Q4-4 Draw a timing diagram for a read operation on a bus in which the read includes two wait states.

Q4-5 Draw a timing diagram for a write operation on a bus in which the write takes two wait states.

Q4-6 Draw a timing diagram for a burst write operation that writes four locations.

Q4-7 Draw a UML state diagram for a burst read operation with wait states. One state diagram is for the bus master and the other is for the device being read.

Q4-8 Draw a UML sequence diagram for a burst read operation with wait states.

Q4-9 Draw timing diagrams for

 a. A device becoming bus master.
 b. The device returning control of the bus to the CPU.

Q4-10 Draw a timing diagram that shows a complete DMA operation, including handing off the bus to the DMA controller, performing the DMA transfer, and returning bus control back to the CPU.

Q4-11 Draw UML state diagrams for a bus mastership transaction in which one side shows the CPU as the default bus master and the other shows the device that can request bus mastership.

Q4-12 Draw a UML sequence diagram for a bus mastership request, grant, and return.

Q4-13 Draw a UML sequence diagram for a complete DMA transaction, including the DMA controller requesting the bus, the DMA transaction itself, and returning control of the bus to the CPU.

Q4-14 Draw a UML sequence diagram showing a read operation across a bus bridge.

Q4-15 Draw a UML sequence diagram showing a write operation with wait states across a bus bridge.

Q4-16 If you have a choice among several DRAMs of the same capacity but with different data widths, when would you want to use a narrower memory? When would you want to use a taller memory?

Q4-17 Draw a UML sequence diagram for a read transaction that includes a DRAM refresh operation. The sequence diagram should include the CPU, the DRAM interface, and the DRAM internals to show the refresh itself.

Q4-18 Design the logic required to build a 64 M \times 32-bit memory out of 16 M \times 32 memories.

Q4-19 Design the logic required to build a 512 M \times 16 memory out of 256 M \times 4 memories.

Q4-20 Design the logic required to build a 1G \times 16 memory out of 256 M \times 4 memories.

Q4-21 Draw a UML class diagram that describes a hardware timer/counter. The device can be loaded with a count value. It can decrement the count down to zero based either on a bus signal or by counting some multiple of clock cycles.

Q4-22 Draw a UML class diagram for an analog/digital converter.

Q4-23 Draw a UML class diagram for a digital/analog converter.

Q4-24 Write ARM assembly language code that handles a breakpoint. It should save the necessary registers, call a subroutine to communicate with the host, and upon return from the host, cause the breakpointed instruction to be properly executed.

Q4-25 Assume an A/D converter is supplying samples at 44.1 kHz.

 a. How much time is available per sample for CPU operations?

 b. If the interrupt handler executes 100 instructions obtaining the sample and passing it onto the application routine, how many instructions can be executed on a 20 MHz RISC processor that executes 1 instruction per cycle?

Q4-26 If an interrupt handler executes for too long and the next interrupt occurs before the last call to the handler has finished, what happens?

Q4-27 Consider a system in which an interrupt handler passes on samples to an FIR filter program that runs in the background.

 a. If the interrupt handler takes too long, how does the FIR filter's output change?

 b. If the FIR filter code takes too long, how does its output change?

Q4-28 Assume that your microprocessor implements an ICE instruction that asserts a bus signal that causes a microprocessor in-circuit emulator to start. Also assume that the microprocessor allows all internal registers to be observed and controlled through a boundary scan chain. Draw a UML sequence diagram of the ICE operation, including execution of the ICE instruction, uploading the microprocessor state to the ICE, and returning control to the microprocessor's program. The sequence diagram should include the microprocessor, the microprocessor in-circuit emulator, and the user.

Q4-29 We are given a 1-word wide bus that supports single-word and burst transfers. The overhead of the single-word transfer is 2 clock cycles. Plot the breakeven point between single-word and burst transfers for several values of burst overhead—for each value of overhead, plot the length of burst transfer at which the burst-transfer is as fast as a series of single-word transfers. Plot breakeven for burst overhead values of 0, 1, 2, and 3 cycles.

Q4-30 You are designing a bus-based computer system: The input device I1 sends its data to program P1; P1 sends its output to output device O1. Is there any way to overlap bus transfers and computations in this system?

LAB EXERCISES

L4-1 Use an instruction-based simulator to simulate a program. How fast was the simulator? Did you have to make any adjustments to your program in order to make it simulate properly?

L4-2 Use a logic analyzer to view system activity on your bus.

L4-3 If your logic analyzer is capable of on-the-fly disassembly, use it to display bus activity in the form of instructions, rather than simply 1s and 0s.

L4-4 Attach LEDs to your system bus so that you can monitor its activity. For example, use an LED to monitor the read/write line on the bus.

L4-5 Design logic to interface an I/O device to your microprocessor.

L4-6 Have someone else deliberately introduce a bug into one of your programs, and then use the appropriate debugging tools to find and correct the bug.

Program Design and Analysis

- Some useful components for embedded software.

- Models of programs, such as data flow and control flow graphs.

- An introduction to compilation methods.

- Analyzing and optimizing programs for performance, size, and power consumption.

- How to test programs to verify their correctness.

- A software modem.

INTRODUCTION

In this chapter we study in detail the process of programming embedded processors. The creation of embedded programs is at the heart of embedded system design. If you are reading this book, you almost certainly have an understanding of programming, but designing and implementing embedded programs is different and more challenging than writing typical workstation or PC programs. Embedded code must not only provide rich functionality, it must also often run at a required rate to meet system deadlines, fit into the allowed amount of memory, and meet power consumption requirements. Designing code that simultaneously meets multiple design constraints is a considerable challenge, but luckily there are techniques and tools that we can use to help us through the design process. Making sure that the program works is also a challenge, but once again methods and tools come to our aid.

Throughout the discussion we concentrate on high-level programming languages, specifically C. High-level languages were once shunned as too inefficient for embedded microcontrollers, but better compilers, more compiler-friendly architectures, and faster processors and memory have made high-level language programs common. Some sections of a program may still need to be written in assembly language if the compiler doesn't give sufficiently good results, but even when coding in assembly language it is often helpful to think about the program's functionality in high-level form. Many of the analysis and optimization techniques that we study in this chapter are equally applicable to programs written in assembly language.

The next section talks about some software components that are commonly used in embedded software. Section 5.2 introduces the control/data flow graph as a model for high-level language programs (which can also be applied to programs

209

written originally in assembly language). Section 5.3 reviews the assembly and linking process and Section 5.4 reviews as background the basic steps in compilation. Section 5.5 discusses code optimization. We talk about optimization techniques specific to embedded computing in the next three sections: performance in Section 5.6, energy consumption in Section 5.8, and size in Section 5.9. Section 5.6 discusses the analysis of software performance while Section 5.7 introduces techniques to optimize software performance. Section 5.8 discusses energy and power optimization while Section 5.9 talks about optimizing programs for size. In Section 5.10, we discuss techniques for ensuring that the programs you write are correct. We close with a software modem as a design example in Section 5.11.

5.1 COMPONENTS FOR EMBEDDED PROGRAMS

In this section, we consider code for three structures or components that are commonly used in embedded software: the state machine, the circular buffer, and the queue. State machines are well suited to **reactive systems** such as user interfaces; circular buffers and queues are useful in digital signal processing.

5.1.1 State Machines

When inputs appear intermittently rather than as periodic samples, it is often convenient to think of the system as reacting to those inputs. The reaction of most systems can be characterized in terms of the input received and the current state of the system. This leads naturally to a **finite-state machine** style of describing the reactive system's behavior. Moreover, if the behavior is specified in that way, it is natural to write the program implementing that behavior in a state machine style. The state machine style of programming is also an efficient implementation of such computations. Finite-state machines are usually first encountered in the context of hardware design. Programming Example 5.1 shows how to write a finite-state machine in a high-level programming language.

Programming Example 5.1

A software state machine

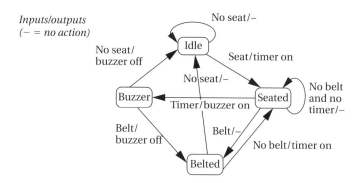

The behavior we want to implement is a simple seat belt controller [Chi94]. The controller's job is to turn on a buzzer if a person sits in a seat and does not fasten the seat belt within a fixed amount of time. This system has three inputs and one output. The inputs are a sensor for the seat to know when a person has sat down, a seat belt sensor that tells when the belt is fastened, and a timer that goes off when the required time interval has elapsed. The output is the buzzer. Appearing below is a state diagram that describes the seat belt controller's behavior.

The idle state is in force when there is no person in the seat. When the person sits down, the machine goes into the seated state and turns on the timer. If the timer goes off before the seat belt is fastened, the machine goes into the buzzer state. If the seat belt goes on first, it enters the belted state. When the person leaves the seat, the machine goes back to idle.

To write this behavior in C, we will assume that we have loaded the current values of all three inputs (seat, belt, timer) into variables and will similarly hold the outputs in variables temporarily (timer_on, buzzer_on). We will use a variable named state to hold the current state of the machine and a switch statement to determine what action to take in each state. The code follows:

```
#define IDLE 0
#define SEATED 1
#define BELTED 2
#define BUZZER 3

switch (state) { /* check the current state */
        case IDLE:
                if (seat) { state = SEATED; timer_on = TRUE; }
                /* default case is self-loop */
                break;
        case SEATED:
                if (belt) state = BELTED; /* won't hear the
                                        buzzer */
                else if (timer) state = BUZZER; /* didn't put on
                                        belt in time */
                /* default is self-loop */
                break;
        case BELTED:
                if (!seat) state = IDLE; /* person left */
                else if (!belt) state = SEATED; /* person still
                                        in seat */
                break;
        case BUZZER:
                if (belt) state = BELTED; /* belt is on—turn off
                                        buzzer */
                else if (!seat) state = IDLE; /* no one in
                                        seat—turn off buzzer */
                break;
    }
```

This code takes advantage of the fact that the state will remain the same unless explicitly changed; this makes self-loops back to the same state easy to implement. This state machine may be executed forever in a while (TRUE) loop or periodically called by some other code. In either case, the code must be executed regularly so that it can check on the current value of the inputs and, if necessary, go into a new state.

5.1.2 Stream-Oriented Programming and Circular Buffers

The data stream style makes sense for data that comes in regularly and must be processed on the fly. The FIR filter of Example 2.5 is a classic example of stream-oriented processing. For each sample, the filter must emit one output that depends on the values of the last n inputs. In a typical workstation application, we would process the samples over a given interval by reading them all in from a file and then computing the results all at once in a batch process. In an embedded system we must not only emit outputs in real time, but we must also do so using a minimum amount of memory.

The circular buffer is a data structure that lets us handle streaming data in an efficient way. Figure 5.1 illustrates how a circular buffer stores a subset of the data stream. At each point in time, the algorithm needs a subset of the data stream that forms a window into the stream. The window slides with time as we throw out old values no longer needed and add new values. Since the size of the window does not

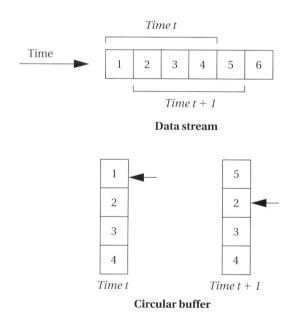

FIGURE 5.1

A circular buffer for streaming data.

change, we can use a fixed-size buffer to hold the current data. To avoid constantly copying data within the buffer, we will move the head of the buffer in time. The buffer points to the location at which the next sample will be placed; every time we add a sample, we automatically overwrite the oldest sample, which is the one that needs to be thrown out. When the pointer gets to the end of the buffer, it wraps around to the top. Programming Example 5.2 provides an efficient implementation of a circular buffer.

Programming Example 5.2

A circular buffer implementation of an FIR filter

Appearing below are the declarations for the circular buffer and filter coefficients, assuming that N, the number of taps in the filter, has been previously defined.

```
int circ_buffer[N]; /* circular buffer for data */
int circ_buffer_head = 0; /* current head of the buffer */
int c[N]; /* filter coefficients (constants) */
```

To write C code for a circular buffer-based FIR filter, we need to modify the original loop slightly. Because the 0th element of data may not be in the 0th element of the circular buffer, we have to change the way in which we access the data. One of the implications of this is that we need separate loop indices for the circular buffer and coefficients.

```
int f, /* loop counter */
    ibuf, /* loop index for the circular buffer */
    ic; /* loop index for the coefficient array */
for (f = 0, ibuf = circ_buffer_head, ic = 0;
    ic < N;
    ibuf = (ibuf == (N - 1) ? 0 : ibuf++),ic++)
    f = f + c[ic] * circ_buffer[ibuf];
```

The above code assumes that some other code, such as an interrupt handler, is replacing the last element of the circular buffer at the appropriate times. The statement ibuf = (ibuf == $(N - 1)$? 0 : ibuf++) is a shorthand C way of incrementing ibuf such that it returns to 0 after reaching the end of the circular buffer array.

5.1.3 Queues

Queues are also used in signal processing and event processing. Queues are used whenever data may arrive and depart at somewhat unpredictable times or when variable amounts of data may arrive. A queue is often referred to as an ***elastic buffer***.

One way to build a queue is with a linked list. This approach allows the queue to grow to an arbitrary size. But in many applications we are unwilling to pay the price of dynamically allocating memory. Another way to design the queue is to use

an array to hold all the data. We used a circular buffer in Example 3.5 to manage interrupt-driven data; here we will develop a non-interrupt version. Programming Example 5.3 gives C code for a queue that is built from an array.

Programming Example 5.3

A buffer-based queue

The first step in designing the queue is to declare the array that we will use for the buffer:

```
#define Q_SIZE 32 /* your queue size may vary */
#define Q_MAX (Q_SIZE-1) /* this is the maximum index value
                            into the array */
int q[Q_SIZE]; /* the array for our queue */
```

We will use two variables to keep track of the state of the queue:

```
int head, tail; /* the position of the head and the tail in
                   the queue */
```

As our initialization code shows, we initialize them to the same position. As we add a value to the tail of the queue, we will increment tail. Similarly, when we remove a value from the head, we will increment head. When we reach the end of the array, we must wrap around these values—for example, when we add a value into the last element of q, the new value of tail becomes the 0th entry of the array.

```
void initialize_queue() {
head = 0;
tail = Q_MAX;
}
```

A useful function adds one to a value with wraparound:

```
Int wrap(int i) { /* increment with wraparound for queue
                     size */
return ((i+1) % Q_SIZE);
}
```

We need to check for two error conditions: removing from an empty queue and adding to a full queue. In the first case, we know the queue is empty if head == wrap(tail). In the second case, we know the queue is full if incrementing tail will cause it to equal head. Testing for fullness, however, is a little harder since we have to worry about wraparound.

Here is the code for adding an element to the tail of the queue, which is known as *enqueueing:*

```
enqueue(int val) {
/* check for a full queue */
if (wrap(wrap(tail) == head) error(ENQUEUE_ERROR);
```

```
/* update the tail */
tail = wrap(tail);
/* add val to the tail of the queue */
q[tail] = val;
}
```

And here is the code for removing an element from the head of the queue, known as
dequeueing:

```
int dequeue() {
        int returnval; /* use this to remember the value that
                               you will return */
/* check for an empty queue */
if (head == wrap(tail)) error(DEQUEUE_ERROR);
/* remove from the head of the queue */
returnval = q[head];
/* update head */

head = wrap(head);
/* return the value */
return returnval;
}
```

5.2 MODELS OF PROGRAMS

In this section, we develop models for programs that are more general than source
code. Why not use the source code directly? First, there are many different types
of source code—assembly languages, C code, and so on—but we can use a single
model to describe all of them. Once we have such a model, we can perform many
useful analyses on the model more easily than we could on the source code.

Our fundamental model for programs is the *control/data flow graph (CDFG)*.
(We can also model hardware behavior with the CDFG.) As the name implies, the
CDFG has constructs that model both data operations (arithmetic and other compu-
tations) and control operations (conditionals). Part of the power of the CDFG comes
from its combination of control and data constructs. To understand the CDFG, we
start with pure data descriptions and then extend the model to control.

5.2.1 Data Flow Graphs

A *data flow graph* is a model of a program with no conditionals. In a high-level
programming language, a code segment with no conditionals—more precisely, with
only one entry and exit point—is known as a basic block. Figure 5.2 shows a simple
basic block. As the C code is executed, we would enter this basic block at the
beginning and execute all the statements.

```
w = a + b;
x = a − c;
y = x + d;
x = a + c;
z = y + e;
```

FIGURE 5.2

A basic block in C.

```
w = a + b;
x1 = a − c;
y = x1 + d;
x2 = a + c;
z = y + e;
```

FIGURE 5.3

The basic block in single-assignment form.

Before we are able to draw the data flow graph for this code we need to modify it slightly. There are two assignments to the variable x—it appears twice on the left side of an assignment. We need to rewrite the code in **single-assignment form**, in which a variable appears only once on the left side. Since our specification is C code, we assume that the statements are executed sequentially, so that any use of a variable refers to its latest assigned value. In this case, x is not reused in this block (presumably it is used elsewhere), so we just have to eliminate the multiple assignment to x. The result is shown in Figure 5.3, where we have used the names $x1$ and $x2$ to distinguish the separate uses of x.

The single-assignment form is important because it allows us to identify a unique location in the code where each named location is computed. As an introduction to the data flow graph, we use two types of nodes in the graph—round nodes denote operators and square nodes represent values. The value nodes may be either inputs to the basic block, such as a and b, or variables assigned to within the block, such as w and $x1$. The data flow graph for our single-assignment code is shown in Figure 5.4. The single-assignment form means that the data flow graph is acyclic—if we assigned to x multiple times, then the second assignment would form a cycle in the graph including x and the operators used to compute x. Keeping the data flow graph acyclic is important in many types of analyses we want to do on the graph. (Of course, it is important to know whether the source code actually assigns to a variable multiple times, because some of those assignments may be mistakes. We consider the analysis of source code for proper use of assignments in Section 5.10.1).

The data flow graph is generally drawn in the form shown in Figure 5.5. Here, the variables are not explicitly represented by nodes. Instead, the edges are labeled with the variables they represent. As a result, a variable can be represented by more

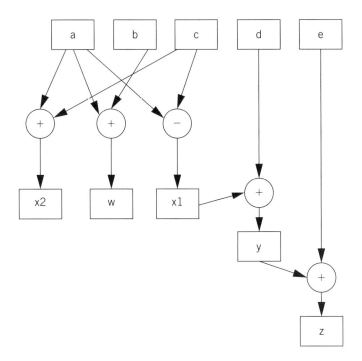

FIGURE 5.4

An extended data flow graph for our sample basic block.

than one edge. However, the edges are directed and all the edges for a variable must come from a single source. We use this form for its simplicity and compactness.

The data flow graph for the code makes the order in which the operations are performed in the C code much less obvious. This is one of the advantages of the data flow graph. We can use it to determine feasible reorderings of the operations, which may help us to reduce pipeline or cache conflicts. We can also use it when the exact order of operations simply doesn't matter. The data flow graph defines a partial ordering of the operations in the basic block. We must ensure that a value is computed before it is used, but generally there are several possible orderings of evaluating expressions that satisfy this requirement.

5.2.2 Control/Data Flow Graphs

A CDFG uses a data flow graph as an element, adding constructs to describe control. In a basic CDFG, we have two types of nodes: ***decision nodes*** and ***data flow nodes***. A data flow node encapsulates a complete data flow graph to represent a basic block. We can use one type of decision node to describe all the types of control in a sequential program. (The jump/branch is, after all, the way we implement all those high-level control constructs.)

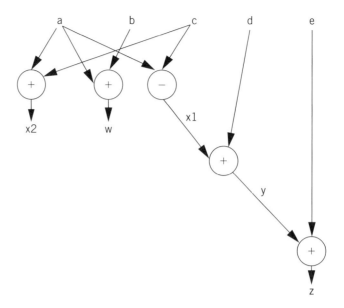

FIGURE 5.5

Standard data flow graph for our sample basic block.

Figure 5.6 shows a bit of C code with control constructs and the CDFG con-structed from it. The rectangular nodes in the graph represent the basic blocks. The basic blocks in the C code have been represented by function calls for simplic-ity. The diamond-shaped nodes represent the conditionals. The node's condition is given by the label, and the edges are labeled with the possible outcomes of evaluating the condition.

Building a CDFG for a while loop is straightforward, as shown in Figure 5.7. The while loop consists of both a test and a loop body, each of which we know how to represent in a CDFG. We can represent for loops by remembering that, in C, a for loop is defined in terms of a while loop. The following for loop

```
for (i = 0; i < N; i++) {
loop_body();
}
```

is equivalent to

```
i = 0;
while (i < N) {
loop_body();
i++;
}
```

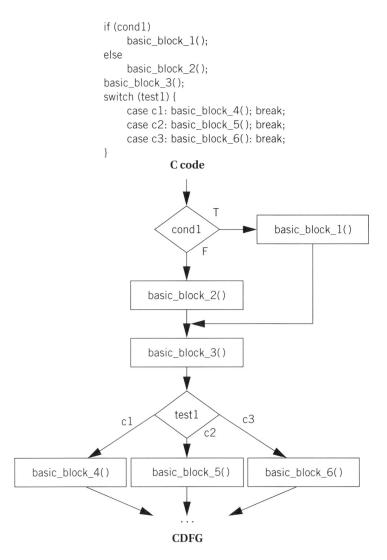

```
if (cond1)
      basic_block_1();
else
      basic_block_2();
basic_block_3();
switch (test1) {
      case c1: basic_block_4(); break;
      case c2: basic_block_5(); break;
      case c3: basic_block_6(): break;
}
```

C code

CDFG

FIGURE 5.6

C code and its CDFG.

For a complete CDFG model, we can use a data flow graph to model each data flow node. Thus, the CDFG is a hierarchical representation—a data flow CDFG can be expanded to reveal a complete data flow graph.

An execution model for a CDFG is very much like the execution of the program it represents. The CDFG does not require explicit declaration of variables, but we assume that the implementation has sufficient memory for all the variables.

```
while (a < b) {
        a = proc1(a,b);
        b = proc2(a,b);
}
```

C code

CDFG

FIGURE 5.7

CDFG for a while loop.

We can define a state variable that represents a program counter in a CPU. (When studying a drawing of a CDFG, a finger works well for keeping track of the program counter state.) As we execute the program, we either execute the data flow node or compute the decision in the decision node and follow the appropriate edge, depending on the type of node the program counter points on. Even though the data flow nodes may specify only a partial ordering on the data flow computations, the CDFG is a sequential representation of the program. There is only one program counter in our execution model of the CDFG, and operations are not executed in parallel.

The CDFG is not necessarily tied to high-level language control structures. We can also build a CDFG for an assembly language program. A jump instruction corresponds to a nonlocal edge in the CDFG. Some architectures, such as ARM and many VLIW processors, support predicated execution of instructions, which may be represented by special constructs in the CDFG.

5.3 ASSEMBLY, LINKING, AND LOADING

Assembly and linking are the last steps in the compilation process—they turn a list of instructions into an image of the program's bits in memory. Loading actually puts the program in memory so that it can be executed. In this section, we survey the basic techniques required for assembly linking to help us understand the complete compilation and loading process.

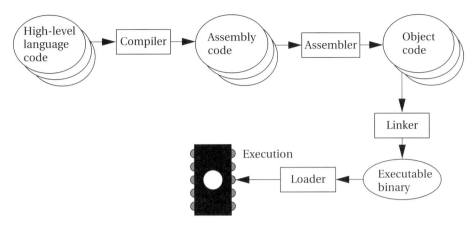

FIGURE 5.8

Program generation from compilation through loading.

Figure 5.8 highlights the role of assemblers and linkers in the compilation process. This process is often hidden from us by compilation commands that do everything required to generate an executable program. As the figure shows, most compilers do not directly generate machine code, but instead create the instruction-level program in the form of human-readable assembly language. Generating assembly language rather than binary instructions frees the compiler writer from details extraneous to the compilation process, which includes the instruction format as well as the exact addresses of instructions and data. The assembler's job is to translate symbolic assembly language statements into bit-level representations of instructions known as ***object code***. The assembler takes care of instruction formats and does part of the job of translating labels into addresses. However, since the program may be built from many files, the final steps in determining the addresses of instructions and data are performed by the linker, which produces an ***executable binary*** file. That file may not necessarily be located in the CPU's memory, however, unless the linker happens to create the executable directly in RAM. The program that brings the program into memory for execution is called a ***loader***.

The simplest form of the assembler assumes that the starting address of the assembly language program has been specified by the programmer. The addresses in such a program are known as ***absolute addresses***. However, in many cases, particularly when we are creating an executable out of several component files, we do not want to specify the starting addresses for all the modules before assembly—if we did, we would have to determine before assembly not only the length of each program in memory but also the order in which they would be linked into the program. Most assemblers therefore allow us to use ***relative addresses*** by specifying at the start of the file that the origin of the assembly language module is to be computed later. Addresses within the module are then computed relative to the start of the module. The linker is then responsible for translating relative addresses into addresses.

5.3.1 Assemblers

When translating assembly code into object code, the assembler must translate opcodes and format the bits in each instruction, and translate labels into addresses. In this section, we review the translation of assembly language into binary.

Labels make the assembly process more complex, but they are the most important abstraction provided by the assembler. Labels let the programmer (a human programmer or a compiler generating assembly code) avoid worrying about the locations of instructions and data. Label processing requires making two passes through the assembly source code as follows:

1. The first pass scans the code to determine the address of each label.

2. The second pass assembles the instructions using the label values computed in the first pass.

As shown in Figure 5.9, the name of each symbol and its address is stored in a **symbol table** that is built during the first pass. The symbol table is built by scanning from the first instruction to the last. (For the moment, we assume that we know the address of the first instruction in the program; we consider the general case in Section 5.3.2.) During scanning, the current location in memory is kept in a **program location counter (PLC)**. Despite the similarity in name to a program counter, the PLC is not used to execute the program, only to assign memory locations to labels. For example, the PLC always makes exactly one pass through the program, whereas the program counter makes many passes over code in a loop. Thus, at the start of the first pass, the PLC is set to the program's starting address and the assembler looks at the first line. After examining the line, the assembler updates the PLC to the next location (since ARM instructions are four bytes long, the PLC would be incremented by four) and looks at the next instruction. If the instruction begins with a label, a new entry is made in the symbol table, which includes the label name and its value. The value of the label is equal to the current value of the PLC. At the end of the first pass, the assembler rewinds to the beginning of the assembly language file to make the second pass. During the second pass, when a label name is found, the label is looked up in the symbol table and its value substituted into the appropriate place in the instruction.

But how do we know the starting value of the PLC? The simplest case is absolute addressing. In this case, one of the first statements in the assembly language program

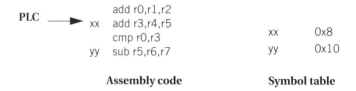

FIGURE 5.9

Symbol table processing during assembly.

is a pseudo-op that specifies the ***origin*** of the program, that is, the location of the first address in the program. A common name for this pseudo-op (e.g., the one used for the ARM) is the ORG statement

```
ORG 2000
```

which puts the start of the program at location 2000. This pseudo-op accomplishes this by setting the PLC's value to its argument's value, 2000 in this case. Assemblers generally allow a program to have many ORG statements in case instructions or data must be spread around various spots in memory. Example 5.1 illustrates the use of the PLC in generating the symbol table.

Example 5.1

Generating a symbol table

Let's use the following simple example of ARM assembly code:

```
        ORG 100
label1 ADR r4,c
        LDR r0,[r4]
label2 ADR r4,d
        LDR r1,[r4]
label3 SUB r0,r0,r1
```

The initial ORG statement tells us the starting address of the program. To begin, let's initialize the symbol table to an empty state and put the PLC at the initial ORG statement.

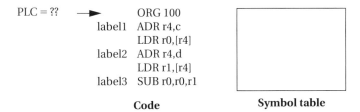

The PLC value shown is at the beginning of this step, before we have processed the ORG statement. The ORG tells us to set the PLC value to 100.

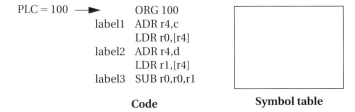

To process the next statement, we move the PLC to point to the next statement. But because the last statement was a pseudo-op that generates no memory values, the PLC value remains at 100.

```
                    ORG 100
PLC = 100  ──▶  label1   ADR r4,c
                    LDR r0,[r4]
            label2   ADR r4,d
                    LDR r1,[r4]
            label3   SUB r0,r0,r1
```
Code **Symbol table**

Because there is a label in this statement, we add it to the symbol table, taking its value from the current PLC value.

```
                    ORG 100
PLC = 100  ──▶  label1   ADR r4,c
                    LDR r0,[r4]
            label2   ADR r4,d
                    LDR r1,[r4]
            label3   SUB r0,r0,r1
```
Code **Symbol table**

| label1 | 100 |

To process the next statement, we advance the PLC to point to the next line of the program and increment its value by the length in memory of the last line, namely, 4.

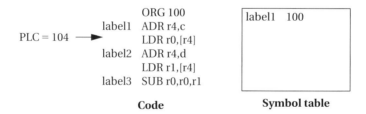

Code **Symbol table**

We continue this process as we scan the program until we reach the end, at which the state of the PLC and symbol table are as shown below.

Code **Symbol table**

Assemblers allow labels to be added to the symbol table without occupying space in the program memory. A typical name of this pseudo-op is EQU for equate. For example, in the code

```
        ADD  r0,r1,r2
FOO EQU 5
BAZ SUB  r3,r4,#FOO
```

the EQU pseudo-op adds a label named FOO with the value 5 to the symbol table. The value of the BAZ label is the same as if the EQU pseudo-op were not present, since EQU does not advance the PLC. The new label is used in the subsequent SUB instruction as the name for a constant. EQUs can be used to define symbolic values to help make the assembly code more structured.

The ARM assembler supports one pseudo-op that is particular to the ARM instruction set. In other architectures, an address would be loaded into a register (e.g., for an indirect access) by reading it from a memory location. ARM does not have an instruction that can load an effective address, so the assembler supplies the ADR pseudo-op to create the address in the register. It does so by using ADD or SUB instructions to generate the address. The address to be loaded can be register relative, program relative, or numeric, but it must assemble to a single instruction. More complicated address calculations must be explicitly programmed.

The assembler produces an object file that describes the instructions and data in binary format. A commonly used object file format, originally developed for Unix but now used in other environments as well, is known as COFF (common object file format). The object file must describe the instructions, data, and any addressing information and also usually carries along the symbol table for later use in debugging.

Generating relative code rather than absolute code introduces some new challenges to the assembly language process. Rather than using an ORG statement to provide the starting address, the assembly code uses a pseudo-op to indicate that the code is in fact relocatable. (Relative code is the default for the ARM assembler.) Similarly, we must mark the output object file as being relative code. We can initialize the PLC to 0 to denote that addresses are relative to the start of the file. However, when we generate code that makes use of those labels, we must be careful, since we do not yet know the actual value that must be put into the bits. We must instead generate relocatable code. We use extra bits in the object file format to mark the relevant fields as relocatable and then insert the label's relative value into the field. The linker must therefore modify the generated code—when it finds a field marked as relative, it uses the addresses that it has generated to replace the relative value with a correct, value for the address. To understand the details of turning relocatable code into executable code, we must understand the linking process described in the next section.

5.3.2 Linking

Many assembly language programs are written as several smaller pieces rather than as a single large file. Breaking a large program into smaller files helps delineate

program modularity. If the program uses library routines, those will already be preassembled, and assembly language source code for the libraries may not be available for purchase. A ***linker*** allows a program to be stitched together out of several smaller pieces. The linker operates on the object files created by the assembler and modifies the assembled code to make the necessary links between files.

Some labels will be both defined and used in the same file. Other labels will be defined in a single file but used elsewhere as illustrated in Figure 5.10. The place in the file where a label is defined is known as an ***entry point***. The place in the file where the label is used is called an ***external reference***. The main job of the loader is to *resolve* external references based on available entry points. As a result of the need to know how definitions and references connect, the assembler passes to the linker not only the object file but also the symbol table. Even if the entire symbol table is not kept for later debugging purposes, it must at least pass the entry points. External references are identified in the object code by their relative symbol identifiers.

The linker proceeds in two phases. First, it determines the address of the start of each object file. The order in which object files are to be loaded is given by the user, either by specifying parameters when the loader is run or by creating a ***load map*** file that gives the order in which files are to be placed in memory. Given the order in which files are to be placed in memory and the length of each object file, it is easy to compute the starting address of each file. At the start of the

label1	LDR r0,[r1]	label2	ADR var1

	ADR a		B label3

	B label2	x	% 1
	...	y	% 1
var1	% 1	a	% 10

External references	Entry points
a	label1
label2	var1

File 1

External references	Entry points
var1	label2
label3	x
	y
	a

File 2

FIGURE 5.10

External references and entry points.

second phase, the loader merges all symbol tables from the object files into a single, large table. It then edits the object files to change relative addresses into addresses. This is typically performed by having the assembler write extra bits into the object file to identify the instructions and fields that refer to labels. If a label cannot be found in the merged symbol table, it is undefined and an error message is sent to the user.

Controlling where code modules are loaded into memory is important in embedded systems. Some data structures and instructions, such as those used to manage interrupts, must be put at precise memory locations for them to work. In other cases, different types of memory may be installed at different address ranges. For example, if we have EPROM in some locations and DRAM in others, we want to make sure that locations to be written are put in the DRAM locations.

Workstations and PCs provide ***dynamically linked libraries***, and some embedded computing environments may provide them as well. Rather than link a separate copy of commonly used routines such as I/O to every executable program on the system, dynamically linked libraries allow them to be linked in at the start of program execution. A brief linking process is run just before execution of the program begins; the dynamic linker uses code libraries to link in the required routines. This not only saves storage space but also allows programs that use those libraries to be easily updated. However, it does introduce a delay before the program starts executing.

5.4 BASIC COMPILATION TECHNIQUES

It is useful to understand how a high-level language program is translated into instructions. Since implementing an embedded computing system often requires controlling the instruction sequences used to handle interrupts, placement of data and instructions in memory, and so forth, understanding how the compiler works can help you know when you cannot rely on the compiler. Next, because many applications are also performance sensitive, understanding how code is generated can help you meet your performance goals, either by writing high-level code that gets compiled into the instructions you want or by recognizing when you must write your own assembly code. Compilation combines translation and optimization. The high-level language program is translated into the lower-level form of instructions; optimizations try to generate better instruction sequences than would be possible if the brute force technique of independently translating source code statements were used. Optimization techniques focus on more of the program to ensure that compilation decisions that appear to be good for one statement are not unnecessarily problematic for other parts of the program.

The compilation process is summarized in Figure 5.11. Compilation begins with high-level language code such as C and generally produces assembly code. (Directly producing object code simply duplicates the functions of an assembler,

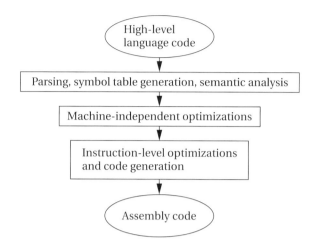

FIGURE 5.11

The compilation process.

which is a very desirable stand-alone program to have.) The high-level language program is parsed to break it into statements and expressions. In addition, a symbol table is generated, which includes all the named objects in the program. Some compilers may then perform higher-level optimizations that can be viewed as modifying the high-level language program input without reference to instructions.

Simplifying arithmetic expressions is one example of a machine-independent optimization. Not all compilers do such optimizations, and compilers can vary widely regarding which combinations of machine-independent optimizations they do perform. Instruction-level optimizations are aimed at generating code. They may work directly on real instructions or on a pseudo-instruction format that is later mapped onto the instructions of the target CPU. This level of optimization also helps modularize the compiler by allowing code generation to create simpler code that is later optimized. For example, consider the following array access code:

```
x[i] = c*x[i];
```

A simple code generator would generate the address for x[i] twice, once for each appearance in the statement. The later optimization phases can recognize this as an example of common expressions that need not be duplicated. While in this simple case it would be possible to create a code generator that never generated the redundant expression, taking into account every such optimization at code generation time is very difficult. We get better code and more reliable compilers by generating simple code first and then optimizing it.

5.4.1 Statement Translation

In this section, we consider the basic job of translating the high-level language program with little or no optimization. Let's first consider how to translate an expression. A large amount of the code in a typical application consists of arithmetic and logical expressions. Understanding how to compile a single expression, as described in Example 5.2, is a good first step in understanding the entire compilation process.

Example 5.2

Compiling an arithmetic expression

In the following arithmetic expression,

```
a*b + 5*(c - d)
```

the variable is written in terms of program variables. In some machines we may be able to perform memory-to-memory arithmetic directly on the locations corresponding to those variables. However, in many machines, such as the ARM, we must first load the variables into registers. This requires choosing which registers receive not only the named variables but also intermediate results such as $(c - d)$.

The code for the expression can be built by walking the data flow graph. The data flow graph for the expression appears on page 230.

The temporary variables for the intermediate values and final result have been named w, x, y, and z. To generate code, we walk from the tree's root (where z, the final result, is generated) by traversing the nodes in post order. During the walk, we generate instructions to cover the operation at every node. The path is presented below.

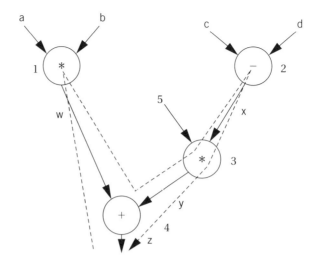

The nodes are numbered in the order in which code is generated. Since every node in the data flow graph corresponds to an operation that is directly supported by the instruction set, we simply generate an instruction at every node. Since we are making an arbitrary register assignment, we can use up the registers in order starting with r1. The resulting ARM code follows:

```
; operator 1 (+)
ADR r4,a          ; get address for a
MOV r1,[r4]       ; load a
ADR r4,b          ; get address for b
MOV r2,[r4]       ; load b
ADD r3,r1,r2      ; put w into r3
; operator 2 (-)
ADR r4,c          ; get address for c
MOV r4,[r4]       ; load c
ADR r4,d          ; get address for d
MOV r5,[r4]       ; load d
SUB r6,r4,r5      ; put x into r6
; operator 3 (*)
MUL r7,r6,#5      ; operator 3, puts y into r7
; operator 4 (+)
ADD r8,r7,r3      ; operator 4, puts z into r8
```

One obvious optimization is to reuse a register whose value is no longer needed. In the case of the intermediate values w, x, and y, we know that they cannot be used after the end of the expression (e.g., in another expression) since they have no name in the C program. However, the final result z may in fact be used in a C assignment and the value reused later in the program. In this case we would need to know when the register is no longer needed to determine its best use.

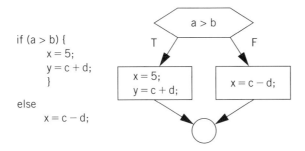

```
if (a > b) {
        x = 5;
        y = c + d;
        }

else
        x = c - d;
```

FIGURE 5.12

Flow of control in C and control flow diagrams.

In the previous example, we made an arbitrary allocation of variables to registers for simplicity. When we have large programs with multiple expressions, we must allocate registers more carefully since CPUs have a limited number of registers. We will consider register allocation in Section 5.5.5.

We also need to be able to translate control structures. Since conditionals are controlled by expressions, the code generation techniques of the last example can be used for those expressions, leaving us with the task of generating code for the flow of control itself. Figure 5.12 shows a simple example of changing flow of control in C—an if statement, in which the condition controls whether the true or false branch of the if is taken. Figure 5.12 also shows the control flow diagram for the if statement.

Example 5.3 illustrates how to implement conditionals in assembly language.

Example 5.3

Generating code for a conditional

Consider the following C statement:

```
if (a + b > 0)
        x = 5;
else
        x = 7;
```

The CDFG for this statement is:

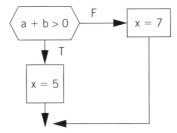

We know how to generate the code for the expressions. We can generate the control flow code by walking the CDFG. One ordered walk through the CDFG is:

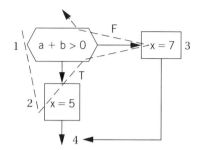

To generate code, we must assign a label to the first instruction at the end of a directed edge and create a branch for each edge that does not go to the following instruction. The exact steps to be taken at the branch points depend on the target architecture. On some machines, evaluating expressions generates condition codes that we can test in subsequent branches, and on other machines we must use test-and-branch instructions. ARM allows us to test condition codes, so we get the following ARM code for the 1-2-3 walk:

```
          ADR  r5,a      ; get address for a
          LDR  r1,[r5]   ; load a
          ADR  r5,b      ; get address for b
          LDR  r2,b      ; load b
          ADD  r3,r1,r2
          BLE  label3    ; true condition falls through branch ;
       ; true case
          LDR  r3,#5     ; load constant
          ADR  r5,x
          STR  r3, [r5]  ; store value into x
          B stmtend      ; done with the true case
       ; false case
       label3 LDR r3,#7  ; load constant
          ADR  r5,x      ; get address of x
          STR  r3,[r5]   ; store value into x
       stmtend    ...
```

The 1-2 and 3-4 edges do not require a branch and label because they are straight-line code. In contrast, the 1-3 and 2-4 edges do require a branch and a label for the target.

Since expressions are generally created as straight-line code, they typically require careful consideration of the order in which the operations are executed. We have much more freedom when generating conditional code because the branches ensure that the flow of control goes to the right block of code. If we walk the CDFG in a different order and lay out the code blocks in a different order in memory, we still get valid code as long as we properly place branches.

Drawing a control flow graph based on the while form of the loop helps us understand how to translate it into instructions.

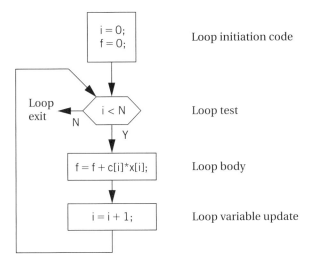

$i = 0;$
$f = 0;$ — Loop initiation code

$i < N$ — Loop test

Loop exit N Y

$f = f + c[i]*x[i];$ — Loop body

$i = i + 1;$ — Loop variable update

C compilers can generate (using the -s flag) assembler source, which some compilers intersperse with the C code. Such code is a very good way to learn about both assembly language programming and compilation.

5.4.2 Procedures

Another major code generation problem is the creation of procedures. Generating code for procedures is relatively straightforward once we know the procedure linkage appropriate for the CPU. At the procedure definition, we generate the code to handle the procedure call and return. At each call of the procedure, we set up the procedure parameters and make the call.

The CPU's subroutine call mechanism is usually not sufficient to directly support procedures in modern programming languages. We introduced the procedure stack and procedure linkages in Section 2.2.3. The linkage mechanism provides a way for the program to pass parameters into the program and for the procedure to return a value. It also provides help in restoring the values of registers that the procedure has modified. All procedures in a given programming language use the same linkage mechanism (although different languages may use different linkages). The mechanism can also be used to call handwritten assembly language routines from compiled code.

Procedure stacks are typically built to grow down from high addresses. A **stack pointer (sp)** defines the end of the current frame, while a **frame pointer (fp)** defines the end of the last frame. (The fp is technically necessary only if the stack frame can be grown by the procedure during execution.) The procedure can refer

to an element in the frame by addressing relative to sp. When a new procedure is called, the sp and fp are modified to push another frame onto the stack.

The **_ARM Procedure Call Standard (APCS)_** is a good illustration of a typical procedure linkage mechanism. Although the stack frames are in main memory, understanding how registers are used is key to understanding the mechanism, as explained below.

- $r0 - r3$ are used to pass parameters into the procedure. $r0$ is also used to hold the return value. If more than four parameters are required, they are put on the stack frame.

- $r4 - r7$ hold register variables.

- $r11$ is the frame pointer and $r13$ is the stack pointer.

- $r10$ holds the limiting address on stack size, which is used to check for stack overflows.

Other registers have additional uses in the protocol.

5.4.3 Data Structures

The compiler must also translate references to data structures into references to raw memories. In general, this requires address computations. Some of these computations can be done at compile time while others must be done at run time.

Arrays are interesting because the address of an array element must in general be computed at run time, since the array index may change. Let us first consider one-dimensional arrays:

```
a[i]
```

The layout of the array in memory is shown in Figure 5.13. The zeroth element is stored as the first element of the array, the first element directly below, and so on.

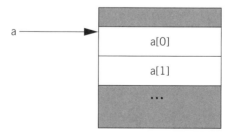

FIGURE 5.13

Layout of a one-dimensional array in memory.

FIGURE 5.14

Memory layout for two-dimensional arrays.

We can create a pointer for the array that points to the array's head, namely, $a[0]$. If we call that pointer aptr for convenience, then we can rewrite the reading of $a[i]$ as

```
*(aptr + i)
```

Two-dimensional arrays are more challenging. There are multiple possible ways to lay out a two-dimensional array in memory, as shown in Figure 5.14. In this form, which is known as **row major**, the inner variable of the array (j in a$[i, j]$) varies most quickly. (Fortran uses a different organization known as column major.) Two-dimensional arrays also require more sophisticated addressing—in particular, we must know the size of the array. Let us consider the row-major form. If the $a[\]$ array is of size $N \times M$, then we can turn the two-dimensional array access into a one-dimensional array access. Thus,

```
a[i,j]
becomes
a[i*M + j]
```

where the maximum value for j is $M - 1$.

A C struct is easier to address. As shown in Figure 5.15, a structure is implemented as a contiguous block of memory. Fields in the structure can be accessed using constant offsets to the base address of the structure. In this example, if field1 is four bytes long, then field2 can be accessed as

```
*(aptr + 4)
```

This addition can usually be done at compile time, requiring only the indirection itself to fetch the memory location during execution.

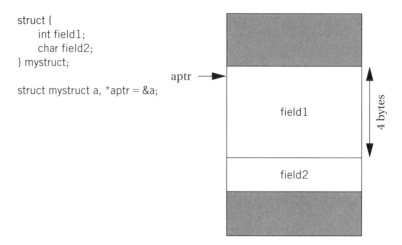

```
struct {
    int field1;
    char field2;
} mystruct;

struct mystruct a, *aptr = &a;
```

FIGURE 5.15

C structure layout and access.

5.5 PROGRAM OPTIMIZATION

Now that we understand something about how programs are created, we can start to understand how to optimize programs. If we want to write programs in a high-level language, then we need to understand how to optimize them without rewriting them in assembly language. This first requires creating the proper source code that causes the compiler to do what we want. Hopefully, the compiler can optimize our program by recognizing features of the code and taking the proper action.

5.5.1 Expression Simplification

Expression simplification is a useful area for machine-independent transformations. We can use the laws of algebra to simplify expressions. Consider the following expression:

```
a*b + a*c
```

We can use the distributive law to rewrite the expression as

```
a*(b + c)
```

Since the new expression has only two operations rather than three for the original form, it is almost certainly cheaper, because it is both faster and smaller. Such transformations make some broad assumptions about the relative cost of operations. In some cases, simple generalizations about the cost of operations may be misleading. For example, a CPU with a multiply-and-accumulate instruction may be

able to do a multiply and addition as cheaply as it can do an addition. However, such situations can often be taken care of in code generation.

We can also use the laws of arithmetic to further simplify expressions on constants. Consider the following C statement:

```
for (i = 0; i < 8 + 1; i++)
```

We can simplify 8 + 1 to 9 at compile time—there is no need to perform that arithmetic while the program is executing. Why would a program ever contain expressions that evaluate to constants? Using named constants rather than numbers is good programming practice and often leads to constant expression. The original form of the for statement could have been

```
for (i = 0; i < NOPS + 1; i++)
```

where, for example, the added 1 takes care of a trailing null character.

5.5.2 Dead Code Elimination

Code that will never be executed can be safely removed from the program. The general problem of identifying code that will never be executed is difficult, but there are some important special cases where it can be done.

Programmers will intentionally introduce dead code in certain situations. Consider this C code fragment:

```
#define DEBUG 0
...
if (DEBUG) print_debug_stuff();
```

In the above case, the print_debug_stuff() function is never executed, but the code allows the programmer to override the preprocessor variable definition (perhaps with a compile-time flag) to enable the debugging code. This case is easy to analyze because the condition is the constant 0, which C uses for the false condition. Since there is no else clause in the if statement, the compiler can totally eliminate the if statement, rewriting the CDFG to provide a direct edge between the statements before and after the if.

Some dead code may be introduced by the compiler. For example, certain optimizations introduce copy statements that copy one variable to another. If uses of the first variable can be replaced by references to the second one, then the copy statement becomes dead code that can be eliminated.

5.5.3 Procedure Inlining

Another machine-independent transformation that requires a little more evaluation is procedure inlining. An inlined procedure does not have a separate procedure body and procedure linkage; rather, the body of the procedure is substituted in place for the procedure call. Figure 5.16 shows an example of function inlining in C.

$$\text{int foo(a,b,c) \{ return a + b - c ; \}}$$

Function definition

$$z = \text{foo(w,x,y)};$$

Function call

$$z = w + x - y;$$

Inlining result

FIGURE 5.16

Function inlining in C.

The C++ programming language provides an inline construct that tells the compiler to generate inline code for a function. In this case, an inlined procedure is generated in expanded form whenever possible. However, inlining is not always the best thing to do. Although it does eliminate the procedure linkage instructions, when a cache is present, having multiple copies of the function body may actually slow down the fetches of these instructions. Inlining also increases code size, and memory may be precious.

5.5.4 Loop Transformations

Loops are important program structures—although they are compactly described in the source code, they often use a large fraction of the computation time. Many techniques have been designed to optimize loops.

A simple but useful transformation is known as ***loop unrolling***, which is illustrated in Example 5.4. Loop unrolling is important because it helps expose parallelism that can be used by later stages of the compiler.

Example 5.4

Loop unrolling

A simple loop in C follows:

```
for (i = 0; i < N; i++) {
        a[i] = b[i]*c[i];
}
```

This loop is executed a fixed number of times, namely, N. A straightforward implementation of the loop would create and initialize the loop variable i, update its value on every iteration, and test it to see whether to exit the loop. However, since the loop is executed a fixed number of times, we can generate more direct code.

If we let $N = 4$, then we can substitute the above C code for the following loop:

```
a[0] = b[0]*c[0];
a[1] = b[1]*c[1];
```

```
a[2] = b[2]*c[2];
a[3] = b[3]*c[3];
```

This unrolled code has no loop overhead code at all, that is, no iteration variable and no tests. But the unrolled loop has the same problems as the inlined procedure—it may interfere with the cache and expands the amount of code required.

We do not, of course, have to fully unroll loops. Rather than unroll the above loop four times, we could unroll it twice. The following code results:

```
for (i = 0; i < 2; i++) {
        a[i*2] = b[i*2]*c[i*2];
        a[i*2 + 1] = b[i*2 + 1]*c[i*2 + 1];
        }
```

In this case, since all operations in the two lines of the loop body are independent, later stages of the compiler may be able to generate code that allows them to be executed efficiently on the CPU's pipeline.

Loop fusion combines two or more loops into a single loop. For this transformation to be legal, two conditions must be satisfied. First, the loops must iterate over the same values. Second, the loop bodies must not have dependencies that would be violated if they are executed together—for example, if the second loop's ith iteration depends on the results of the $I + 1$th iteration of the first loop, the two loops cannot be combined. *Loop distribution* is the opposite of loop fusion, that is, decomposing a single loop into multiple loops.

Loop tiling breaks up a loop into a set of nested loops, with each inner loop performing the operations on a subset of the data. An example is shown in Figure 5.17. Here, each loop is broken up into tiles of size two. Each loop is split into two loops—for example, the inner *ii* loop iterates within the tile and the outer *i* loop iterates across the tiles. The result is that the pattern of accesses across the *a* array is drastically different—rather than walking across one row in its entirety, the code walks through rows and columns following the tile structure. Loop tiling changes the order in which array elements are accessed, thereby allowing us to better control the behavior of the cache during loop execution.

We can also modify the arrays being indexed in loops. *Array padding* adds dummy data elements to a loop in order to change the layout of the array in the cache. Although these array locations will not be used, they do change how the useful array elements fall into cache lines. Judicious padding can in some cases significantly reduce the number of cache conflicts during loop execution.

5.5.5 Register Allocation

Register allocation is a very important compilation phase. Given a block of code, we want to choose assignments of variables (both declared and temporary) to

Code

for (i = 0 ; i < N; i++) for (i = 0 ; i < N; i += 2)
 for (j = 0 ; j < N; j++) for (j = 0 ; j < N; j += 2)
 c[i] = a [i,j] * b[i]; for (ii = i; ii < min(i + 2 ,N); i++)
 for (jj = j; jj < min(j + 2 ,N); j++)
 c[ii] = a [ii,jj] * b[ii];

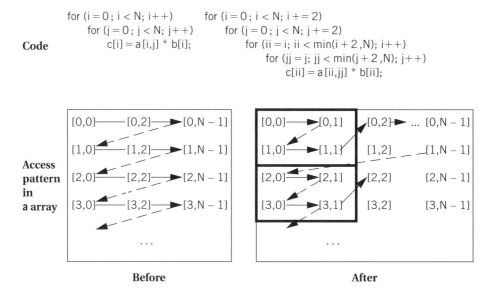

Access pattern in a array

Before **After**

FIGURE 5.17

Loop tiling.

registers to minimize the total number of required registers. Example 5.5 illustrates the importance of proper register allocation.

Example 5.5

Register allocation

To keep the example small, we assume that we can use only four of the ARM's registers. In fact, such a restriction is not unthinkable—programming conventions can reserve certain registers for special purposes and significantly reduce the number of general-purpose registers available.

Consider the following C code:

```
w = a + b; /* statement 1 */
x = c + w; /* statement 2 */
y = c + d; /* statement 3 */
```

A naive register allocation, assigning each variable to a separate register, would require seven registers for the seven variables in the above code. However, we can do much better by reusing a register once the value stored in the register is no longer needed. To understand how to do this, we can draw a **lifetime graph** that shows the statements on which each statement is used. Appearing below is a lifetime graph in which the x-axis is the statement number in the C code and the y-axis shows the variables.

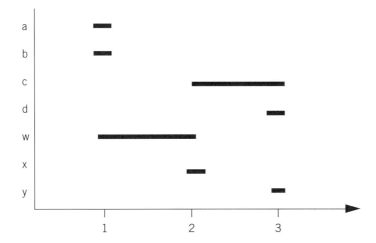

A horizontal line stretches from the first statement where the variable is used to the last use of the variable; a variable is said to be *live* during this interval. At each statement, we can determine every variable currently in use. The maximum number of variables in use at any statement determines the maximum number of registers required. In this case, statement two requires three registers: *c*, *w*, and *x*. This fits within the four registers limitation. By reusing registers once their current values are no longer needed, we can write code that requires no more than four registers. Appearing below is one register assignment.

A	r0
B	r1
C	r2
D	r0
W	r3
X	r0
Y	r3

The ARM assembly code that uses the above register assignment follows:

```
LDR r0,[p_a]    ; load a into r0 using pointer to a (p_a)
LDR r1,[p_b]    ; load b into r1
ADD r3,r0,r1    ; compute a + b
STR r3,[p_w]    ; w = a + b
LDR r2,[p_c]    ; load c into r2
ADD r0,r2,r3    ; compute c + w, reusing r0 for x
STR r0,[p_x]    ; x = c + w
LDR r0,[p_d]    ; load d into r0
ADD r3,r2,r0    ; compute c + d, reusing r3 for y
STR r3,[p_y]    ; y = c + d
```

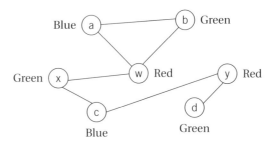

FIGURE 5.18

Using graph coloring to solve the problem of Example 5.5.

If a section of code requires more registers than are available, we must ***spill*** some of the values out to memory temporarily. After computing some values, we write the values to temporary memory locations, reuse those registers in other computations, and then reread the old values from the temporary locations to resume work. Spilling registers is problematic in several respects. For example, it requires extra CPU time and uses up both instruction and data memory. Putting effort into register allocation to avoid unnecessary register spills is worth your time.

We can solve register allocation problems by building a ***conflict graph*** and solving a graph coloring problem. As shown in Figure 5.18, each variable in the high-level language code is represented by a node. An edge is added between two nodes if they are both live at the same time. The graph coloring problem is to use the smallest number of distinct colors to color all the nodes such that no two nodes are directly connected by an edge of the same color. The figure shows a satisfying coloring that uses three colors. Graph coloring is NP-complete, but there are efficient heuristic algorithms that can give good results on typical register allocation problems.

Lifetime analysis assumes that we have already determined the order in which we will evaluate operations. In many cases, we have freedom in the order in which we do things. Consider the following expression:

```
(a + b) * (c - d)
```

We have to do the multiplication last, but we can do either the addition or the subtraction first. Different orders of loads, stores, and arithmetic operations may also result in different execution times on pipelined machines. If we can keep values in registers without having to reread them from main memory, we can save execution time and reduce code size as well. Example 5.6 illustrates how proper ***operator scheduling*** can improve register allocation.

Example 5.6

Operator scheduling for register allocation

Here is sample C code fragment:

```
w = a + b;  /* statement 1 */
x = c + d;  /* statement 2 */
y = x + e;  /* statement 3 */
z = a - b;  /* statement 4 */
```

If we compile the statements in the order in which they were written, we get the register

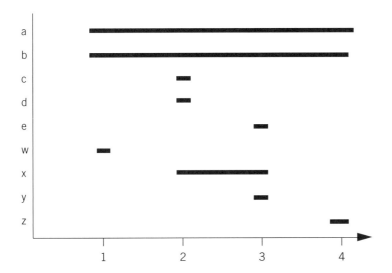

Since *w* is needed until the last statement, we need five registers at statement 3, even though only three registers are needed for the statement at line 3. If we swap statements 3 and 4 (renumbering them 39 and 49), we reduce our requirements to three registers. The modified C code follows:

```
w = a + b;  /* statement 1 */
z = a - b;  /* statement 29 */
x = c + d;  /* statement 39 */
y = x + e;  /* statement 49 */
```

The lifetime graph for the new code appears below.

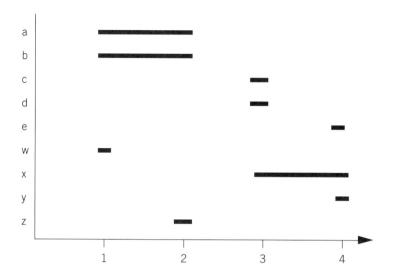

Compare the ARM assembly code for the two code fragments. We have written both assuming that we have only four free registers. In the *before* version, we do not have to write out any values, but we must read *a* and *b* twice. The *after* version allows us to retain all values in registers as long as we need them.

Before version	*After version*
LDR r0,a	LDR r0,a
LDR r1,b	LDR r1,b
ADD r2,r0,r1	ADD r2,r1,r0
STR r2,w ; w = a + b	STR r2,w ; w = a + b
LDRr r0,c	SUB r2,r0,r1
LDR r1,d	STR r2,z ; z = a - b
ADD r2,r0,r1	LDR r0,c
STR r2,x ; x = c + d	LDR r1,d
LDR r1,e	ADD r2,r1,r0
ADD r0,r1,r2	STR r2,x ; x = c + d
STR r0,y ; y = x + e	LDR r1,e
LDR r0,a ; reload a	ADD r0,r1,r2
LDR r1,b ; reload b	STR r0,y ; y = x + e
SUB r2,r1,r0	
STR r2,z ; z = a - b	

5.5.6 Scheduling

We have some freedom to choose the order in which operations will be performed. We can use this to our advantage—for example, we may be able to improve the

register allocation by changing the order in which operations are performed, thereby changing the lifetimes of the variables.

We can solve scheduling problems by keeping track of resource utilization over time. We do not have to know the exact microarchitecture of the CPU—all we have to know is that, for example, instruction types 1 and 2 both use resource A while instruction types 3 and 4 use resource B. CPU manufacturers generally disclose enough information about the microarchitecture to allow us to schedule instructions even when they do not provide a detailed description of the CPU's internals.

We can keep track of CPU resources during instruction scheduling using a ***reservation table*** [Kog81]. As illustrated in Figure 5.19, rows in the table represent instruction execution time slots and columns represent resources that must be scheduled. Before scheduling an instruction to be executed at a particular time, we check the reservation table to determine whether all resources needed by the instruction are available at that time. Upon scheduling the instruction, we update the table to note all resources used by that instruction. Various algorithms can be used for the scheduling itself, depending on the types of resources and instructions involved, but the reservation table provides a good summary of the state of an instruction scheduling problem in progress.

We can also schedule instructions to maximize performance. As we know from Section 3.5, when an instruction that takes more cycles than normal to finish is in the pipeline, pipeline bubbles appear that reduce performance. ***Software pipelining*** is a technique for reordering instructions across several loop iterations to reduce pipeline bubbles. Some instructions take several cycles to complete; if the value produced by one of these instructions is needed by other instructions in the loop iteration, then they must wait for that value to be produced. Rather than pad the loop with no-ops, we can start instructions from the next iteration. The loop body then contains instructions that manipulate values from several different loop iterations—some of the instructions are working on the early part of iteration $n + 1$, others are working on iteration n, and still others are finishing iteration $n - 1$.

Time	Resource A	Resource B
t	X	
t + 1	X	X
t + 2	X	
t + 3		X

FIGURE 5.19

A reservation table for instruction scheduling.

5.5.7 Instruction Selection

Selecting the instructions to use to implement each operation is not trivial. There may be several different instructions that can be used to accomplish the same goal, but they may have different execution times. Moreover, using one instruction for one part of the program may affect the instructions that can be used in adjacent code. Although we cannot discuss all the problems and methods for code generation here, a little bit of knowledge helps us envision what the compiler is doing.

One useful technique for generating code is **template matching**, illustrated in Figure 5.20. We have a DAG that represents the expression for which we want to generate code. In order to be able to match up instructions and operations, we represent instructions using the same DAG representation. We shaded the instruction template nodes to distinguish them from code nodes. Each node has a cost, which may be simply the execution time of the instruction or may include factors for size, power consumption, and so on. In this case, we have shown that each instruction takes the same amount of time, and thus all have a cost of 1. Our goal is to cover all nodes in the code DAG with instruction DAGs—until we have covered the code DAG we have not generated code for all the operations in the expression. In this

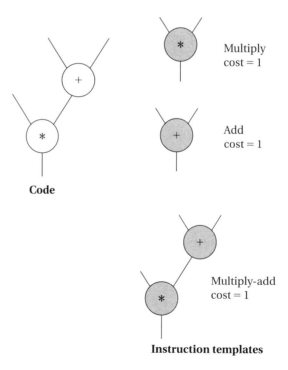

FIGURE 5.20

Code generation by template matching.

case, the lowest-cost covering uses the multiply-add instruction to cover both nodes. If we first tried to cover the bottom node with the multiply instruction, we would find ourselves blocked from using the multiply-add instruction. Dynamic programming can be used to efficiently find the lowest-cost covering of trees, and heuristics can extend the technique to DAGs.

5.5.8 Understanding and Using Your Compiler

Clearly, the compiler can vastly transform your program during the creation of assembly language. But compilers are also substantially different in terms of the optimizations they perform. Understanding your compiler can help you get the best code out of it.

Studying the assembly language output of the compiler is a good way to learn about what the compiler does. Some compilers will annotate sections of code to help you make the correspondence between the source and assembler output. Starting with small examples that exercise only a few types of statements will help. You can experiment with different optimization levels (the -O flag on most C compilers). You can also try writing the same algorithm in several ways to see how the compiler's output changes.

If you cannot get your compiler to generate the code you want, you may need to write your own assembly language. You can do this by writing it from scratch or modifying the output of the compiler. If you write your own assembly code, you must ensure that it conforms to all compiler conventions, such as procedure call linkage. If you modify the compiler output, you should be sure that you have the algorithm right before you start writing code so that you don't have to repeatedly edit the compiler's assembly language output. You also need to clearly document the fact that the high-level language source is, in fact, not the code used in the system.

5.5.9 Interpreters and JIT Compilers

Programs are not always compiled and then separately executed. In some cases, it may make sense to translate the program into instructions during execution. Two well-known techniques for on-the-fly translation are ***interpretation*** and ***just-in-time (JIT) compilation***. The trade-offs for both techniques are similar. Interpretation or JIT compilation adds overhead—both time and memory—to execution. However, that overhead may be more than made up for in some circumstances. For example, if only parts of the program are executed over some period of time, interpretation or JIT compilation may save memory, even taking overhead into account. Interpretation and JIT compilation also provide added security when programs arrive over the network.

An ***interpreter*** translates program statements one at a time. The program may be expressed in a high-level language, with Forth being a prime example of an embedded language that is interpreted. An interpreter may also interpret instructions in

FIGURE 5.21

Structure of a program interpretation system.

some abstract machine language. As illustrated in Figure 5.21, the interpreter sits between the program and the machine. It translates one statement of the program at a time. The interpreter may or may not generate an explicit piece of code to represent the statement. Because the interpreter translates only a very small piece of the program at any given time, a small amount of memory is used to hold intermediate representations of the program. In many cases, a Forth program plus the Forth interpreter are smaller than the equivalent native machine code.

Just-in-time compilers have been used for many years, but are best known today for their use in Java environments [Cra97]. A JIT compiler is somewhere between an interpreter and a stand-alone compiler. A JIT compiler produces executable code segments for pieces of the program. However, it compiles a section of the program (such as a function) only when it knows it will be executed. Unlike an interpreter, it saves the compiled version of the code so that the code does not have to be retranslated the next time it is executed. A JIT compiler saves some execution time overhead relative to an interpreter because it does not translate the same piece of code multiple times, but it also uses more memory for the intermediate representation. The JIT compiler usually generates machine code directly rather than building intermediate program representation data structures such as the CDFG. A JIT compiler also usually performs only simple optimizations as compared to a stand-alone compiler.

5.6 PROGRAM-LEVEL PERFORMANCE ANALYSIS

Because embedded systems must perform functions in real time, we often need to know how fast a program runs. The techniques we use to analyze program execution time are also helpful in analyzing properties such as power consumption. In this

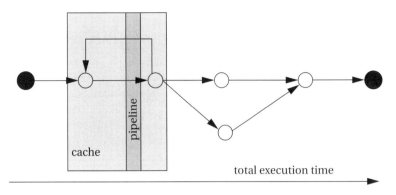

FIGURE 5.22

Execution time is a global property of a program.

section, we study how to analyze programs to estimate their run times. We also examine how to optimize programs to improve their execution times; of course, optimization relies on analysis.

It is important to keep in mind that CPU performance is not judged in the same way as program performance. Certainly, CPU clock rate is a very unreliable metric for program performance. But more importantly, the fact that the CPU executes part of our program quickly does not mean that it will execute the entire program at the rate we desire. As illustrated in Figure 5.22, the CPU pipeline and cache act as windows into our program. In order to understand the total execution time of our program, we must look at execution paths, which in general are far longer than the pipeline and cache windows. The pipeline and cache influence execution time, but execution time is a global property of the program.

While we might hope that the execution time of programs could be precisely determined, this is in fact difficult to do in practice:

- The execution time of a program often varies with the input data values because those values select different execution paths in the program. For example, loops may be executed a varying number of times, and different branches may execute blocks of varying complexity.

- The cache has a major effect on program performance, and once again, the cache's behavior depends in part on the data values input to the program.

- Execution times may vary even at the instruction level. Floating-point operations are the most sensitive to data values, but the normal integer execution pipeline can also introduce data-dependent variations. In general, the execution time of an instruction in a pipeline depends not only on that instruction but on the instructions around it in the pipeline.

We can measure program performance in several ways:

- Some microprocessor manufacturers supply simulators for their CPUs: The simulator runs on a workstation or PC, takes as input an executable for the microprocessor along with input data, and simulates the execution of that program. Some of these simulators go beyond functional simulation to measure the execution time of the program. Simulation is clearly slower than executing the program on the actual microprocessor, but it also provides much greater visibility during execution. Be careful—some microprocessor performance simulators are not 100% accurate, and simulation of I/O-intensive code may be difficult.

- A timer connected to the microprocessor bus can be used to measure performance of executing sections of code. The code to be measured would reset and start the timer at its start and stop the timer at the end of execution. The length of the program that can be measured is limited by the accuracy of the timer.

- A logic analyzer can be connected to the microprocessor bus to measure the start and stop times of a code segment. This technique relies on the code being able to produce identifiable events on the bus to identify the start and stop of execution. The length of code that can be measured is limited by the size of the logic analyzer's buffer.

We are interested in the following three different types of performance measures on programs:

- *Average-case execution time* This is the typical execution time we would expect for typical data. Clearly, the first challenge is defining typical inputs.

- *Worst-case execution time* The longest time that the program can spend on any input sequence is clearly important for systems that must meet deadlines. In some cases, the input set that causes the worst-case execution time is obvious, but in many cases it is not.

- *Best-case execution time* This measure can be important in multirate real-time systems, as seen in Chapter 6.

First, we look at the fundamentals of program performance in more detail. We then consider trace-driven performance based on executing the program and observing its behavior.

5.6.1 Elements of Program Performance

The key to evaluating execution time is breaking the performance problem into parts. Program execution time [Sha89] can be seen as

$$execution\ time = program\ path + instruction\ timing$$

The path is the sequence of instructions executed by the program (or its equivalent in the high-level language representation of the program). The instruction timing is determined based on the sequence of instructions traced by the program path, which takes into account data dependencies, pipeline behavior, and caching. Luckily, these two problems can be solved relatively independently.

Although we can trace the execution path of a program through its high-level language specification, it is hard to get accurate estimates of total execution time from a high-level language program. This is because there is not, as we saw in Section 5.4, a direct correspondence between program statements and instructions. The number of memory locations and variables must be estimated, and results may be either saved for reuse or recomputed on the fly, among other effects. These problems become more challenging as the compiler puts more and more effort into optimizing the program. However, some aspects of program performance can be estimated by looking directly at the C program. For example, if a program contains a loop with a large, fixed iteration bound or if one branch of a conditional is much longer than another, we can get at least a rough idea that these are more time-consuming segments of the program.

Of course, a precise estimate of performance also relies on the instructions to be executed, since different instructions take different amounts of time. (In addition, to make life even more difficult, the execution time of one instruction can depend on the instructions executed before and after it.) Example 5.7 illustrates data-dependent program paths.

Example 5.7

Data-dependent paths in if **statements**

Here is a set of nested if statements:

```
if (a) { /* test 1 */
        if (b) { /* test 2 */
                x = r * s + t; /* assignment 1 */
                }
        else    {
                y = r + s; /* assignment 2 */
                }
        z = r + s + u; /* assignment 3 */
        }
else {
        if (c) { /* test 3 */
                y = r - t; /* assignment 4 */
                }
        }
```

The conditional tests and assignments are labeled within each if statement to make it easier to identify paths. What execution paths may be exercised? One way to enumerate all the paths

is to create a truth table–like structure. The paths are controlled by the variables in the if conditions, namely, a, b, and c. For any given combination of values of those variables, we can trace through the program to see which branch is taken at each if and which assignments are performed. For example, when $a = 1$, $b = 0$, and $c = 1$, then test 1 is true and test 2 is true. This means we first perform assignment 1 and then assignment 3.

Results for all the controlling variable values follow:

a	b	c	Path
0	0	0	test 1 false, test 3 false: no assignments
0	0	1	test 1 false, test 3 true: assignment 4
0	1	0	test 1 false, test 3 false: no assignments
0	1	1	test 1 false, test 3 true: assignment 4
1	0	0	test 1 true, test 2 false: assignments 2, 3
1	0	1	test 1 true, test 2 false: assignments 2, 3
1	1	0	test 1 true, test 2 true: assignments 1, 3
1	1	1	test 1 true, test 2 true: assignments 1, 3

Notice that there are only four distinct cases: no assignment, assignment 4, assignments 2 and 3, or assignments 1 and 3. These correspond to the possible paths through the nested ifs; the table adds value by telling us which variable values exercise each of these paths.

Enumerating the paths through a fixed-iteration for loop is seemingly simple. In the code below,

```
for (i = 0; i < N; i++)
        a[i] = b[i]*c[i];
```

the assignment in the loop is performed exactly N times. However, we can't forget the code executed to set up the loop and to test the iteration variable. Example 5.8 illustrates how to determine the path through a loop.

Example 5.8

Paths in a loop

Here is the loop code for the FIR filter of Example 2.5:

```
for (i = 0, f = 0; i < N; i++)
        f = f + c[i] * x[i];
```

By examining the CDFG for the code we can more easily determine how many times various statements are executed. Here is the CDFG once again:

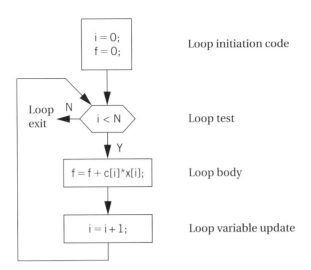

i = 0;
f = 0;
Loop initiation code

Loop exit N i < N Loop test

Y

f = f + c[i]*x[i]; Loop body

i = i + 1; Loop variable update

The CDFG makes it clear that the loop initiation block is executed once, the test is executed $N + 1$ times, and the body and loop variable update are each executed N times.

To measure the longest path length, we must find the longest path through the optimized CDFG since the compiler may change the structure of the control and data flow to optimize the program's implementation. It is important to keep in mind that choosing the longest path through a CDFG as measured by the number of nodes or edges touched may not correspond to the longest execution time. Since the execution time of a node in the CDFG will vary greatly depending on the instructions represented by that node, we must keep in mind that the longest path through the CDFG depends on the execution times of the nodes. In general, it is good policy to choose several of what we estimate are the longest paths through the program and measure the lengths of all of them in sufficient detail to be sure that we have in fact captured the longest path.

Once we know the execution path of the program, we have to measure the execution time of the instructions executed along that path. The simplest estimate is to assume that every instruction takes the same number of clock cycles, which means we need only count the instructions and multiply by the per-instruction execution time to obtain the program's total execution time. However, even ignoring cache effects, this technique is simplistic for the reasons summarized below.

■ *Not all instructions take the same amount of time.* Although RISC architectures tend to provide uniform instruction execution times in order to keep the CPU's pipeline full, even many RISC architectures take different amounts of time to execute certain instructions. Multiple load-store instructions are examples of longer-executing instructions in the ARM architecture. Floating-point instructions show especially wide variations in execution time—while

basic multiply and add operations are fast, some transcendental functions can take thousands of cycles to execute.

- *Execution times of instructions are not independent.* The execution time of one instruction depends on the instructions around it. For example, many CPUs use register bypassing to speed up instruction sequences when the result of one instruction is used in the next instruction. As a result, the execution time of an instruction may depend on whether its destination register is used as a source for the next operation (or vice versa).

- *The execution time of an instruction may depend on operand values.* This is clearly true of floating-point instructions in which a different number of iterations may be required to calculate the result. Other specialized instructions can, for example, perform a data-dependent number of integer operations.

We can handle the first two problems more easily than the third. We can look up instruction execution time in a table; the table will be indexed by opcode and possibly by other parameter values such as the registers used. To handle interdependent execution times, we can add columns to the table to consider the effects of nearby instructions. Since these effects are generally limited by the size of the CPU pipeline, we know that we need to consider a relatively small window of instructions to handle such effects. Handling variations due to operand values is difficult to do without actually executing the program using a variety of data values, given the large number of factors that can affect value-dependent instruction timing. Luckily, these effects are often small. Even in floating-point programs, most of the operations are typically additions and multiplications whose execution times have small variances.

Thus far we have not considered the effect of the cache. Because the access time for main memory can be 10–100 times larger than the cache access time, caching can have huge effects on instruction execution time by changing both the instruction and data access times. Caching performance inherently depends on the program's execution path since the cache's contents depend on the history of accesses.

5.6.2 Measurement-Driven Performance Analysis

The most direct way to determine the execution time of a program is by measuring it. This approach is appealing, but it does have some drawbacks. First, in order to cause the program to execute its worst-case execution path, we have to provide the proper inputs to it. Determining the set of inputs that will guarantee the worst-case execution path is infeasible. Furthermore, in order to measure the program's performance on a particular type of CPU, we need the CPU or its simulator.

Despite these drawbacks, measurement is the most commonly used way to determine the execution time of embedded software. Worst-case execution time analysis algorithms have been used successfully in some areas, such as flight control software, but many system design projects determine the execution time of their programs by measurement.

Most methods of measuring program performance combine the determination of the execution path and the timing of that path: as the program executes, it chooses a path and we observe the execution time along that path. We refer to the record of the execution path of a program as a ***program trace*** (or more succinctly, a ***trace***). Traces can be valuable for other purposes, such as analyzing the cache behavior of the program.

Perhaps the biggest problem in measuring program performance is figuring out a useful set of inputs to provide to the program. This problem has two aspects. First, we have to determine the actual input values. We may be able to use benchmark data sets or data captured from a running system to help us generate typical values. For simple programs, we may be able to analyze the algorithm to determine the inputs that cause the worst-case execution time. The software testing methods of Section 5.10 can help us generate some test values and determine how thoroughly we have exercised the program.

The other problem with input data is the ***software scaffolding*** that we may need to feed data into the program and get data out. When we are designing a large system, it may be difficult to extract out part of the software and test it independently of the other parts of the system. We may need to add new testing modules to the system software to help us introduce testing values and to observe testing outputs.

We can measure program performance either directly on the hardware or by using a simulator. Each method has its advantages and disadvantages.

Physical measurement requires some sort of hardware instrumentation. The most direct method of measuring the performance of a program would be to watch the program counter's value: start a timer when the PC reaches the program's start, stop the timer when it reaches the program's end. Unfortunately, it generally isn't possible to directly observe the program counter. However, it is possible in many cases to modify the program so that it starts a timer at the beginning of execution and stops the timer at the end. While this doesn't give us direct information about the program trace, it does give us execution time. If we have several timers available, we can use them to measure the execution time of different parts of the program.

A logic analyzer or an oscilloscope can be used to watch for signals that mark various points in the execution of the program. However, because logic analyzers have a limited amount of memory, this approach doesn't work well for programs with extremely long execution times.

Some CPUs have hardware facilities for automatically generating trace information. For example, the Pentium family microprocessors generate a special bus cycle, a branch trace message, that shows the source and/or destination address of a branch [Col97]. If we record only traces, we can reconstruct the instructions executed within the basic blocks while greatly reducing the amount of memory required to hold the trace.

The alternative to physical measurement of execution time is simulation. A CPU simulator is a program that takes as input a memory image for a CPU and performs the operations on that memory image that the actual CPU would perform, leaving

the results in the modified memory image. For purposes of performance analysis, the most important type of CPU simulator is the **cycle-accurate simulator**, which performs a sufficiently detailed simulation of the processor's internals so that it can determine the exact number of clock cycles required for execution. A cycle-accurate simulator is built with detailed knowledge of how the processor works, so that it can take into account all the possible behaviors of the microarchitecture that may affect execution time. Cycle-accurate simulators are slower than the processor itself, but a variety of techniques can be used to make them surprisingly fast, running only hundreds of times slower than the hardware itself.

A cycle-accurate simulator has a complete model of the processor, including the cache. It can therefore provide valuable information about why the program runs too slowly. The next example discusses a simulator that can be used to model many different processors.

Example 5.9

Cycle-accurate simulation

SimpleScalar (http://www.simplescalar.com) is a framework for building cycle-accurate CPU models. Some aspects of the processor can be configured easily at run time. For more complex changes, we can use the SimpleScalar toolkit to write our own simulator.

We can use SimpleScalar to simulate the FIR filter code. SimpleScalar can model a number of different processors; we will use a standard ARM model here.

We want to include the data as part of the program so that the execution time doesn't include file I/O. File I/O is slow and the time it takes to read or write data can change substantially from one execution to another. We get around this problem by setting up an array that holds the FIR data. And since the test program will include some initialization and other miscellaneous code, we execute the FIR filter many times in a row using a simple loop. Here is the complete test program:

```
#define COUNT 100
#define N 12

int x[N] = {8,17,3,122,5,93,44,2,201,11,74,75};
int c[N] = {1,2,4,7,3,4,2,2,5,8,5,1};

main() {
  int i, k, f;

  for (k=0; k<COUNT; k++) { /* run the filter */
      for (i=0; i<N; i++)
          f += c[i]*x[i];
  }
}
```

To start the simulation process, we compile our test program using a special compiler:

```
% arm-linux-gcc firtest.c
```

This gives us an executable program (by default, a.out) that we use to simulate our program:

```
% arm-outorder a.out
```

SimpleScalar produces a large output file with a great deal of information about the program's execution. Since this is a simple example, the most useful piece of data is the total number of simulated clock cycles required to execute the program:

```
 sim_cycle              25854 # total simulation time in cycles
```

To make sure that we can ignore the effects of program overhead, we will execute the FIR filter for several different values of N and compare. This run used $N = 100$; when we also run $N = 1,000$ and $N = 10,000$, we get these results:

N	Total simulation time in cycles	Simulation time for one filter execution
100	25854	259
1000	155759	156
10000	1451840	145

Because the FIR filter is so simple and ran in so few cycles, we had to execute it a number of times to wash out all the other overhead of program execution. However, the time for 1,000 and 10,000 filter executions are within 10% of each other, so those values are reasonably close to the actual execution time of the FIR filter itself.

5.7 SOFTWARE PERFORMANCE OPTIMIZATION

In this section we will look at several techniques for optimizing software performance.

5.7.1 Loop Optimizations

Loops are important targets for optimization because programs with loops tend to spend a lot of time executing those loops. There are three important techniques in optimizing loops: *code motion, induction variable elimination,* and *strength reduction*.

 Code motion lets us move unnecessary code out of a loop. If a computation's result does not depend on operations performed in the loop body, then we can safely move it out of the loop. Code motion opportunities can arise because programmers may find some computations clearer and more concise when put in the loop body,

even though they are not strictly dependent on the loop iterations. A simple example of code motion is also common. Consider the following loop:

```
for (i = 0; i < N*M; i++) {
    z[i] = a[i] + b[i];
    }
```

The code motion opportunity becomes more obvious when we draw the loop's CDFG as shown in Figure 5.23. The loop bound computation is performed on every iteration during the loop test, even though the result never changes. We can avoid $N \times M - 1$ unnecessary executions of this statement by moving it before the loop, as shown in the figure.

An ***induction variable*** is a variable whose value is derived from the loop iteration variable's value. The compiler often introduces induction variables to help it implement the loop. Properly transformed, we may be able to eliminate some variables and apply strength reduction to others.

A nested loop is a good example of the use of induction variables. Here is a simple nested loop:

```
for (i = 0; i < N; i++)
    for (j = 0; j < M; j++)
        z[i][j] = b[i][j];
```

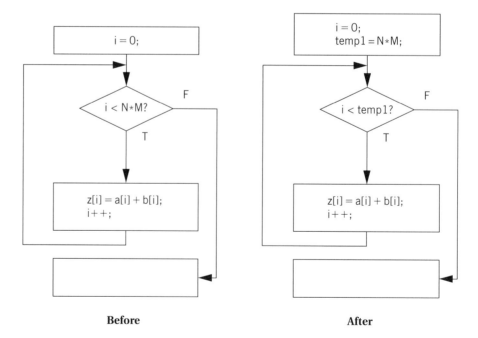

Before **After**

FIGURE 5.23

Code motion in a loop.

The compiler uses induction variables to help it address the arrays. Let us rewrite the loop in C using induction variables and pointers. (Later, we use a common induction variable for the two arrays, even though the compiler would probably introduce separate induction variables and then merge them.)

```
for (i = 0; i < N; i++)
    for (j = 0; j < M; j++) {
        zbinduct = i*M + j;
        *(zptr + zbinduct) = *(bptr + zbinduct);
    }
```

In the above code, zptr and bptr are pointers to the heads of the z and b arrays and zbinduct is the shared induction variable. However, we do not need to compute zbinduct afresh each time. Since we are stepping through the arrays sequentially, we can simply add the update value to the induction variable:

```
zbinduct = 0;
for (i = 0; i < N; i++) {
    for (j = 0; j < M; j++) {
        *(zptr + zbinduct) = *(bptr + zbinduct);
        zbinduct++;
    }
}
```

This is a form of strength reduction since we have eliminated the multiplication from the induction variable computation.

Strength reduction helps us reduce the cost of a loop iteration. Consider the following assignment:

```
y = x * 2;
```

In integer arithmetic, we can use a left shift rather than a multiplication by 2 (as long as we properly keep track of overflows). If the shift is faster than the multiply, we probably want to perform the substitution. This optimization can often be used with induction variables because loops are often indexed with simple expressions. Strength reduction can often be performed with simple substitution rules since there are relatively few interactions between the possible substitutions.

Cache Optimizations

A *loop nest* is a set of loops, one inside the other. Loop nests occur when we process arrays. A large body of techniques has been developed for optimizing loop nests. Rewriting a loop nest changes the order in which array elements are accessed. This can expose new parallelism opportunities that can be exploited by later stages of the compiler, and it can also improve cache performance. In this section we concentrate on the analysis of loop nests for cache performance.

Example 5.10

Data realignment and array padding

Assume we want to optimize the cache behavior of the following code:

```
for (j = 0; j < M; j++)
    for (i = 0; i < N; i++)
        a[j][i] = b[j][i] * c;
```

Let us also assume that the a and b arrays are sized with M at 265 and N at 4 and a 256-line, four-way set-associative cache with four words per line. Even though this code does not reuse any data elements, cache conflicts can cause serious performance problems because they interfere with spatial reuse at the cache line level.

Assume that the starting location for $a[]$ is 1024 and the starting location for $b[]$ is 4099. Although $a[0][0]$ and $b[0][0]$ do not map to the same word in the cache, they do map to the same block.

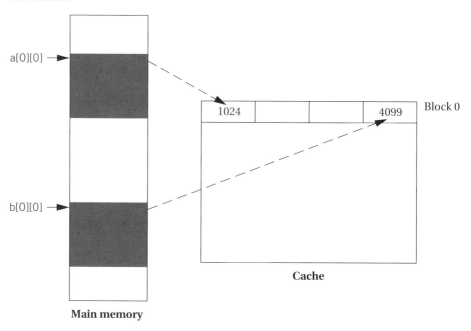

As a result, we see the following scenario in execution:

- The access to $a[0][0]$ brings in the first four words of $a[]$.

- The access to $b[0][0]$ replaces $a[0][0]$ through $a[0][3]$ with $b[0][3]$ and the contents of the three locations before $b[]$.

- When $a[0][1]$ is accessed, the same cache line is again replaced with the first four elements of $a[]$.

Once the *a*[0][1] access brings that line into the cache, it remains there for the *a*[0][2] and *a*[0][3] accesses since the *b*[] accesses are now on the next line. However, the scenario repeats itself at *a*[1][0] and every four iterations of the cache.

One way to eliminate the cache conflicts is to move one of the arrays. We do not have to move it far. If we move *b*'s start to 4100, we eliminate the cache conflicts.

However, that fix won't work in more complex situations. Moving one array may only introduce cache conflicts with another array. In such cases, we can use another technique called padding. If we extend each of the rows of the arrays to have four elements rather than three, with the padding word placed at the beginning of the row, we eliminate the cache conflicts. In this case, *b*[0][0] is located at 4100 by the padding. Although padding wastes memory, it substantially improves memory performance. In complex situations with multiple arrays and sophisticated access patterns, we have to use a combination of techniques—relocating arrays and padding them—to be able to minimize cache conflicts.

5.7.2 Performance Optimization Strategies

Let's look more generally at how to improve program execution time. First, make sure that the code really needs to be accelerated. If you are dealing with a large program, the part of the program using the most time may not be obvious. **Profiling** the program will help you find hot spots. A profiler does not measure execution time—instead, it counts the number of times that procedures or basic blocks in the program are executed. There are two major ways to profile a program: We can modify the executable program by adding instructions that increment a location every time the program passes that point in the program; or we can sample the program counter during execution and keep track of the distribution of PC values. Profiling adds relatively little overhead to the program and it gives us some useful information about where the program spends most of its time.

You may be able to redesign your algorithm to improve efficiency. Examining asymptotic performance is often a good guide to efficiency. Doing fewer operations is usually the key to performance. In a few cases, however, brute force may provide a better implementation. A seemingly simple high-level language statement may in fact hide a very long sequence of operations that slows down the algorithm. Using dynamically allocated memory is one example, since managing the heap takes time but is hidden from the programmer. For example, a sophisticated algorithm that uses dynamic storage may be slower in practice than an algorithm that performs more operations on statically allocated memory.

Finally, you can look at the implementation of the program itself. A few hints on program implementation are summarized below.

- *Try to use registers efficiently*. Group accesses to a value together so that the value can be brought into a register and kept there.

- *Make use of page mode accesses in the memory system whenever possible*. Page mode reads and writes eliminate one step in the memory access. You

can increase use of page mode by rearranging your variables so that more can be referenced contiguously.

■ *Analyze cache behavior to find major cache conflicts.* Restructure the code to eliminate as many of these as you can as follows:

—For instruction conflicts, if the offending code segment is small, try to rewrite the segment to make it as small as possible so that it better fits into the cache. Writing in assembly language may be necessary. For conflicts across larger spans of code, try moving the instructions or padding with NOPs.

—For scalar data conflicts, move the data values to different locations to reduce conflicts.

—For array data conflicts, consider either moving the arrays or changing your array access patterns to reduce conflicts.

5.8 PROGRAM-LEVEL ENERGY AND POWER ANALYSIS AND OPTIMIZATION

Power consumption is a particularly important design metric for battery-powered systems because the battery has a very limited lifetime. However, power consumption is increasingly important in systems that run off the power grid. Fast chips run hot, and controlling power consumption is an important element of increasing reliability and reducing system cost.

How much control do we have over power consumption? Ultimately, we must consume the energy required to perform necessary computations. However, there are opportunities for saving power. Examples appear below.

■ We may be able to replace the algorithms with others that do things in clever ways that consume less power.

■ Memory accesses are a major component of power consumption in many applications. By optimizing memory accesses we may be able to significantly reduce power.

■ We may be able to turn off parts of the system—such as subsystems of the CPU, chips in the system, and so on—when we do not need them in order to save power.

The first step in optimizing a program's energy consumption is knowing how much energy the program consumes. It is possible to measure power consumption for an instruction or a small code fragment [Tiw94]. The technique, illustrated in Figure 5.24, executes the code under test over and over in a loop. By measuring the current flowing into the CPU, we are measuring the power consumption of the complete loop, including both the body and other code. By separately measuring the power consumption of a loop with no body (making sure, of course, that the

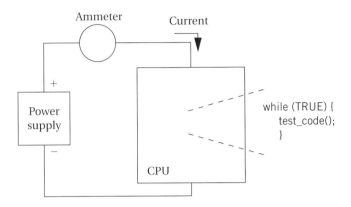

FIGURE 5.24

Measuring energy consumption for a piece of code.

compiler hasn't optimized away the empty loop), we can calculate the power consumption of the loop body code as the difference between the full loop and the bare loop energy cost of an instruction.

Several factors contribute to the energy consumption of the program.

- Energy consumption varies somewhat from instruction to instruction.
- The sequence of instructions has some influence.
- The opcode and the locations of the operands also matter.

Choosing which instructions to use can make some difference in a program's energy consumption, but concentrating on the instruction opcodes has limited payoffs in most CPUs. The program has to do a certain amount of computation to perform its function. While there may be some clever ways to perform that computation, the energy cost of the basic computation will change only a fairly small amount compared to the total system energy consumption, and usually only after a great deal of effort. We are further hampered in our ability to optimize instruction-level energy consumption because most manufacturers do not provide detailed, instruction-level energy consumption figures for their processors.

In many applications, the biggest payoff in energy reduction for a given amount of designer effort comes from concentrating on the memory system. Catthoor *et al*. [Cat98] showed that memory transfers are by far the most expensive type of operation performed by a CPU—in their studies, a memory transfer takes 33 times more energy than does an addition. As a result, the biggest payoffs in energy optimization come from properly organizing instructions and data in memory. Accesses to registers are the most energy efficient; cache accesses are more energy efficient than main memory accesses.

Caches are an important factor in energy consumption. On the one hand, a cache hit saves a costly main memory access, and on the other, the cache itself is relatively

power hungry because it is built from SRAM, not DRAM. If we can control the size of the cache, we want to choose the smallest cache that provides us with the necessary performance. Li and Henkel [Li98] measured the influence of caches on energy consumption in detail. Figure 5.25 breaks down the energy consumption of a computer running MPEG (a video encoder) into several components: software running on the CPU, main memory, data cache, and instruction cache.

As the instruction cache size increases, the energy cost of the software on the CPU declines, but the instruction cache comes to dominate the energy consumption. Experiments like this on several benchmarks show that many programs have sweet spots in energy consumption. If the cache is too small, the program runs slowly and the system consumes a lot of power due to the high cost of main memory accesses. If the cache is too large, the power consumption is high without a corresponding payoff in performance. At intermediate values, the execution time and power consumption are both good.

How can we optimize a program for low power consumption? The best overall advice is that *high performance = low power*. Generally speaking, making the program run faster also reduces energy consumption.

Clearly, the biggest factor that can be reasonably well controlled by the programmer is the memory access patterns. If the program can be modified to reduce instruction or data cache conflicts, for example, the energy required by the memory system can be significantly reduced. The effectiveness of changes such as reordering instructions or selecting different instructions depends on the processor involved, but they are generally less effective than cache optimizations.

A few optimizations mentioned previously for performance are also often useful for improving energy consumption:

- *Try to use registers efficiently.* Group accesses to a value together so that the value can be brought into a register and kept there.

- *Analyze cache behavior to find major cache conflicts.* Restructure the code to eliminate as many of these as you can:

 —For instruction conflicts, if the offending code segment is small, try to rewrite the segment to make it as small as possible so that it better fits into the cache. Writing in assembly language may be necessary. For conflicts across larger spans of code, try moving the instructions or padding with NOPs.

 —For scalar data conflicts, move the data values to different locations to reduce conflicts.

 —For array data conflicts, consider either moving the arrays or changing your array access patterns to reduce conflicts.

- *Make use of page mode accesses in the memory system whenever possible.* Page mode reads and writes eliminate one step in the memory access, saving a considerable amount of power.

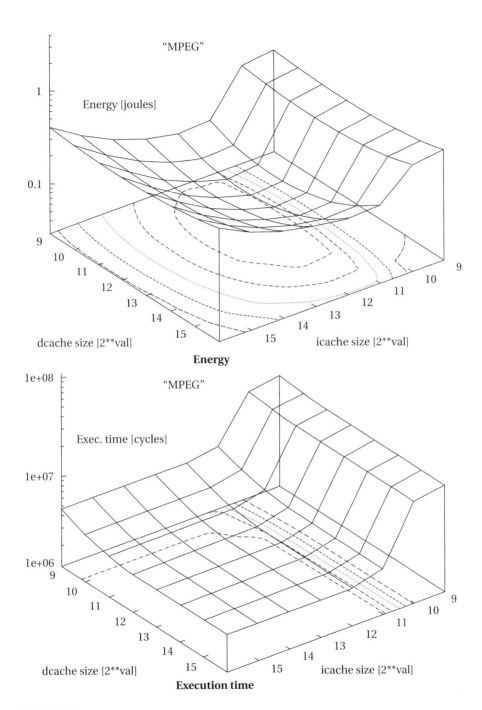

FIGURE 5.25

Energy and execution time vs. instruction/data cache size for a benchmark program [Li98].

Metha et al. [Met97] present some additional observations about energy optimization as follows:

- Moderate loop unrolling eliminates some loop control overhead. However, when the loop is unrolled too much, power increases due to the lower hit rates of straight-line code.

- Software pipelining reduces pipeline stalls, thereby reducing the average energy per instruction.

- Eliminating recursive procedure calls where possible saves power by getting rid of function call overhead. Tail recursion can often be eliminated; some compilers do this automatically.

5.9 ANALYSIS AND OPTIMIZATION OF PROGRAM SIZE

The memory footprint of a program is determined by the size of its data and instructions. Both must be considered to minimize program size.

Data provide an excellent opportunity for minimizing size because the data are most highly dependent on programming style. Because inefficient programs often keep several copies of data, identifying and eliminating duplications can lead to significant memory savings usually with little performance penalty. Buffers should be sized carefully—rather than defining a data array to a large size that the program will never attain, determine the actual maximum amount of data held in the buffer and allocate the array accordingly. Data can sometimes be packed, such as by storing several flags in a single word and extracting them by using bit-level operations.

A very low-level technique for minimizing data is to reuse values. For instance, if several constants happen to have the same value, they can be mapped to the same location. Data buffers can often be reused at several different points in the program. This technique must be used with extreme caution, however, since subsequent versions of the program may not use the same values for the constants. A more generally applicable technique is to generate data on the fly rather than store it. Of course, the code required to generate the data takes up space in the program, but when complex data structures are involved there may be some net space savings from using code to generate data.

Minimizing the size of the instruction text of a program requires a mix of high-level program transformations and careful instruction selection. Encapsulating functions in subroutines can reduce program size when done carefully. Because subroutines have overhead for parameter passing that is not obvious from the high-level language code, there is a minimum-size function body for which a subroutine makes sense. Architectures that have variable-size instruction lengths are particularly good candidates for careful coding to minimize program size, which may require assembly

language coding of key program segments. There may also be cases in which one or a sequence of instructions is much smaller than alternative implementations—for example, a multiply-accumulate instruction may be both smaller and faster than separate arithmetic operations.

When reducing the number of instructions in a program, one important technique is the proper use of subroutines. If the program performs identical operations repeatedly, these operations are natural candidates for subroutines. Even if the operations vary somewhat, you may be able to construct a properly parameterized subroutine that saves space. Of course, when considering the code size savings, the subroutine linkage code must be counted into the equation. There is extra code not only in the subroutine body but also in each call to the subroutine that handles parameters. In some cases, proper instruction selection may reduce code size; this is particularly true in CPUs that use variable-length instructions.

Some microprocessor architectures support ***dense instruction sets***, specially designed instruction sets that use shorter instruction formats to encode the instructions. The ARM Thumb instruction set and the MIPS-16 instruction set for the MIPS architecture are two examples of this type of instruction set. In many cases, a microprocessor that supports the dense instruction set also supports the normal instruction set, although it is possible to build a microprocessor that executes only the dense instruction set. Special compilation modes produce the program in terms of the dense instruction set. Program size of course varies with the type of program, but programs using the dense instruction set are often 70 to 80% of the size of the standard instruction set equivalents.

5.10 PROGRAM VALIDATION AND TESTING

Complex systems need testing to ensure that they work as they are intended. But bugs can be subtle, particularly in embedded systems, where specialized hardware and real-time responsiveness make programming more challenging. Fortunately, there are many available techniques for software testing that can help us generate a comprehensive set of tests to ensure that our system works properly. We examine the role of validation in the overall design methodology in Section 9.5. In this section, we concentrate on nuts-and-bolts techniques for creating a good set of tests for a given program.

The first question we must ask ourselves is how much testing is enough. Clearly, we cannot test the program for every possible combination of inputs. Because we cannot implement an infinite number of tests, we naturally ask ourselves what a reasonable standard of thoroughness is. One of the major contributions of software testing is to provide us with standards of thoroughness that make sense. Following these standards does not guarantee that we will find all bugs. But by breaking the testing problem into subproblems and analyzing each subproblem,

we can identify testing methods that provide reasonable amounts of testing while keeping the testing time within reasonable bounds.

The two major types of testing strategies:

- **Black-box** methods generate tests without looking at the internal structure of the program.

- **Clear-box** (also known as **white-box**) methods generate tests based on the program structure.

In this section we cover both types of tests, which complement each other by exercising programs in very different ways.

5.10.1 Clear-Box Testing

The control/data flow graph extracted from a program's source code is an important tool in developing clear-box tests for the program. To adequately test the program, we must exercise both its control and data operations.

In order to execute and evaluate these tests, we must be able to control variables in the program and observe the results of computations, much as in manufacturing testing. In general, we may need to modify the program to make it more testable. By adding new inputs and outputs, we can usually substantially reduce the effort required to find and execute the test. Example 5.11 illustrates the importance of observability and controllability in software testing.

No matter what we are testing, we must accomplish the following three things in a test:

- Provide the program with inputs that exercise the test we are interested in.

- Execute the program to perform the test.

- Examine the outputs to determine whether the test was successful.

Example 5.11

Controlling and observing programs

Let's first consider controllability by examining the following FIR filter with a limiter:

```
firout = 0.0; /* initialize filter output */
/* compute buff*c in bottom part of circular buffer */
for (j = curr, k = 0; j < N; j++, k++)
    firout += buff[j] * c[k];
/* compute buff*c in top part of circular buffer */
for (j = 0; j < curr; j++, k++)
    firout += buff[j] * c[k];
/* limit output value */
```

```
if (firout > 100.0) firout = 100.0;
if (firout < 100.0) firout = -100.0;
```

The above code computes the output of an FIR filter from a circular buffer of values and then limits the maximum filter output (much as an overloaded speaker will hit a range limit). If we want to test whether the limiting code works, we must be able to generate two out-of-range values for firout: positive and negative. To do that, we must fill the FIR filter's circular buffer with N values in the proper range. Although there are many sets of values that will work, it will still take time for us to properly set up the filter output for each test.

This code also illustrates an observability problem. If we want to test the FIR filter itself, we look at the value of firout before the limiting code executes. We could use a debugger to set breakpoints in the code, but this is an awkward way to perform a large number of tests. If we want to test the FIR code independent of the limiting code, we would have to add a mechanism for observing firout independently.

Being able to perform this process for a large number of tests entails some amount of drudgery, but that drudgery can be alleviated with good program design that simplifies controllability and observability.

The next task is to determine the set of tests to be performed. We need to perform many different types of tests to be confident that we have identified a large fraction of the existing bugs. Even if we thoroughly test the program using one criterion, that criterion ignores other aspects of the program. Over the next few pages we will describe several very different criteria for program testing.

The most fundamental concept in clear-box testing is the path of execution through a program. Previously, we considered paths for performance analysis; we are now concerned with making sure that a path is covered and determining how to ensure that the path is in fact executed. We want to test the program by forcing the program to execute along chosen paths. We force the execution of a path by giving it inputs that cause it to take the appropriate branches. Execution of a path exercises both the control and data aspects of the program. The control is exercised as we take branches; both the computations leading up to the branch decision and other computations performed along the path exercise the data aspects.

Is it possible to execute every complete path in an arbitrary program? The answer is no, since the program may contain a while loop that is not guaranteed to terminate. The same is true for any program that operates on a continuous stream of data, since we cannot arbitrarily define the beginning and end of the data stream. If the program always terminates, then there are indeed a finite number of complete paths that can be enumerated from the path graph. This leads us to the next question: Does it make sense to exercise every path? The answer to this question is no for most programs, since the number of paths, especially for any program with a loop, is extremely large. However, the choice of an appropriate subset of paths to test requires some thought. Example 5.12 illustrates the consequences of two different choices of testing strategies.

Example 5.12

Choosing the paths to test

Two reasonable choices for a set of paths to test follow:

- Execute every statement at least once.
- Execute every direction of a branch at least once.

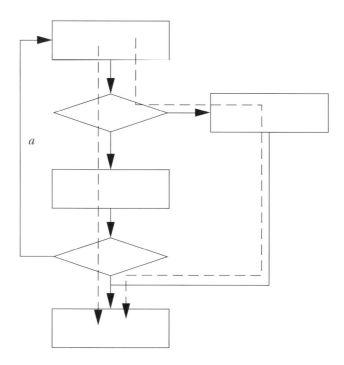

These conditions are equivalent for structured programming languages without gotos, but are not the same for unstructured code. Most assembly language is unstructured, and state machines may be coded in high-level languages with gotos.

To understand the difference between statement and branch coverage, consider the CDFG below. We can execute every statement at least once by executing the program along two distinct paths.

However, this leaves branch *a* out of the lower conditional uncovered. To ensure that we have executed along every edge in the CDFG, we must execute a third path through the program. This path does not test any new statements, but it does cause *a* to be exercised.

How do we choose a set of paths that adequately covers the program's behavior? Intuition tells us that a relatively small number of paths should be able to cover most practical programs. Graph theory helps us get a quantitative handle on the

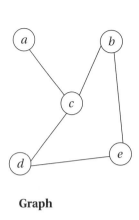

$$
\begin{array}{c}
\quad\quad a\ b\ c\ d\ e \\
\begin{array}{c}
a \\
b \\
c \\
d \\
e
\end{array}
\left[
\begin{array}{ccccc}
0 & 0 & 1 & 0 & 0 \\
0 & 0 & 1 & 0 & 1 \\
1 & 1 & 0 & 1 & 0 \\
0 & 0 & 1 & 0 & 1 \\
0 & 1 & 0 & 1 & 0
\end{array}
\right]
\end{array}
$$

Incidence matrix

$$
\begin{array}{c}
a \\
b \\
c \\
d \\
e
\end{array}
\left[
\begin{array}{ccccc}
1 & 0 & 0 & 0 & 0 \\
0 & 1 & 0 & 0 & 0 \\
0 & 0 & 1 & 0 & 0 \\
0 & 0 & 0 & 1 & 0 \\
0 & 0 & 0 & 0 & 1
\end{array}
\right]
$$

Basis set

Graph

FIGURE 5.26

The matrix representation of a graph and its basis set.

different paths required. In an undirected graph, we can form any path through the graph from combinations of **basis paths**. (Unfortunately, this property does not strictly hold for directed graphs such as CDFGs, but this formulation still helps us understand the nature of selecting a set of covering paths through a program.) The term "basis set" comes from linear algebra. Figure 5.26 shows how to evaluate the basis set of a graph. The graph is represented as an **incidence matrix**. Each row and column represents a node; a 1 is entered for each node pair connected by an edge. We can use standard linear algebra techniques to identify the basis set of the graph. Each vector in the basis set represents a primitive path. We can form new paths by adding the vectors modulo 2. Generally, there is more than one basis set for a graph.

The basis set property provides a metric for test coverage. If we cover all the basis paths, we can consider the control flow adequately covered. Although the basis set measure is not entirely accurate since the directed edges of the CDFG may make some combinations of paths infeasible, it does provide a reasonable and justifiable measure of test coverage.

There is a simple measure, **cyclomatic complexity** [McC76], which allows us to measure the control complexity of a program. Cyclomatic complexity is an upper bound on the size of the basis set that we found in Section 5.6.1. If e is the number of edges in the flow graph, n the number of nodes, and p the number of components in the graph, then the cyclomatic complexity is given by

$$
M = e - n + 2p. \tag{5.1}
$$

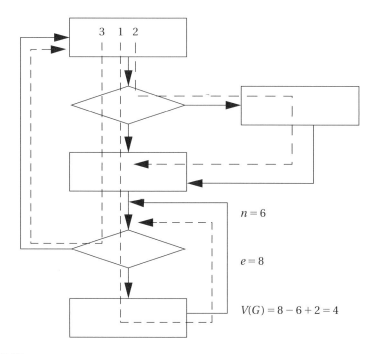

FIGURE 5.27

Cyclomatic complexity.

For a structured program, M can be computed by counting the number of binary decisions in the flow graph and adding 1. If the CDFG has higher-order branch nodes, add $b-1$ for each b-way branch. In the example of Figure 5.27, the cyclomatic complexity evaluates to 4. Because there are actually only three distinct paths in the graph, cyclomatic complexity in this case is an overly conservative bound.

Another way of looking at control flow-oriented testing is to analyze the conditions that control the conditional statements. Consider this if statement:

```
if ((a == b) || (c >= d)) { ... }
```

This complex condition can be exercised in several different ways. If we want to truly exercise the paths through this condition, it is prudent to exercise the conditional's elements in ways related to their own structure, not just the structure of the paths through them. A simple condition testing strategy is known as **branch testing** [Mye79]. This strategy requires the true and false branches of a conditional and every simple condition in the conditional's expression to be tested at least once. Example 5.13 illustrates branch testing.

Example 5.13

Condition testing with the branch testing strategy

Assume that the code below is what we meant to write.

```
if (a || (b >= c)) { printf("OK\n"); }
```

The code that we mistakenly wrote instead follows:

```
if (a && (b >= c)) { printf("OK\n"); }
```

If we apply branch testing to the code we wrote, one of the tests will use these values: $a = 0$, $b = 3$, $c = 2$ (making a false and $b >= c$ true). In this case, the code should print the OK term [0 || (3 >= 2) is true] but instead doesn't print [0 && (3 >= 2) evaluates to false]. That test picks up the error.

Let's consider another more subtle error that is nonetheless all too common in C. The code we meant to write follows:

```
if ((x == good_pointer) && (x->field1 == 3))
    { printf("got the value\n"); }
```

Here is the bad code we actually wrote:

```
if ((x = good_pointer) && (x->field1 == 3))
    { printf("got the value\n"); }
```

The problem here is that we typed = rather than ==, creating an assignment rather than a test. The code x = good_pointer first assigns the value good_pointer to x and then, because assignments are also expressions in C, returns good_pointer as the result of evaluating this expression.

If we apply the principles of branch testing, one of the tests we want to use will contain x != good_pointer and x->field1 == 3. Whether this test catches the error depends on the state of the record pointed to by good_pointer. If it is equal to 3 at the time of the test, the message will be printed erroneously. Although this test is not guaranteed to uncover the bug, it has a reasonable chance of success. One of the reasons to use many different types of tests is to maximize the chance that supposedly unrelated elements will cooperate to reveal the error in a particular situation.

Another more sophisticated strategy for testing conditionals is known as ***domain testing*** [How82], illustrated in Figure 5.28. Domain testing concentrates on linear inequalities. In the figure, the inequality the program should use for the test is $j <= i + 1$. We test the inequality with three test points—two on the boundary of the valid region and a third outside the region but between the i values of the other two points. When we make some common mistakes in typing the inequality, these three tests are sufficient to uncover them, as shown in the figure.

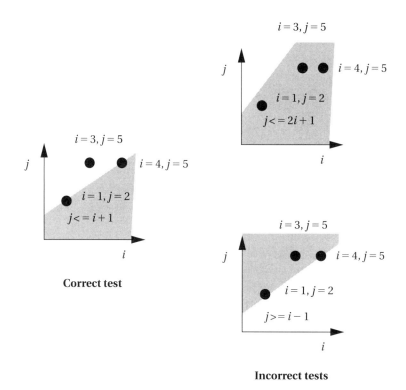

FIGURE 5.28

Domain testing for a pair of variables.

A potential problem with path coverage is that the paths chosen to cover the CDFG may not have any important relationship to the program's function. Another testing strategy known as **data flow testing** makes use of **def-use analysis** (short for definition-use analysis). It selects paths that have some relationship to the program's function.

The terms def and use come from compilers, which use def-use analysis for optimization [Aho06]. A variable's value is **defined** when an assignment is made to the variable; it is **used** when it appears on the right side of an assignment (sometimes called a **c-use** for computation use) or in a conditional expression (sometimes called **p-use** for predicate use). A **def-use pair** is a definition of a variable's value and a use of that value. Figure 5.29 shows a code fragment and all the def-use pairs for the first assignment to *a*. Def-use analysis can be performed on a program using iterative algorithms. Data flow testing chooses tests that exercise chosen def-use pairs. The test first causes a certain value to be assigned at the definition and then observes the result at the use point to be sure that the desired value arrived there. Frankl and

```
a = mypointer;
if (c > 5){
        while (a->field1 != val1)
            a = a->next;
}
if (a->field2 == val2)
        someproc(a,b);
```

FIGURE 5.29

Definitions and uses of variables.

Weyuker [Fra88] have defined criteria for choosing which def-use pairs to exercise to satisfy a well-behaved adequacy criterion.

We can write some specialized tests for loops. Since loops are common and often perform important steps in the program, it is worth developing loop-centric testing methods. If the number of iterations is fixed, then testing is relatively simple. However, many loops have bounds that are executed at run time. Consider first the case of a single loop:

```
for (i = 0; i < terminate(); i++)
        proc(i,array);
```

It would be too expensive to evaluate the above loop for all possible termination conditions. However, there are several important cases that we should try at a minimum:

1. Skipping the loop entirely [if possible, such as when terminate() returns 0 on its first call].

2. One loop iteration.

3. Two loop iterations.

4. If there is an upper bound n on the number of loop iterations (which may come from the maximum size of an array), a value that is significantly below that maximum number of iterations.

5. Tests near the upper bound on the number of loop iterations, that is, $n-1, n$, and $n + 1$.

We can also have nested loops like this:

```
for (i = 0; i < terminate1(); i++)
        for (j = 0; j < terminate2(); j++)
                for (k = 0; k < terminate3(); k++)
                        proc(i,j,k,array);
```

There are many possible strategies for testing nested loops. One thing to keep in mind is which loops have fixed vs. variable numbers of iterations. Beizer [Bei90] suggests an inside-out strategy for testing loops with multiple variable iteration bounds. First, concentrate on testing the innermost loop as above—the outer loops should be controlled to their minimum numbers of iterations. After the inner loop has been thoroughly tested, the next outer loop can be tested more thoroughly, with the inner loop executing a typical number of iterations. This strategy can be repeated until the entire loop nest has been tested. Clearly, nested loops can require a large number of tests. It may be worthwhile to insert testing code to allow greater control over the loop nest for testing.

5.10.2 Black-Box Testing

Black-box tests are generated without knowledge of the code being tested. When used alone, black-box tests have a low probability of finding all the bugs in a program. But when used in conjunction with clear-box tests they help provide a well-rounded test set, since black-box tests are likely to uncover errors that are unlikely to be found by tests extracted from the code structure. Black-box tests can really work. For instance, when asked to test an instrument whose front panel was run by a microcontroller, one acquaintance of the author used his hand to depress all the buttons simultaneously. The front panel immediately locked up. This situation could occur in practice if the instrument were placed face-down on a table, but discovery of this bug would be very unlikely via clear-box tests.

One important technique is to take tests directly from the specification for the code under design. The specification should state which outputs are expected for certain inputs. Tests should be created that provide specified outputs and evaluate whether the results also satisfy the inputs.

We can't test every possible input combination, but some rules of thumb help us select reasonable sets of inputs. When an input can range across a set of values, it is a very good idea to test at the ends of the range. For example, if an input must be between 1 and 10, 0, 1, 10, and 11 are all important values to test. We should be sure to consider tests both within and outside the range, such as, testing values within the range and outside the range. We may want to consider tests well outside the valid range as well as boundary-condition tests.

Random tests form one category of black-box test. Random values are generated with a given distribution. The expected values are computed independently of the system, and then the test inputs are applied. A large number of tests must be applied for the results to be statistically significant, but the tests are easy to generate.

Another scenario is to test certain types of data values. For example, integer-valued inputs can be generated at interesting values such as 0, 1, and values near the maximum end of the data range. Illegal values can be tested as well.

Regression tests form an extremely important category of tests. When tests are created during earlier stages in the system design or for previous versions of the system, those tests should be saved to apply to the later versions of the

system. Clearly, unless the system specification changed, the new system should be able to pass old tests. In some cases old bugs can creep back into systems, such as when an old version of a software module is inadvertently installed. In other cases regression tests simply exercise the code in different ways than would be done for the current version of the code and therefore possibly exercise different bugs.

Some embedded systems, particularly digital signal processing systems, lend themselves to numerical analysis. Signal processing algorithms are frequently implemented with limited-range arithmetic to save hardware costs. Aggressive data sets can be generated to stress the numerical accuracy of the system. These tests can often be generated from the original formulas without reference to the source code.

5.10.3 Evaluating Function Tests

How much testing is enough? Horgan and Mathur [Hor96] evaluated the coverage of two well-known programs, *TeX* and *awk*. They used functional tests for these programs that had been developed over several years of extensive testing. Upon applying those functional tests to the programs, they obtained the code coverage statistics shown in Figure 5.30. The columns refer to various types of test coverage: *block* refers to basic blocks, *decision* to conditionals, *p-use* to a use of a variable in a predicate (decision), and *c-use* to variable use in a nonpredicate computation. These results are at least suggestive that functional testing does not fully exercise the code and that techniques that explicitly generate tests for various pieces of code are necessary to obtain adequate levels of code coverage.

Methodological techniques are important for understanding the quality of your tests. For example, if you keep track of the number of bugs tested each day, the data you collect over time should show you some trends on the number of errors per page of code to expect on the average, how many bugs are caught by certain kinds of tests, and so on. We address methodological approaches to quality control in more detail in Section 9.5.

One interesting method for analyzing the coverage of your tests is **error injection**. First, take your existing code and add bugs to it, keeping track of where the bugs were added. Then run your existing tests on the modified program. By counting the number of added bugs your tests found, you can get an idea of how effective

	Block	Decision	P-use	C-use
TeX	85%	72%	53%	48%
awk	70%	59%	48%	55%

FIGURE 5.30

Code coverage of functional tests for TeX and awk (after Horgan and Mathur [Hor96]).

the tests are in uncovering the bugs you haven't yet found. This method assumes that you can deliberately inject bugs that are of similar varieties to those created naturally by programming errors.

If the bugs are too easy or too difficult to find or simply require different types of tests, then bug injection's results will not be relevant. Of course, it is essential that you finally use the correct code, not the code with added bugs.

5.11 SOFTWARE MODEM

In this section we design a modem. Low-cost modems generally use specialized chips, but some PCs implement the modem functions in software. Before jumping into the modem design itself, we discuss principles of how to transmit digital data over a telephone line. We will then go through a specification and discuss architecture, module design, and testing.

5.11.1 Theory of Operation and Requirements

The modem will use *frequency-shift keying (FSK)*, a technique used in 1200-baud modems. Keying alludes to Morse code—style keying. As shown in Figure 5.31, the FSK scheme transmits sinusoidal tones, with 0 and 1 assigned to different frequencies. Sinusoidal tones are much better suited to transmission over analog phone lines than are the traditional high and low voltages of digital circuits. The 01 bit patterns create the chirping sound characteristic of modems. (Higher-speed modems

FIGURE 5.31

Frequency-shift keying.

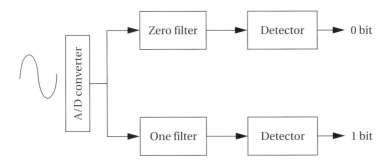

FIGURE 5.32

The FSK detection scheme.

are backward compatible with the 1200-baud FSK scheme and begin a transmission with a protocol to determine which speed and protocol should be used.)

The scheme used to translate the audio input into a bit stream is illustrated in Figure 5.32. The analog input is sampled and the resulting stream is sent to two digital filters (such as an FIR filter). One filter passes frequencies in the range that represents a 0 and rejects the 1-band frequencies, and the other filter does the converse. The outputs of the filters are sent to detectors, which compute the average value of the signal over the past n samples. When the energy goes above a threshold value, the appropriate bit is detected.

We will send data in units of 8-bit bytes. The transmitting and receiving modems agree in advance on the length of time during which a bit will be transmitted (otherwise known as the baud rate). But the transmitter and receiver are physically separated and therefore are not synchronized in any way. The receiving modem does not know when the transmitter has started to send a byte. Furthermore, even when the receiver does detect a transmission, the clock rates of the transmitter and receiver may vary somewhat, causing them to fall out of sync. In both cases, we can reduce the chances for error by sending the waveforms for a longer time.

The receiving process is illustrated in Figure 5.33. The receiver will detect the start of a byte by looking for a start bit, which is always 0. By measuring the length of the start bit, the receiver knows where to look for the start of the first bit. However, since the receiver may have slightly misjudged the start of the bit, it does not immediately try to detect the bit. Instead, it runs the detection algorithm at the predicted middle of the bit.

The modem will not implement a hardware interface to a telephone line or software for dialing a phone number. We will assume that we have analog audio inputs and outputs for sending and receiving. We will also run at a much slower bit rate than 1200 baud to simplify the implementation. Next, we will not implement a serial interface to a host, but rather put the transmitter's message in memory and save the receiver's result in memory as well. Given those understandings, let's fill out the requirements table.

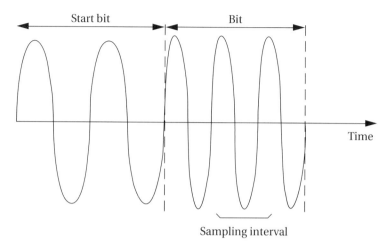

FIGURE 5.33

Receiving bits in the modem.

Name	Modem.
Purpose	A fixed baud rate frequency-shift keyed modem.
Inputs	Analog sound input, reset button.
Outputs	Analog sound output, LED bit display.
Functions	Transmitter: Sends data stored in microprocessor memory in 8-bit bytes. Sends start bit for each byte equal in length to one bit.
	Receiver: Automatically detects bytes and stores results in main memory. Displays currently received bit on LED.
Performance	1200 baud.
Manufacturing cost	Dominated by microprocessor and analog I/O.
Power	Powered by AC through a standard power supply.
Physical size and weight	Small and light enough to fit on a desktop.

5.11.2 Specification

The basic classes for the modem are shown in Figure 5.34.

5.11.3 System Architecture

The modem consists of one small subsystem (the interrupt handlers for the samples) and two major subsystems (transmitter and receiver). Two sample interrupt handlers are required, one for input and another for output, but they are very simple. The transmitter is simpler, so let's consider its software architecture first.

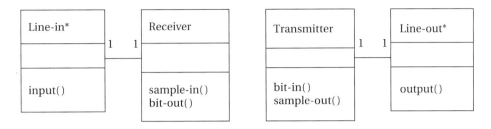

FIGURE 5.34

Class diagram for the modem.

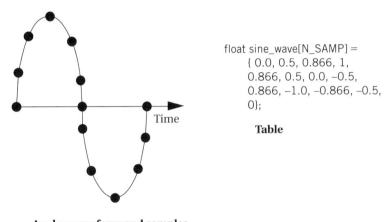

```
float sine_wave[N_SAMP] =
    { 0.0, 0.5, 0.866, 1,
    0.866, 0.5, 0.0, –0.5,
    0.866, –1.0, –0.866, –0.5,
    0};
```

Table

Analog waveform and samples

FIGURE 5.35

Waveform generation by table lookup.

The best way to generate waveforms that retain the proper shape over long intervals is *table lookup*. Software oscillators can be used to generate periodic signals, but numerical problems limit their accuracy. Figure 5.35 shows an analog waveform with sample points and the C code for these samples. Table lookup can be combined with interpolation to generate high-resolution waveforms without excessive memory costs, which is more accurate than oscillators because no feedback is involved. The required number of samples for the modem can be found by experimentation with the analog/digital converter and the sampling code.

The structure of the receiver is considerably more complex. The filters and detectors of Figure 5.33 can be implemented with circular buffers. But that module must feed a state machine that recognizes the bits. The recognizer state machine must use a timer to determine when to start and stop computing the filter output average based on the starting point of the bit. It must then determine the nature of the bit at the proper interval. It must also detect the start bit and measure it using the

counter. The receiver sample interrupt handler is a natural candidate to double as the receiver timer since the receiver's time points are relative to samples.

The hardware architecture is relatively simple. In addition to the analog/digital and digital/analog converters, a timer is required. The amount of memory required to implement the algorithms is relatively small.

5.11.4 Component Design and Testing

The transmitter and receiver can be tested relatively thoroughly on the host platform since the timing-critical code only delivers data samples. The transmitter's output is relatively easy to verify, particularly if the data are plotted. A testbench can be constructed to feed the receiver code sinusoidal inputs and test its bit recognition rate. It is a good idea to test the bit detectors first before testing the complete receiver operation. One potential problem in host-based testing of the receiver is encountered when library code is used for the receiver function. If a DSP library for the target processor is used to implement the filters, then a substitute must be found or built for the host processor testing. The receiver must then be retested when moved to the target system to ensure that it still functions properly with the library code.

Care must be taken to ensure that the receiver does not run too long and miss its deadline. Since the bulk of the computation is in the filters, it is relatively simple to estimate the total computation time early in the implementation process.

5.11.5 System Integration and Testing

There are two ways to test the modem system: by having the modem's transmitter send bits to its receiver, and or by connecting two different modems. The ultimate test is to connect two different modems, particularly modems designed by different people to be sure that incompatible assumptions or errors were not made. But single-unit testing, called ***loop-back*** testing in the telecommunications industry, is simpler and a good first step. Loop-back can be performed in two ways. First, a shared variable can be used to directly pass data from the transmitter to the receiver. Second, an audio cable can be used to plug the analog output to the analog input. In this case it is also possible to inject analog noise to test the resiliency of the detection algorithm.

SUMMARY

The program is a very fundamental unit of embedded system design and it usually contains tightly interacting code. Because we care about more than just functionality, we need to understand how programs are created. Because today's compilers do not take directives such as "compile this to run in <1 μs," we have to be able to optimize the programs ourselves for speed, power, and space. Our earlier understanding of computer architecture is critical to our ability to perform these optimizations.

We also need to test programs to make sure they do what we want. Some of our testing techniques can also be useful in exercising the programs for performance optimization.

What We Learned

- We can use data flow graphs to model straight-line code and CDFGs to model complete programs.

- Compilers perform numerous tasks, such as generating control flow, assigning variables to registers, creating procedure linkages, and so on.

- Remember the performance optimization equation: *execution time = program path + instruction timing*.

- Memory and cache optimizations are very important to performance optimization.

- Optimizing for power consumption often goes hand in hand with performance optimization.

- Optimizing programs for size is possible, but don't expect miracles.

- Programs can be tested as black boxes (without knowing the code) or as clear boxes (by examining the code structure).

FURTHER READING

Aho, Sethi, and Ullman [Aho06] wrote a classic text on compilers, and Muchnick [Muc97] describes advanced compiler techniques in detail. A paper on the ATOM system [Sri94] provides a good description of instrumenting programs for gathering traces. Cramer et al. [Cra97] describe the Java JIT compiler. Li and Malik [Li97] describe a method for statically analyzing program performance. Banerjee [Ban93, Ban94] describes loop transformations. Two books by Beizer, one on fundamental functional and structural testing techniques [Bei90] and the other on system-level testing [Bei84], provide comprehensive introductions to software testing and, as a bonus, are well written. Lyu [Lyu96] provides a good advanced survey of software reliability. Walsh [Wal97] describes a software modem implemented on an ARM processor.

QUESTIONS

Q5-1 Write C code for a state machine that implements a four-cycle handshake.

Q5-2 Write C code for a program that takes two values from an input circular buffer and puts the sum of those two values into a separate output circular buffer.

Q5-3 Write C code for a producer/consumer program that takes one value from one input queue, another value from another input queue, and puts the sum of those two values into a separate queue.

Q5-4 For each basic block given below, rewrite it in single-assignment form, and then draw the data flow graph for that form.

 a. x = a + b;
 y = c + d;
 z = x + e;

 b. r = a + b - c;
 s = 2 * r;
 t = b - d;
 r = d + e;

 c. a = q - r;
 b = a + t;
 a = r + s;
 c = t - u;

 d. w = a - b + c;
 x = w - d;
 y = x - 2;
 w = a + b - c;
 z = y + d;
 y = b * c;

Q5-5 Draw the CDFG for the following code fragments:

 a. if (y == 2) {r = a + b; s = c - d;}
 else r = a - c

 b. x = 1; if (y == 2) { r = a + b; s = c - d; }
 else { r = a - c; }

 c. x = 2;
 while (x < 40) {
 x = foo[x];
 }

 d. for (i = 0; i < N; i++)
 x[i] = a[i]*b[i];

e.
```
for (i = 0; i < N; i++) {
        if (a[i] == 0)
                x[i] = 5;
        else
                x[i] = a[i]*b[i];
}
```

Q5-6 Show the contents of the assembler's symbol table at the end of code generation for each line of the following programs:

a.
```
        ORG 200
p1      ADR r4,a
        LDR r0,[r4]
        ADR r4,e
        LDR r1,[r4]
        ADD r0,r0,r1
        CMP r0,r1
        BNE q1
p2      ADR r4,e
```

b.
```
        ORG 100
p1      CMP r0,r1
        BEQ x1
p2      CMP r0,r2
        BEQ x2
p3      CMP r0,r3
        BEQ x3
```

Q5-7 Your linker uses a single pass through the set of given object files to find and resolve external references. Each object file is processed in the order given, all external references are found, and then the previously loaded files are searched for labels that resolve those references. Will this linker be able to successfully load a program with these external references and entry points?

Object file	Entry points	External references
o1	a, b, c, d	s, t
o2	r, s, t	w, y, d
o3	w, x, y, z	a, c, d

Q5-8 Provide the required order of execution of operations in these data flow graphs. If several operations can be performed in arbitrary order, show them as a set: $\{a + b, c - d\}$.

a.

b.

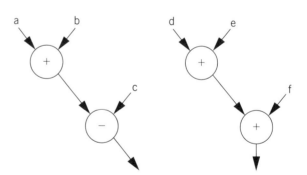

Q5-9 Draw the CDFG for the following C code before and after applying dead code elimination to the if statement:

```c
#define DEBUG 0
proc1();
if (DEBUG) debug_stuff();
switch (foo) {
        case A: a_case();
        case B: b_case();
        default: default_case();
        }
```

Q5-10 Unroll the loop below

 a. two times

 b. three times

```
for (i = 0; i < 32; i++)
        x[i] = a[i] * c[i];
```

Q5-11 Can you apply code motion to the following example? Explain.

```
for (i = 0; i < N; i++)
      for (j = 0; j < M; j++)
            z[i][j] = a[i] * b[i][j];
```

Q5-12 For each of the basic blocks of question Q5-4, determine the minimum number of registers required to perform the operations when they are executed in the order shown in the code. (You can assume that all computed values are used outside the basic blocks, so that no assignments can be eliminated.)

Q5-13 For each of the basic blocks of question Q5-4, determine the order of execution of operations that gives the smallest number of required registers. Next, state the number of registers required in each case. (You can assume that all computed values are used outside the basic blocks, so that no assignments can be eliminated.)

Q5-14 Draw a data flow graph for the code fragment of Example 5.5. Assign an order of execution to the nodes in the graph so that no more than four registers are required. Explain how you arrived at your solution using the structure of the data flow graph.

Q5-15 Determine the longest path through each code fragment, assuming that all statements can be executed in equal time and that all branch directions are equally probable.

 a.
```
if (i < CONST1) { x = a + b; }
else { x = c - d; y = e + f; }
```

 b.
```
for (i = 0; i < 32; i++)
        if (a[i] < CONST2)
                x[i] = a[i] * c[i];
```

 c.
```
if (a < CONST3) {
        if (b < CONST4)
                w = r + s;
        else {
                w = r - s;
```

```
                             x = s + t;
                    }
            } else {
                    if (c > CONST5) {
                            w = r + t;
                            x = r - s;
                            y = s + u;
                    }
            }
```

Q5-16 For each of the code fragments of question Q5-14, determine the shortest path through each code fragment, assuming that all statements can be executed in equal time and that all branch directions are equally probable.

Q5-17 The loop appearing below is executed on a machine that has a 1K word data cache with four words per cache line.

 a. How must x and a be placed relative to each other in memory to produce a conflict miss every time the inner loop's body is executed?

 b. How must x and a be placed relative to each other in memory to produce a conflict miss one out of every four times the inner loop's body is executed?

 c. How must x and a be placed relative to each other in memory to produce no conflict misses?

```
for (i = 0; i < 50; i++)
    for (j = 0; j < 4; j++)
        x[i][j] = a[i][j] * c[i];
```

Q5-18 Explain why the person generating clear-box program tests should not be the person who wrote the code being tested.

Q5-19 Find the cyclomatic complexity of the CDFGs for each of the code fragments given below.

 a.
```
if (a < b) {
if (c < d)
x = 1;
else
x = 2;
} else {
if (e < f)
x = 3;
```

```
    else
    x = 4;
    }
```

b.
```
switch (state) {
        case A:
                if (x = 1) { r = a + b; state = B; }
                else { s = a - b; state = C; }
                break;
        case B:
                s = c + d;
                state = A;
                break;
        case C:
                if (x < 5) { r = a - f; state = D; }
                else if (x == 5) { r = b + d; state = A; }
                else { r = c + e; state = D; }
                break;
        case D:
                r = r + 1;
                state = D;
                break;
    }
```

c.
```
for (i = 0; i < M; i++)
        for (j = 0; j < N; j++)
                x[i][j] = a[i][j] * c[i];
```

Q5-20 Use the branch condition testing strategy to determine a set of tests for each of the following statements.

a.
```
if (a < b || ptr1 == NULL) proc1();
else proc2();
```

b.
```
switch (x) {
    case 0: proc1(); break;
    case 1: proc2(); break;
    case 2: proc3(); break;
    case 3: proc4(); break;
    default; dproc(); break;
    }
```

 c. `if (a < 5 && b > 7) proc1();`
 `else if (a < 5) proc2();`
 `else if (b > 7) proc3();`
 `else proc4();`

Q5-21 Find all the def-use pairs for each code fragment given below.

 a.
```
x = a + b;
if (x < 20) proc1();
else {
        y ⁻ c + d;
        while (y < 10)
                y = y + e;
}
```

 b.
```
r = 10;
s = a - b;
for (i = 0; i < 10; i++)
        x[i] = a[i] * b[s];
```

 c.
```
x = a - b;
y = c - d;
z = e - f;
if (x < 10) {
        q = y + e;
        z = e + f;
}
if (z < y) proc1();
```

Q5-22 For each of the code fragments of question Q5-21, determine values for the variables that will cause each def-use pair to be exercised at least once.

Q5-23 Assume you want to use random tests on an FIR filter program. How would you know when the program under test is executing correctly?

Q5-24 Generate a set of functional tests for a moderate-size program. Evaluate your test coverage in one of two ways: Have someone else independently identify bugs and see how many of those bugs your tests catch (and how many tests they catch that were not found by the human inspector); or inject bugs into the code and see how many of those are caught by your tests.

LAB EXERCISES

L5-1 Compare the source code and assembly code for a moderate-size program. (Most C compilers will provide an assembly language listing with the -s flag.) Can you trace the high-level language statements in the assembly code? Can you see any optimizations that can be done on the assembly code?

L5-2 Write C code for an FIR filter. Measure the execution time of the filter, either using a simulator or by measuring the time on a running microprocessor. Vary the number of taps in the FIR filter and measure execution time as a function of the filter size.

L5-3 Generate a trace for a program using software techniques. Use the trace to analyze the program's cache behavior.

L5-4 Use a cycle-accurate CPU simulator to determine the execution time of a program.

L5-5 Measure the power consumption of your microprocessor on a simple block of code.

L5-6 Use software testing techniques to determine how well your input sequences to the cycle-accurate simulator exercise of your program.

Processes and Operating Systems

- The process abstraction.

- Switching contexts between programs.

- Real-time operating systems (RTOSs).

- Interprocess communication.

- Task-level performance analysis and power consumption.

- A telephone answering machine design.

INTRODUCTION

Although simple applications can be programmed on a microprocessor by writing a single piece of code, many applications are sophisticated enough that writing one large program does not suffice. When multiple operations must be performed at widely varying times, a single program can easily become too complex and unwieldy. The result is spaghetti code that is too difficult to verify for either performance or functionality.

This chapter studies the two fundamental abstractions that allow us to build complex applications on microprocessors: the **process** and the **operating system (OS)**. Together, these two abstractions let us switch the state of the processor between multiple tasks. The process cleanly defines the state of an executing program, while the OS provides the mechanism for switching execution between the processes.

These two mechanisms together let us build applications with more complex functionality and much greater flexibility to satisfy timing requirements. The need to satisfy complex timing requirements—events happening at very different rates, intermittent events, and so on—causes us to use processes and OSs to build embedded software. Satisfying complex timing tasks can introduce extremely complex control into programs. Using processes to compartmentalize functions and encapsulating in the OS the control required to switch between processes make it much easier to satisfy timing requirements with relatively clean control within the processes.

We are particularly interested in ***real-time operating systems (RTOSs)***, which are OSs that provide facilities for satisfying real-time requirements. A RTOS allocates resources using algorithms that take real time into account. General-purpose OSs, in contrast, generally allocate resources using other criteria like fairness. Trying to allocate the CPU equally to all processes without regard to time can easily cause processes to miss their deadlines.

In the next section, we will introduce the concepts of task and process. Section 6.2 looks at how the RTOS implements processes. Section 6.3 develops algorithms for scheduling those processes to meet real-time requirements. Section 6.4 introduces some basic concepts in interprocess communication. Section 6.5 considers the performance of RTOSs while Section 6.6 looks at power consumption. Section 6.7 walks through the design of a telephone answering machine.

6.1 MULTIPLE TASKS AND MULTIPLE PROCESSES

Most embedded systems require functionality and timing that is too complex to embody in a single program. We break the system into multiple tasks in order to manage when things happen. In this section we will develop the basic abstractions that will be manipulated by the RTOS to build multirate systems.

6.1.1 Tasks and Processes

Many (if not most) embedded computing systems do more than one thing—that is, the environment can cause mode changes that in turn cause the embedded system to behave quite differently. For example, when designing a telephone answering machine, we can define recording a phone call and operating the user's control panel as distinct tasks, because they perform logically distinct operations and they must be performed at very different rates. These different **tasks** are part of the system's functionality, but that application-level organization of functionality is often reflected in the structure of the program as well.

A ***process*** is a single execution of a program. If we run the same program two different times, we have created two different processes. Each process has its own state that includes not only its registers but all of its memory. In some OSs, the memory management unit is used to keep each process in a separate address space. In others, particularly lightweight RTOSs, the processes run in the same address space. Processes that share the same address space are often called ***threads***.

In this book, we will use the terms tasks and processes somewhat interchangeably, as do many people in the field. To be more precise, task can be composed of several processes or threads; it is also true that a task is primarily an implementation concept and process more of an implementation concept.

To understand why the separation of an application into tasks may be reflected in the program structure, consider how we would build a stand-alone compression unit based on the compression algorithm we implemented in Section 3.7. As shown in Figure 6.1, this device is connected to serial ports on both ends. The input to the box is an uncompressed stream of bytes. The box emits a compressed string of bits on the output serial line, based on a predefined compression table. Such a box may be used, for example, to compress data being sent to a modem.

The program's need to receive and send data at different rates—for example, the program may emit 2 bits for the first byte and then 7 bits for the second byte— will obviously find itself reflected in the structure of the code. It is easy to create irregular, ungainly code to solve this problem; a more elegant solution is to create a queue of output bits, with those bits being removed from the queue and sent to the serial port in 8-bit sets. But beyond the need to create a clean data structure that simplifies the control structure of the code, we must also ensure that we process the inputs and outputs at the proper rates. For example, if we spend too much time in packaging and emitting output characters, we may drop an input character. Solving timing problems is a more challenging problem.

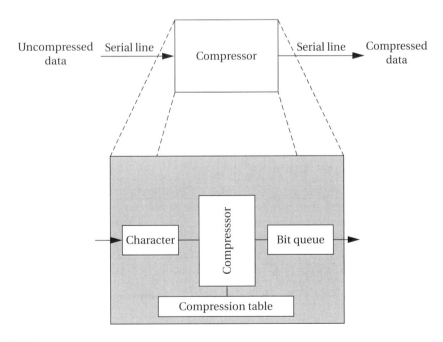

FIGURE 6.1

An on-the-fly compression box.

The text compression box provides a simple example of rate control problems. A control panel on a machine provides an example of a different type of rate control problem, the **asynchronous input**. The control panel of the compression box may, for example, include a compression mode button that disables or enables compression, so that the input text is passed through unchanged when compression is disabled. We certainly do not know when the user will push the compression mode button—the button may be depressed asynchronously relative to the arrival of characters for compression.

We do know, however, that the button will be depressed at a much lower rate than characters will be received, since it is not physically possible for a person to repeatedly depress a button at even slow serial line rates. Keeping up with the input and output data while checking on the button can introduce some very complex control code into the program. Sampling the button's state too slowly can cause the machine to miss a button depression entirely, but sampling it too frequently and duplicating a data value can cause the machine to incorrectly compress data. One solution is to introduce a counter into the main compression loop, so that a subroutine to check the input button is called once every n times the compression loop is executed. But this solution does not work when either the compression loop or the button-handling routine has highly variable execution times—if the execution time of either varies significantly, it will cause the other to execute later than expected, possibly causing data to be lost. We need to be able to keep track of these two different tasks separately, applying different timing requirements to each. This is the sort of control that processes allow.

The above two examples illustrate how requirements on timing and execution rate can create major problems in programming. When code is written to satisfy several different timing requirements at once, the control structures necessary to get any sort of solution become very complex very quickly. Worse, such complex control is usually quite difficult to verify for either functional or timing properties.

6.1.2 Multirate Systems

Implementing code that satisfies timing requirements is even more complex when multiple rates of computation must be handled. **Multirate** embedded computing systems are very common, including automobile engines, printers, and cell phones. In all these systems, certain operations must be executed periodically, and each operation is executed at its own rate. Application Example 6.1 describes why automobile engines require multirate control.

Application Example 6.1

Automotive engine control

The simplest automotive engine controllers, such as the ignition controller for a basic motorcycle engine, perform only one task—timing the firing of the spark plug, which takes the place

of a mechanical distributor. The spark plug must be fired at a certain point in the combustion cycle, but to obtain better performance, the phase relationship between the piston's movement and the spark should change as a function of engine speed. Using a microcontroller that senses the engine crankshaft position allows the spark timing to vary with engine speed. Firing the spark plug is a periodic process (but note that the period depends on the engine's operating speed).

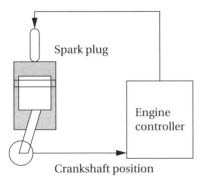

The control algorithm for a modern automobile engine is much more complex, making the need for microprocessors that much greater. Automobile engines must meet strict requirements (mandated by law in the United States) on both emissions and fuel economy. On the other hand, the engines must still satisfy customers not only in terms of performance but also in terms of ease of starting in extreme cold and heat, low maintenance, and so on.

Automobile engine controllers use additional sensors, including the gas pedal position and an oxygen sensor used to control emissions. They also use a multimode control scheme. For example, one mode may be used for engine warm-up, another for cruise, and yet another for climbing steep hills, and so forth. The larger number of sensors and modes increases the number of discrete tasks that must be performed. The highest-rate task is still firing the spark plugs. The throttle setting must be sampled and acted upon regularly, although not as frequently as the crankshaft setting and the spark plugs. The oxygen sensor responds much more slowly than the throttle, so adjustments to the fuel/air mixture suggested by the oxygen sensor can be computed at a much lower rate.

The engine controller takes a variety of inputs that determine the state of the engine. It then controls two basic engine parameters: the spark plug firings and the fuel/air mixture. The engine control is computed periodically, but the periods of the different inputs and outputs range over several orders of magnitude of time. An early paper on automotive electronics by Marley [Mar78] described the rates at which engine inputs and outputs must be handled.

Variable	Time to move full range (ms)	Update period (ms)
Engine spark timing	300	2
Throttle	40	2
Airflow	30	4
Battery voltage	80	4
Fuel flow	250	10
Recycled exhaust gas	500	25
Set of status switches	100	50
Air temperature	seconds	500
Barometric pressure	seconds	1000
Spark/dwell	10	1
Fuel adjustments	80	4
Carburetor adjustments	500	25
Mode actuators	100	100

6.1.3 Timing Requirements on Processes

Processes can have several different types of timing requirements imposed on them by the application. The timing requirements on a set of processes strongly influence the type of scheduling that is appropriate. A scheduling policy must define the timing requirements that it uses to determine whether a schedule is valid. Before studying scheduling proper, we outline the types of process timing requirements that are useful in embedded system design.

Figure 6.2 illustrates different ways in which we can define two important requirements on processes: *release time* and *deadline*. The release time is the time at which the process becomes ready to execute; this is not necessarily the time at which it actually takes control of the CPU and starts to run. An aperiodic process is by definition initiated by an event, such as external data arriving or data computed by another process. The release time is generally measured from that event, although the system may want to make the process ready at some interval after the event itself. For a periodically executed process, there are two common possibilities. In simpler systems, the process may become ready at the beginning of the period. More sophisticated systems, such as those with data dependencies between processes, may set the release time at the arrival time of certain data, at a time after the start of the period.

A deadline specifies when a computation must be finished. The deadline for an aperiodic process is generally measured from the release time, since that is the only reasonable time reference. The deadline for a periodic process may in general occur at some time other than the end of the period. As seen in Section 6.3.1, some scheduling policies make the simplifying assumption that the deadline occurs at the end of the period.

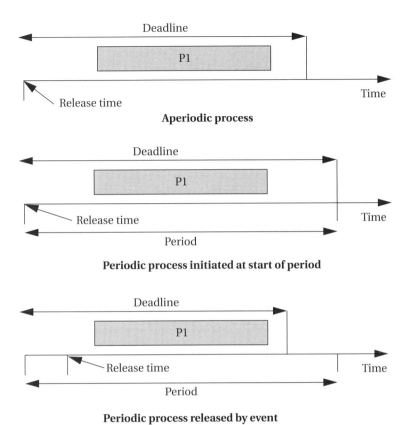

FIGURE 6.2

Example definitions of release times and deadlines.

Rate requirements are also fairly common. A rate requirement specifies how quickly processes must be initiated. The **period** of a process is the time between successive executions. For example, the period of a digital filter is defined by the time interval between successive input samples. The process's **rate** is the inverse of its period. In a multirate system, each process executes at its own distinct rate. The most common case for periodic processes is for the initiation interval to be equal to the period. However, pipelined execution of processes allows the initiation interval to be less than the period. Figure 6.3 illustrates process execution in a system with four CPUs. The various execution instances of program P1 have been subscripted to distinguish their initiation times. In this case, the initiation interval is equal to one-fourth of the period. It is possible for a process to have an initiation rate less than the period even in single-CPU systems. If the process execution time is significantly less than the period, it may be possible to initiate multiple copies of a program at slightly offset times.

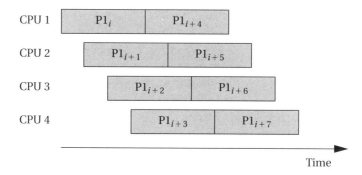

FIGURE 6.3

A sequence of processes with a high initiation rate.

What happens when a process misses a deadline? The practical effects of a timing violation depend on the application—the results can be catastrophic in an automotive control system, whereas a missed deadline in a multimedia system may cause an audio or video glitch. The system can be designed to take a variety of actions when a deadline is missed. Safety-critical systems may try to take compensatory measures such as approximating data or switching into a special safety mode. Systems for which safety is not as important may take simple measures to avoid propagating bad data, such as inserting silence in a phone line, or may completely ignore the failure.

Even if the modules are functionally correct, their timing improper behavior can introduce major execution errors. Application Example 6.2 describes a timing problem in space shuttle software that caused the delay of the first launch of the shuttle.

Application Example 6.2

A space shuttle software error

Garman [Gar81] describes a software problem that delayed the first launch of the U.S. space shuttle. No one was hurt and the launch proceeded after the computers were reset. However, this bug was serious and unanticipated.

The shuttle's primary control system was known as the Primary Avionics Software System (PASS). It used four computers to monitor events, with the four machines voting to ensure fault tolerance. Four computers allowed one machine to fail while still leaving three operating machines to vote, such that a majority vote would still be possible to determine operating procedures. If at least two machines failed, control was to be turned over to a fifth computer called the Backup Flight Control System (BFS). The BFS used the same computer, requirements, programming language, and compiler, but it was developed by a different organization than the one that built the PASS to ensure that methodological errors did not cause simultaneous failure of both systems. The switchover from PASS to BFS was controlled by the astronauts.

During normal operation, the BFS would listen to the operation of the PASS computers so that it could keep track of the state of the shuttle. However, BFS would stop listening when it thought that PASS was compromising data fetching. This would prevent PASS failures from inadvertently destroying the state of the BFS. PASS used an asynchronous, priority-driven software architecture. If high-priority processes take too much time, the OS can skip or delay lower-priority processing. The BFS, in contrast, used a time-slot system that allocated a fixed amount of time to each process. Since the BFS monitored the PASS, it could get confused by temporary overloads on the primary system. As a result, the PASS was changed late in the design cycle to make its behavior more amenable to the backup system.

On the morning of the launch attempt, the BFS failed to synchronize itself with the primary system. It saw the events on the PASS system as inconsistent and therefore stopped listening to PASS behavior. It turned out that all PASS and BFS processing had been running late relative to telemetry data. This occurred because the system incorrectly calculated its start time.

After much analysis of system traces and software, it was determined that a few minor changes to the software had caused the problem. First, about 2 years before the incident, a subroutine used to initialize the data bus was modified. Since this routine was run prior to calculating the start time, it introduced an additional, unnoticed delay into that computation. About a year later, a constant was changed in an attempt to fix that problem. As a result of these changes, there was a 1 in 67 probability for a timing problem. When this occurred, almost all computations on the computers would occur a cycle late, leading to the observed failure. The problems were difficult to detect in testing since they required running through all the initialization code; many tests start with a known configuration to save the time required to run the setup code. The changes to the programs were also not obviously related to the final changes in timing.

The order of execution of processes may be constrained when the processes pass data between each other. Figure 6.4 shows a set of processes with data dependencies among them. Before a process can become ready, all the processes on which it depends must complete and send their data to it. The data dependencies define a partial ordering on process execution—P1 and P2 can execute in any order (or in interleaved fashion) but must both complete before P3, and P3 must complete before P4. All processes must finish before the end of the period. The data dependencies must form a directed acyclic graph (DAG)—a cycle in the data dependencies is difficult to interpret in a periodically executed system.

A set of processes with data dependencies is known as a ***task graph***. Although the terminology for elements of a task graph varies from author to author, we will consider a component of the task graph (a set of nodes connected by data dependencies) as a ***task*** and the complete graph as the ***task set***. Figure 6.4 also shows a second task with two processes. The two tasks ({P1, P2, P3, P4} and {P5, P6}) have no timing relationships between them.

Communication among processes that run at different rates cannot be represented by data dependencies because there is no one-to-one relationship between data coming out of the source process and going into the destination process.

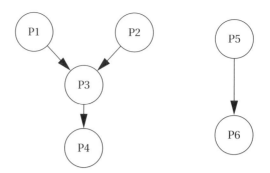

FIGURE 6.4

Data dependencies among processes.

FIGURE 6.5

Communication among processes at different rates.

Nevertheless, communication among processes of different rates is very common. Figure 6.5 illustrates the communication required among three elements of an MPEG audio/video decoder. Data come into the decoder in the system format, which multiplexes audio and video data. The system decoder process demultiplexes the audio and video data and distributes it to the appropriate processes. Multirate communication is necessarily one way—for example, the system process writes data to the video process, but a separate communication mechanism must be provided for communication from the video process back to the system process.

6.1.4 CPU Metrics

We also need some terminology to describe how the process actually executes. The ***initiation time*** is the time at which a process actually starts executing on the CPU. The ***completion time*** is the time at which the process finishes its work.

The most basic measure of work is the amount of ***CPU time*** expended by a process. The CPU time of process i is called C_i. Note that the CPU time is not equal to the completion time minus initiation time; several other processes may interrupt execution. The total CPU time consumed by a set of processes is

$$T = \sum_{1 \le i \le n} T_i. \tag{6.1}$$

We need a basic measure of the efficiency with which we use the CPU. The simplest and most direct measure is **utilization**:

$$U = \frac{\text{CPU time for useful work}}{\text{total available CPU time}}. \tag{6.2}$$

Utilization is the ratio of the CPU time that is being used for useful computations to the total available CPU time. This ratio ranges between 0 and 1, with 1 meaning that all of the available CPU time is being used for system purposes. The utilization is often expressed as a percentage. If we measure the total execution time of all processes over an interval of time t, then the CPU utilization is

$$U = \frac{T}{t}. \tag{6.3}$$

6.1.5 Process State and Scheduling

The first job of the OS is to determine that process runs next. The work of choosing the order of running processes is known as scheduling.

The OS considers a process to be in one of three basic **scheduling states**: **waiting**, **ready**, or **executing**. There is at most one process executing on the CPU at any time. (If there is no useful work to be done, an idling process may be used to perform a null operation.) Any process that could execute is in the ready state; the OS chooses among the ready processes to select the next executing process. A process may not, however, always be ready to run. For instance, a process may be waiting for data from an I/O device or another process, or it may be set to run from a timer that has not yet expired. Such processes are in the waiting state. Figure 6.6 shows the possible transitions between states available to a process. A process goes into the waiting state when it needs data that it has not yet received or when it has finished all its work for the current period. A process goes into the ready state when it receives its required data and when it enters a new period. A process can go into the executing state only when it has all its data, is ready to run, and the scheduler selects the process as the next process to run.

6.1.6 Some Scheduling Policies

A **scheduling policy** defines how processes are selected for promotion from the ready state to the running state. Every multitasking OS implements some type of scheduling policy. Choosing the right scheduling policy not only ensures that the system will meet all its timing requirements, but it also has a profound influence on the CPU horsepower required to implement the system's functionality.

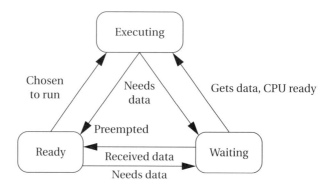

FIGURE 6.6

Scheduling states of a process.

Schedulability means whether there exists a schedule of execution for the processes in a system that satisfies all their timing requirements. In general, we must construct a schedule to show schedulability, but in some cases we can eliminate some sets of processes as unschedulable using some very simple tests. Utilization is one of the key metrics in evaluating a scheduling policy. Our most basic requirement is that CPU utilization be no more than 100% since we can't use the CPU more than 100% of the time.

When we evaluate the utilization of the CPU, we generally do so over a finite period that covers all possible combinations of process executions. For periodic processes, the length of time that must be considered is the ***hyperperiod***, which is the least-common multiple of the periods of all the processes. (The complete schedule for the least-common multiple of the periods is sometimes called the ***unrolled schedule***.) If we evaluate the hyperperiod, we are sure to have considered all possible combinations of the periodic processes. The next example evaluates the utilization of a simple set of processes.

Example 6.1

Utilization of a set of processes

We are given three processes, their execution times, and their periods:

Process	Period	Execution time
P1	1.0×10^{-3}	1.0×10^{-4}
P2	1.0×10^{-3}	2.0×10^{-4}
P3	5.0×10^{-3}	3.0×10^{-4}

The least common multiple of these periods is 5×10^{-3} s.

In order to calculate the utilization, we have to figure out how many times each process is executed in one hyperperiod: P1 and P2 are each executed five times while P3 is executed once.

We can now determine the utilization over the hyperperiod:

$$U = \frac{5.1 \times 10^{-4} + 5.2 \times 10^{-4} + 1.3 \times 10^{-4}}{5 \times 10^{-3}} = 0.36$$

This is well below our maximum utilization of 1.0.

We will see that some types of timing requirements for a set of processes imply that we cannot utilize 100% of the CPU's execution time on useful work, even ignoring context switching overhead. However, some scheduling policies can deliver higher CPU utilizations than others, even for the same timing requirements. The best policy depends on the required timing characteristics of the processes being scheduled.

One very simple scheduling policy is known as *cyclostatic* scheduling or sometimes as *Time Division Multiple Access* scheduling. As illustrated in Figure 6.7, a cyclostatic schedule is divided into equal-sized time slots over an interval equal to the length of the hyperperiod H. Processes always run in the same time slot. Two factors affect utilization: the number of time slots used and the fraction of each time slot that is used for useful work. Depending on the deadlines for some of the processes, we may need to leave some time slots empty. And since the time slots are of equal size, some short processes may have time left over in their time slot. We can use utilization as a schedulability measure: the total CPU time of all the processes must be less than the hyperperiod.

Another scheduling policy that is slightly more sophisticated is *round robin*. As illustrated in Figure 6.8, round robin uses the same hyperperiod as does cyclostatic. It also evaluates the processes in order. But unlike cyclostatic scheduling, if a process

FIGURE 6.7

Cyclostatic scheduling.

FIGURE 6.8

Round-robin scheduling.

does not have any useful work to do, the round-robin scheduler moves on to the next process in order to fill the time slot with useful work. In this example, all three processes execute during the first hyperperiod, but during the second one, P_1 has no useful work and is skipped. The processes are always evaluated in the same order. The last time slot in the hyperperiod is left empty; if we have occasional, non-periodic tasks without deadlines, we can execute them in these empty time slots. Round-robin scheduling is often used in hardware such as buses because it is very simple to implement but it provides some amount of flexibility.

In addition to utilization, we must also consider **scheduling overhead**—the execution time required to choose the next execution process, which is incurred in addition to any context switching overhead. In general, the more sophisticated the scheduling policy, the more CPU time it takes during system operation to implement it. Moreover, we generally achieve higher theoretical CPU utilization by applying more complex scheduling policies with higher overheads. The final decision on a scheduling policy must take into account both theoretical utilization and practical scheduling overhead.

6.1.7 Running Periodic Processes

We need to find a programming technique that allows us to run periodic processes, ideally at different rates. For the moment, let's think of a process as a subroutine; we will call them p1(), p2(), etc. for simplicity. Our goal is to run these subroutines at rates determined by the system designer.

Here is a very simple program that runs our process subroutines repeatedly:

```
while (TRUE) {
    p1();
    p2();
    }
```

This program has several problems. First, it does not control the rate at which the processes execute—the loop runs as quickly as possible, starting a new iteration as soon as the previous iteration has finished. Second, all the processes run at the same rate.

Before worrying about multiple rates, let's first make the processes run at a controlled rate. One could imagine controlling the execution rate by carefully designing the code—by determining the execution time of the instructions executed during an iteration, we could pad the loop with useless operations (NOPs) to make the execution time of an iteration equal to the desired period. Although some video games were designed this way in the 1970s, this technique should be avoided. Modern processors make it hard to accurately determine execution time, as we saw in Chapter 5. Conditionals anywhere in the program make it even harder to be sure that the loop consumes the same amount of execution time on every iteration. Furthermore, if any part of the program is changed, the entire timing scheme must be re-evaluated.

A timer is a much more reliable way to control execution of the loop. We would probably use the timer to generate periodic interrupts. Let's assume for the moment that the pall() function is called by the timer's interrupt handler. Then this code will execute each process once after a timer interrupt:

```
void pall() {
    p1();
    p2();
    }
```

But what happens when a process runs too long? The timer's interrupt will cause the CPU's interrupt system to mask its interrupts, so the interrupt will not occur until after the pall() routine returns. As a result, the next iteration will start late. This is a serious problem, but we will have to wait for further refinements before we can fix it.

Our next problem is to execute different processes at different rates. If we have several timers, we can set each timer to a different rate. We could then use a function to collect all the processes that run at that rate:

```
void pA() {
    /* processes that run at rate A*/
    p1();
    p3();
    }
void pB() {
    /* processes that run at rate B */
    p2();
    p4();
    p5();
    }
```

This works, but it does require multiple timers, and we may not have enough timers to support all the rates required by a system.

An alternative is to use counters to divide the counter rate. If, for example, process p2() must run at 1/3 the rate of p1(), then we can use this code:

```
static int p2count = 0; /* use this to remember count across
                           timer interrupts */

void pall() {
    p1();
    if (p2count >= 2) { /* execute p2() and reset count */
        p2();
        p2count = 0;
        }
    else p2count++; /* just update count in this case */
    }
```

This solution allows us to execute processes at rates that are simple multiples of each other. However, when the rates aren't related by a simple ratio, the counting process becomes more complex and more likely to contain bugs.

We have developed somewhat more reliable code, but this programming style is still limited in capability and prone to bugs. To improve both the capabilities and reliability of our systems, we need to invent the RTOS.

6.2 PREEMPTIVE REAL-TIME OPERATING SYSTEMS

A RTOS executes processes based upon timing constraints provided by the system designer. The most reliable way to meet timing constraints accurately is to build a **preemptive** OS and to use **priorities** to control what process runs at any given time. We will use these two concepts to build up a basic RTOS. We will use as our example OS FreeRTOS.org [Bar07]. This operating system runs on many different platforms.

6.2.1 Preemption

Preemption is an alternative to the C function call as a way to control execution. To be able to take full advantage of the timer, we must change our notion of a process as something more than a function call. We must, in fact, break the assumptions of our high-level programming language. We will create new routines that allow us to jump from one subroutine to another at any point in the program. That, together with the timer, will allow us to move between functions whenever necessary based upon the system's timing constraints.

We want to share the CPU across two processes. The **kernel** is the part of the OS that determines what process is running. The kernel is activated periodically by the timer. The length of the timer period is known as the **time quantum** because it is the smallest increment in which we can control CPU activity. The kernel determines what process will run next and causes that process to run. On the next timer interrupt, the kernel may pick the same process or another process to run.

Note that this use of the timer is very different from our use of the timer in the last section. Before, we used the timer to control loop iterations, with one loop

iteration including the execution of several complete processes. Here, the time quantum is in general smaller than the execution time of any of the processes.

How do we switch between processes before the process is done? We cannot rely on C-level mechanisms to do so. We can, however, use assembly language to switch between processes. The timer interrupt causes control to change from the currently executing process to the kernel; assembly language can be used to save and restore registers. We can similarly use assembly language to restore registers not from the process that was interrupted by the timer but to use registers from any process we want. The set of registers that define a process are known as its **context** and switching from one process's register set to another is known as **context switching**. The data structure that holds the state of the process is known as the **process control block**.

6.2.2 Priorities

How does the kernel determine what process will run next? We want a mechanism that executes quickly so that we don't spend all our time in the kernel and starve out the processes that do the useful work. If we assign each task a numerical **priority**, then the kernel can simply look at the processes and their priorities, see which ones actually want to execute (some may be waiting for data or for some event), and select the highest priority process that is ready to run. This mechanism is both flexible and fast. The priority is a non-negative integer value. The exact value of the priority is not as important as the relative priority of different processes. In this book, we will generally use priority 1 as the highest priority, but it is equally reasonable to use 1 or 0 as the lowest priority value (as FreeRTOS.org does).

Example 6.2 shows how priorities can be used to schedule processes.

Example 6.2

Priority-driven scheduling

For this example, we will adopt the following simple rules:

- Each process has a fixed priority that does not vary during the course of execution. (More sophisticated scheduling schemes do, in fact, change the priorities of processes to control what happens next.)

- The ready process with the highest priority (with 1 as the highest priority of all) is selected for execution.

- A process continues execution until it completes or it is preempted by a higher-priority process.

Let's define a simple system with three processes as seen below.

Process	Priority	Execution time
P1	1	10
P2	2	30
P3	3	20

In addition to describing the properties of the processes in general, we need to know the environmental setup. We assume that P2 is ready to run when the system is started, P1 is released at time 15, and P3 is released at time 18.

Once we know the process properties and the environment, we can use the priorities to determine which process is running throughout the complete execution of the system.

When the system begins execution, P2 is the only ready process, so it is selected for execution. At time 15, P1 becomes ready; it preempts P2 and begins execution since it has a higher priority. Since P1 is the highest-priority process in the system, it is guaranteed to execute until it finishes. P3's data arrive at time 18, but it cannot preempt P1. Even when P1 finishes, P3 is not allowed to run. P2 is still ready and has higher priority than P3. Only after both P1 and P2 finish can P3 execute.

6.2.3 Processes and Context

The best way to understand processes and context is to dive into an RTOS implementation. We will use the FreeRTOS.org kernel as an example; in particular, we will use version 4.7.0 for the ARM7 AT91 platform. A process is known in FreeRTOS.org as a task. Task priorities in FreeRTOS.org are ranked opposite to the convention we use in the rest of the book: higher numbers denote higher priorities and the priority 0 task is the idle task.

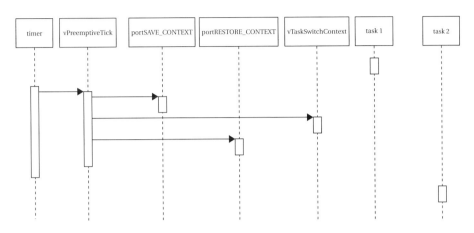

FIGURE 6.9

Sequence diagram for freeRTOS.org context switch.

To understand the basics of a context switch, let's assume that the set of tasks is in steady state: Everything has been initialized, the OS is running, and we are ready for a timer interrupt. Figure 6.9 shows a sequence diagram for a context switch in freeRTOS.org. This diagram shows the application tasks, the hardware timer, and all the functions in the kernel that are involved in the context switch:

- vPreemptiveTick() is called when the timer ticks.

- portSAVE_CONTEXT() swaps out the current task context.

- vTaskSwitchContext() chooses a new task.

- portRESTORE_CONTEXT() swaps in the new context.

Here is the code for vPreemptiveTick() in the file portISR.c:

```
void vPreemptiveTick( void )
{
    /* Save the context of the interrupted task. */
    portSAVE_CONTEXT();

    /* WARNING - Do not use local (stack) variables here.
       Use globals if you must! */
    static volatile unsigned portLONG ulDummy;

    /* Clear tick timer interrupt indication. */
    ulDummy = portTIMER_REG_BASE_PTR->TC_SR;
    /* Increment the RTOS tick count, then look for the
       highest priority task that is ready to run. */
    vTaskIncrementTick();
    vTaskSwitchContext();
```

```
        /* Acknowledge the interrupt at AIC level... */
        AT91C_BASE_AIC->AIC_EOICR = portCLEAR_AIC_INTERRUPT;

        /* Restore the context of the new task. */
        portRESTORE_CONTEXT();
    }
```

vPreemptiveTick() has been declared as a naked function; this means that it does not use the normal procedure entry and exit code that is generated by the compiler. Because the function is **naked**, the registers for the process that was interrupted are still available; vPreemptiveTick() doesn't have to go to the procedure call stack to get their values. This is particularly handy since the procedure mechanism would save only part of the process state, making the state-saving code a little more complex.

The first thing that this routine must do is save the context of the task that was interrupted. To do this, it uses the routine portSAVE_CONTEXT(), which saves all the context of the stack. It then performs some housekeeping, such as incrementing the tick count. The tick count is the internal timer that is used to determine deadlines. After the tick is incremented, some tasks may have become ready as they passed their deadlines.

Next, the OS determines which task to run next using the routine vTaskSwitchContext(). After some more housekeeping, it uses port RESTORE_CONTEXT() to restore the context of the task that was selected by vTaskSwitchContext(). The action of portRESTORE_CONTEXT() causes control to transfer to that task without using the standard C return mechanism.

The code for portSAVE_CONTEXT(), in the file portmacro.h, is defined as a macro and not as a C function. It is structured in this way so that it doesn't disturb the register values that need to be saved. Because it is a macro, it has to be written in a hard-to-read way—all code must be on the same line or end-of-line continuations (back slashes) must be used. Here is the code in more readable form, with the end-of-line continuations removed and the assembly language that is the heart of this routine temporarily removed.:

```
#define portSAVE_CONTEXT()
{
extern volatile void * volatile pxCurrentTCB;
extern volatile unsigned portLONG ulCriticalNesting;

    /* Push R0 as we are going to use the register. */
    asm volatile( /* assembly language code here */ );
    ( void ) ulCriticalNesting;
    ( void ) pxCurrentTCB;
}
```

The asm statement allows assembly language code to be introduced in-line into the C program. The keyword volatile tells the compiler that the assembly language

may change register values, which means that many compiler optimizations cannot be performed across the assembly language code. The code uses ulCriticalNesting and pxCurrentTCB simply to avoid compiler warnings about unused variables—the variables are actually used in the assembly code, but the compiler cannot see that.

The asm statement requires that the assembly language be entered as strings, one string per line, which makes the code hard to read. The fact that the code is included in a #define makes it even harder to read. Here is a cleaned-up version of the assembly language code from the asm volatile() statement:

```
STMDB           SP!, {R0}
/* Set R0 to point to the task stack pointer. */
STMDB           SP, {SP}^
NOP
SUB             SP, SP, #4
LDMIA           SP!,{R0}
/* Push the return address onto the stack. */
STMDB           R0!, {LR}
/* Now we have saved LR we can use it instead of R0. */
MOV             LR, R0
/* Pop R0 so we can save it onto the system mode stack. */
LDMIA           SP!, {R0}
/* Push all the system mode registers onto the task
   stack. */
STMDB           LR,{R0-LR}^
NOP
SUB             LR, LR, #60 /*
Push the SPSR onto the task stack. */
MRS             R0, SPSR
STMDB           LR!, {R0}
LDR             R0, =ulCriticalNesting
LDR             R0, [R0]
STMDB           LR!, {R0}
/*Store the new top of stack for the task. */
LDR             R0, =pxCurrentTCB
LDR             R0, [R0]
STR             LR, [R0]
```

Here is the code for vTaskSwitchContext(), which is defined in the file tasks.c:

```
void vTaskSwitchContext( void )
{
    if( uxSchedulerSuspended != ( unsigned portBASE_TYPE )
        pdFALSE )
```

```
        {
                /* The scheduler is currently suspended - do not
                   allow a context switch. */
                xMissedYield = pdTRUE;
        return;
        }

        /* Find the highest priority queue that contains ready
           tasks. */
        while( listLIST_IS_EMPTY(&( pxReadyTasksLists[
                uxTopReadyPriority ]) ) )
        {
                --uxTopReadyPriority;
        }

        /* listGET_OWNER_OF_NEXT_ENTRY walks through the list,
           so the tasks of the same priority get an equal share
           of the processor time. */
        listGET_OWNER_OF_NEXT_ENTRY( pxCurrentTCB,
        &(pxReadyTasksLists[uxTopReadyPriority ] ) );
        vWriteTraceToBuffer();
}
```

This function is relatively straightforward—it walks down the list of tasks to identify the highest-priority task. This function is designed to deterministically choose the next task to run as long as the selected task is of equal or higher priority to the interrupted task; the list of tasks that is checked is determined by the variable uxTopReadyPriority. Each list contains the set of processes with the same priority; once the proper priority has selected by determining the value of uxTopReadyPriority, the system rotates through processes of equal priority by walking down their list.

The portRESTORE_CONTEXT() routine is also defined in portmacro.h and is implemented as a macro with embedded assembly language. Here is the macro with the line continuations and assembly language code removed:

```
#define portRESTORE_CONTEXT()
{
extern volatile void * volatile
pxCurrentTCB;
extern volatile unsigned portLONG
ulCriticalNesting;

        /* Set the LR to the task stack. */
        asm volatile (/* assembly language code here */);
```

```
        ( void ) ulCriticalNesting;
        ( void ) pxCurrentTCB;
}
```

Here is the assembly language code for portRESTORE_CONTEXT:

```
LDR             R0, =pxCurrentTCB
LDR             R0, [R0]
LDR             LR, [R0]
/* The critical nesting depth is the first item on the
   stack. */
/* Load it into the ulCriticalNesting variable. */
LDR             R0, =ulCriticalNesting
LDMFD           LR!, {R1}
STR             R1, [R0]
/* Get the SPSR from the stack. */
LDMFD           LR!, {R0}
MSR             SPSR, R0
/* Restore all system mode registers for the task. */
LDMFD           LR, {R0-R14}^
NOP
/* Restore the return address. */
LDR             LR, [LR, #+60]
/* And return - correcting the offset in the LR to obtain
   the */
/* correct address. */
SUBS            PC, LR, #4
```

6.2.4 Processes and Object-Oriented Design

We need to design systems with processes as components. In this section, we survey the ways we can describe processes in UML and how to use processes as components in object-oriented design.

UML often refers to processes as ***active objects***, that is, objects that have independent threads of control. The class that defines an active object is known as an ***active class***. Figure 6.10 shows an example of a UML active class. It has all the normal characteristics of a class, including a name, attributes, and operations. It also provides a set of signals that can be used to communicate with the process. A signal is an object that is passed between processes for asynchronous communication. We describe signals in more detail in Section 6.2.4.

We can mix active objects and normal objects when describing a system. Figure 6.11 shows a simple collaboration diagram in which an object is used as an interface between two processes: $p1$ uses the w object to manipulate its data before the data is sent to the *master* process.

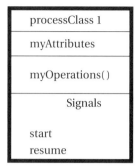

FIGURE 6.10

An active class in UML.

FIGURE 6.11

A collaboration diagram with active and normal objects.

6.3 PRIORITY-BASED SCHEDULING

Now that we have a priority-based context switching mechanism, we have to determine an algorithm by which to assign priorities to processes. After assigning priorities, the OS takes care of the rest by choosing the highest-priority ready process. There are two major ways to assign priorities: ***static*** priorities that do not change during execution and ***dynamic*** priorities that do change. We will look at examples of each in this section.

6.3.1 Rate-Monotonic Scheduling

Rate-monotonic scheduling (RMS), introduced by Liu and Layland [Liu73], was one of the first scheduling policies developed for real-time systems and is still very widely used. RMS is a static scheduling policy. It turns out that these fixed priorities are sufficient to efficiently schedule the processes in many situations.

The theory underlying RMS is known as ***rate-monotonic analysis (RMA)***. This theory, as summarized below, uses a relatively simple model of the system.

- All processes run periodically on a single CPU.
- Context switching time is ignored.

- There are no data dependencies between processes.
- The execution time for a process is constant.
- All deadlines are at the ends of their periods.
- The highest-priority ready process is always selected for execution.

The major result of RMA is that a relatively simple scheduling policy is optimal under certain conditions. Priorities are assigned by rank order of period, with the process with the shortest period being assigned the highest priority. This fixed-priority scheduling policy is the optimum assignment of static priorities to processes, in that it provides the highest CPU utilization while ensuring that all processes meet their deadlines.

Example 6.3 illustrates RMS.

Example 6.3

Rate-monotonic scheduling

Here is a simple set of processes and their characteristics.

Process	Execution time	Period
P1	1	4
P2	2	6
P3	3	12

Applying the principles of RMA, we give P1 the highest priority, P2 the middle priority, and P3 the lowest priority. To understand all the interactions between the periods, we need to construct a time line equal in length to hyperperiod, which is 12 in this case.

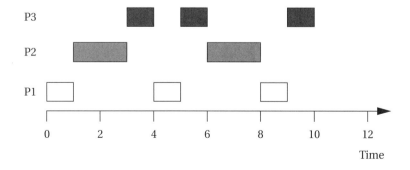

All three periods start at time zero. P1's data arrive first. Since P1 is the highest-priority process, it can start to execute immediately. After one time unit, P1 finishes and goes out of the ready state until the start of its next period. At time 1, P2 starts executing as the

highest-priority ready process. At time 3, P2 finishes and P3 starts executing. P1's next iteration starts at time 4, at which point it interrupts P3. P3 gets one more time unit of execution between the second iterations of P1 and P2, but P3 does not get to finish until after the third iteration of P1.

Consider the following different set of execution times for these processes, keeping the same deadlines.

Process	Execution time	Period
P1	2	4
P2	3	6
P3	3	12

In this case, we can show that there is no feasible assignment of priorities that guarantees scheduling. Even though each process alone has an execution time significantly less than its period, combinations of processes can require more than 100% of the available CPU cycles. For example, during one 12 time-unit interval, we must execute P1 three times, requiring 6 units of CPU time; P2 twice, costing 6 units of CPU time; and P3 one time, requiring 3 units of CPU time. The total of 6 + 6 + 3 = 15 units of CPU time is more than the 12 time units available, clearly exceeding the available CPU capacity.

Liu and Layland [Liu73] proved that the RMA priority assignment is optimal using critical-instant analysis. We define the **response time** of a process as the time at which the process finishes. The **critical** instant for a process is defined as the instant during execution at which the task has the largest response time. It is easy to prove that the critical instant for any process P, under the RMA model, occurs when it is ready and all higher-priority processes are also ready—if we change any higher-priority process to waiting, then P's response time can only go down.

We can use critical-instant analysis to determine whether there is any feasible schedule for the system. In the case of the second set of execution times in Example 6.3, there was no feasible schedule. Critical-instant analysis also implies that priorities should be assigned in order of periods. Let the periods and computation times of two processes P_1 and P_2 be τ_1, τ_2 and T_1, T_2, with $\tau_1 < \tau_2$. We can generalize the result of Example 6.3 to show the total CPU requirements for the two processes in two cases. In the first case, let P_1 have the higher priority. In the worst case we then execute P_2 once during its period and as many iterations of P_1 as fit in the same interval. Since there are $\lfloor \tau_2/\tau_1 \rfloor$ iterations of P_1 during a single period of P_2, the required constraint on CPU time, ignoring context switching overhead, is

$$\left\lfloor \frac{\tau_2}{\tau_1} \right\rfloor T_1 + T_2 \leq \tau_2. \tag{6.4}$$

period comp
 time

If, on the other hand, we give higher priority to P_2, then critical-instant analysis tells us that we must execute all of P_2 and all of P_1 in one of P_1's periods in the worst case:

$$T_1 + T_2 \leq \tau_1. \qquad (6.5)$$

There are cases where the first relationship can be satisfied and the second cannot, but there are no cases where the second relationship can be satisfied and the first cannot. We can inductively show that the process with the shorter period should always be given higher priority for process sets of arbitrary size. It is also possible to prove that RMS always provides a feasible schedule if such a schedule exists.

The bad news is that, although RMS is the optimal static-priority schedule, it does not always allow the system to use 100% of the available CPU cycles. In the RMS framework, the total CPU utilization for a set of n tasks is

$$U = \sum_{i=1}^{n} \frac{T_i}{\tau_i}. \qquad (6.6)$$

The fraction T_i/τ_i is the fraction of time that the CPU spends executing task i. It is possible to show that for a set of two tasks under RMS scheduling, the CPU utilization U will be no greater than $2(2^{1/2} - 1) \cong 0.83$. In other words, the CPU will be idle at least 17% of the time. This idle time is due to the fact that priorities are assigned statically; we see in the next section that more aggressive scheduling policies can improve CPU utilization. When there are m tasks with fixed priorities, the maximum processor utilization is

$$U = m(2^{1/m} - 1). \qquad (6.7)$$

As m approaches infinity, the least upper bound to CPU utilization is $\ln 2 = 0.69$—the CPU will be idle 31% of the time. This does not mean that we can never use 100% of the CPU. If the periods of the tasks are arranged properly, then we can schedule tasks to make use of 100% of the CPU. But the least upper bound of 69% tells us that RMS can in some cases deliver utilizations significantly below 100%.

The implementation of RMS is very simple. Figure 6.12 shows C code for an RMS scheduler run at the OS's timer interrupt. The code merely scans through the list of processes in priority order and selects the highest-priority ready process to run. Because the priorities are static, the processes can be sorted by priority in advance before the system starts executing. As a result, this scheduler has an asymptotic complexity of $O(n)$, where n is the number of processes in the system. (This code assumes that processes are not created dynamically. If dynamic process creation is required, the array can be replaced by a linked list of processes, but the asymptotic complexity remains the same.) The RMS scheduler has both low asymptotic complexity and low actual execution time, which helps minimize the discrepancies between the zero-context-switch assumption of RMA and the actual execution of an RMS system.

```
/* processes[] is an array of process activation records,
   stored in order of priority, with processes[0] being
   the highest-priority process */
Activation_record processes[NPROCESSES];

void RMA(int current) { /* current = currently executing
process */
  int i;
  /* turn off current process (may be turned back on) */
  processes[current].state = READY_STATE;
  /* find process to start executing */
  for (i = 0; i < NPROCESSES; i++)
      if (processes[i].state == READY_STATE) {
          /* make this the running process */
          processes[i].state == EXECUTING_STATE;
          break;
      }
}
```

FIGURE 6.12

C code for rate-monotonic scheduling.

6.3.2 Earliest-Deadline-First Scheduling

Earliest deadline first (EDF) is another well-known scheduling policy that was also studied by Liu and Layland [Liu73]. It is a dynamic priority scheme—it changes process priorities during execution based on initiation times. As a result, it can achieve higher CPU utilizations than RMS.

The EDF policy is also very simple: It assigns priorities in order of deadline. The highest-priority process is the one whose deadline is nearest in time, and the lowest-priority process is the one whose deadline is farthest away. Clearly, priorities must be recalculated at every completion of a process. However, the final step of the OS during the scheduling procedure is the same as for RMS—the highest-priority ready process is chosen for execution.

Example 6.4 illustrates EDF scheduling in practice.

Example 6.4

Earliest-deadline-first scheduling

Consider the following processes:

Process	Execution time	Period
P1	1	3
P2	1	4
P3	2	5

The hyperperiod is 60. In order to be able to see the entire period, we write it as a table:

Time	Running process	Deadlines
0	P1	
1	P2	
2	P3	P1
3	P3	P2
4	P1	P3
5	P2	P1
6	P1	
7	P3	P2
8	P3	P1
9	P1	P3
10	P2	
11	P3	P1, P2
12	P1	
13	P3	
14	P2	P1, P3
15	P1	P2
16	P2	
17	P3	P1
18	P1	
19	P3	P2, P3
20	P2	P1
21	P1	
22	P3	
23	P3	P1, P2
24	P1	P3
25	P2	
26	P3	P1
27	P1	P2
28	P3	
29	P2	P1, P3
30	idle	
31	P1	P2
32	P3	P1
33	P3	
34	P1	P3
35	P2	P1, P2
36	P1	
37	P2	
38	P3	P1
39	P3	P2, P3
40	P1	

(*Continued*)

Time	Running process	Deadlines
41	P2	P1
42	P1	
43	P3	P2
44	P3	P1, P3
45	P1	
46	P2	
47	P3	P1, P2
48	P3	
49	P1	P3
50	P2	P1
51	P1	P2
52	P3	
53	P3	P1
54	P2	P3
55	P1	P2
56	P2	P1
57	P1	
58	P3	
59	P3	P1, P2, P3

There is one time slot left at $t = 30$, giving a CPU utilization of 59/60.

Liu and Layland showed that EDF can achieve 100% utilization. A feasible schedule exists if the CPU utilization (calculated in the same way as for RMA) is ≤ 1. They also showed that when an EDF system is overloaded and misses a deadline, it will run at 100% capacity for a time before the deadline is missed.

The implementation of EDF is more complex than the RMS code. Figure 6.13 outlines one way to implement EDF. The major problem is keeping the processes sorted by time to deadline—since the times to deadlines for the processes change during execution, we cannot presort the processes into an array, as we could for RMS. To avoid resorting the entire set of records at every change, we can build a binary tree to keep the sorted records and incrementally update the sort. At the end of each period, we can move the record to its new place in the sorted list by deleting it from the tree and then adding it back to the tree using standard tree manipulation techniques. We must update process priorities by traversing them in sorted order, so the incremental sorting routines must also update the linked list pointers that let us traverse the records in deadline order. (The linked list lets us avoid traversing the tree to go from one node to another, which would require more time.) After putting in the effort to building the sorted list of records, selecting the next executing process is done in a manner similar to that of RMS. However, the dynamic sorting adds complexity to the entire scheduling process. Each update of the sorted list

Data structure

```
/* linked list, sorted by deadline */
Activation_record *processes;
/* data structure for sorting processes */
Deadline_tree *deadlines;

void expired_deadline(Activation_record *expired){
    remove(expired); /* remove from the deadline-sorted list */
    add(expired,expired->deadline); /* add at new deadline */
}

Void EDF(int current) { /* current = currently executing process */
    int i;
    /* turn off current process (may be turned back on) */
    processes->state = READY_STATE;
    /* find process to start executing */
    for (alink = processes; alink != NULL; alink = alink->next_deadline)
        if (processes->state == READY_STATE) {
            /* make this the running process */
            processes->state == EXECUTING_STATE;
            break;
        }
}
```

Code

FIGURE 6.13

C code for earliest-deadline-first scheduling.

requires $O(\log n)$ steps. The EDF code is also significantly more complex than the RMS code.

6.3.3 RMS vs. EDF

Which scheduling policy is better: RMS or EDF? That depends on your criteria. EDF can extract higher utilization out of the CPU, but it may be difficult to diagnose the possibility of an imminent overload. Because the scheduler does take some overhead to make scheduling decisions, a factor that is ignored in the schedulability analysis of both EDF and RMS, running a scheduler at very high utilizations is somewhat problematic. RMS achieves lower CPU utilization but is easier to ensure that all deadlines

will be satisfied. In some applications, it may be acceptable for some processes to occasionally miss deadlines. For example, a set-top box for video decoding is not a safety-critical application, and the occasional display artifacts caused by missing deadlines may be acceptable in some markets.

What if your set of processes is unschedulable and you need to guarantee that they complete their deadlines? There are several possible ways to solve this problem:

- *Get a faster CPU.* That will reduce execution times without changing the periods, giving you lower utilization. This will require you to redesign the hardware, but this is often feasible because you are rarely using the fastest CPU available.

- *Redesign the processes to take less execution time.* This requires knowledge of the code and may or may not be possible.

- *Rewrite the specification to change the deadlines.* This is unlikely to be feasible, but may be in a few cases where some of the deadlines were initially made tighter than necessary.

6.3.4 A Closer Look at Our Modeling Assumptions

Our analyses of RMS and EDF have made some strong assumptions. These assumptions have made the analyses much more tractable, but the predictions of analysis may not hold up in practice. Since a misprediction may cause a system to miss a critical deadline, it is important to at least understand the consequences of these assumptions.

In all of the above discussions, we have assumed that each process is totally self-contained. However, that is not always the case—for instance, a process may need a system resource, such as an I/O device or the bus, to complete its work. Scheduling the processes without considering the resources those processes require can cause ***priority inversion***, in which a low-priority process blocks execution of a higher-priority process by keeping hold of its resource. Example 6.5 illustrates priority inversion.

Example 6.5

Priority inversion

Consider a system with two processes: the higher-priority P1 and the lower-priority P2. Each uses the microprocessor bus to communicate to peripherals. When P2 executes, it requests the bus from the operating system and receives it. If P1 becomes ready while P2 is using the bus, the OS will preempt P2 for P1, leaving P2 with control of the bus. When P1 requests the bus, it will be denied the bus, since P2 already owns it. Unless P1 has a way to take the bus from P2, the two processes may deadlock.

The most common method for dealing with priority inversion is to promote the priority of any process when it requests a resource from the OS. The priority of the process temporarily becomes higher than that of any other process that may use

the resource. This ensures that the process will continue executing once it has the resource so that it can finish its work with the resource, return it to the OS, and allow other processes to use it. Once the process is finished with the resource, its priority is demoted to its normal value. Several methods have been developed to manage the priority swapping process [Liu00].

Rate-monotonic scheduling assumes that there are no data dependencies between processes. Example 6.6 shows that knowledge of data dependencies can help use the CPU more efficiently.

Example 6.6

Data dependencies and scheduling

Data dependencies imply that certain combinations of processes can never occur. Consider the simple example [Yen98] below.

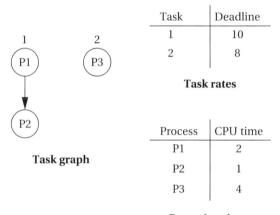

Task	Deadline
1	10
2	8

Task rates

Task graph

Process	CPU time
P1	2
P2	1
P3	4

Execution times

We know that P1 and P2 cannot execute at the same time, since P1 must finish before P2 can begin. Furthermore, we also know that because P3 has a higher priority, it will not preempt both P1 and P2 in a single iteration. If P3 preempts P1, then P3 will complete before P2 begins; if P3 preempts P2, then it will not interfere with P1 in that iteration. Because we know that some combinations of processes cannot be ready at the same time, we know that our worst-case CPU requirements are less than would be required if all processes could be ready simultaneously.

6.4 INTERPROCESS COMMUNICATION MECHANISMS

Processes often need to communicate with each other. ***Interprocess communication mechanisms*** are provided by the operating system as part of the process abstraction.

FIGURE 6.14

Shared memory communication implemented on a bus.

In general, a process can send a communication in one of two ways: ***blocking*** or ***nonblocking***. After sending a blocking communication, the process goes into the waiting state until it receives a response. Nonblocking communication allows the process to continue execution after sending the communication. Both types of communication are useful.

There are two major styles of interprocess communication: ***shared memory*** and ***message passing***. The two are logically equivalent—given one, you can build an interface that implements the other. However, some programs may be easier to write using one rather than the other. In addition, the hardware platform may make one easier to implement or more efficient than the other.

6.4.1 Shared Memory Communication

Figure 6.14 illustrates how shared memory communication works in a bus-based system. Two components, such as a CPU and an I/O device, communicate through a shared memory location. The software on the CPU has been designed to know the address of the shared location; the shared location has also been loaded into the proper register of the I/O device. If, as in the figure, the CPU wants to send data to the device, it writes to the shared location. The I/O device then reads the data from that location. The read and write operations are standard and can be encapsulated in a procedural interface.

Example 6.7 describes the use of shared memory as a practical communication mechanism.

Example 6.7

Elastic buffers as shared memory

The text compressor of Application Example 3.4 provides a good example of a shared memory. As shown below, the text compressor uses the CPU to compress incoming text, which is then sent on a serial line by a UART.

The input data arrive at a constant rate and are easy to manage. But because the output data are consumed at a variable rate, these data require an elastic buffer. The CPU and output UART share a memory area—the CPU writes compressed characters into the buffer and the UART removes them as necessary to fill the serial line. Because the number of bits in the buffer changes constantly, the compression and transmission processes need additional size information. In this case, coordination is simple—the CPU writes at one end of the buffer and the UART reads at the other end. The only challenge is to make sure that the UART does not overrun the buffer.

As an application of shared memory, let us consider the situation of Figure 6.14 in which the CPU and the I/O device want to communicate through a shared memory block. There must be a flag that tells the CPU when the data from the I/O device is ready. The flag, an additional shared data location, has a value of 0 when the data are not ready and 1 when the data are ready. The CPU, for example, would write the data, and then set the flag location to 1. If the flag is used only by the CPU, then the flag can be implemented using a standard memory write operation. If the same flag is used for bidirectional signaling between the CPU and the I/O device, care must be taken. Consider the following scenario:

1. CPU reads the flag location and sees that it is 0.

2. I/O device reads the flag location and sees that it is 0.

3. CPU sets the flag location to 1 and writes data to the shared location.

4. I/O device erroneously sets the flag to 1 and overwrites the data left by the CPU.

The above scenario is caused by a critical timing race between the two programs. To avoid such problems, the microprocessor bus must support an atomic ***test-and-set*** operation, which is available on a number of microprocessors. The test-and-set operation first reads a location and then sets it to a specified value. It returns the result of the test. If the location was already set, then the additional set has no effect but the test-and-set instruction returns a false result. If the location was not set, the

instruction returns true and the location is in fact set. The bus supports this as an *atomic* operation that cannot be interrupted. Programming Example 6.1 describes a test-and-set operation in more detail.

A test-and-set can be used to implement a **semaphore**, which is a language-level synchronization construct. For the moment, let's assume that the system provides one semaphore that is used to guard access to a block of protected memory. Any process that wants to access the memory must use the semaphore to ensure that no other process is actively using it. As shown below, the semaphore names by tradition are P() to gain access to the protected memory and V() to release it.

```
/* some nonprotected operations here */
P(); /* wait for semaphore */
/* do protected work here */
V(); /* release semaphore */
```

The P() operation uses a test-and-set to repeatedly test a location that holds a lock on the memory block. The P() operation does not exit until the lock is available; once it is available, the test-and-set automatically sets the lock. Once past the P() operation, the process can work on the protected memory block. The V() operation resets the lock, allowing other processes access to the region by using the P() function.

Programming Example 6.1

Test-and-set operation

The SWP (swap) instruction is used in the ARM to implement atomic test-and-set:

```
SWP Rd,Rm,Rn
```

The SWP instruction takes three operands—the memory location pointed to by *Rn* is loaded and saved into *Rd*, and the value of *Rm* is then written into the location pointed to by *Rn*. When *Rd* and *Rn* are the same register, the instruction swaps the register's value and the value stored at the address pointed to by *Rd*/*Rn*. For example, consider this code sequence:

```
ADR r0, SEMAPHORE        ; get semaphore address
LDR r1, #1
GETFLAG SWP r1,r1,[r0]   ; test-and-set the flag
BNZ GETFLAG              ; no flag yet, try again
HASFLAG ...
```

The program first loads the constant 1 into *r1* and the address of the semaphore FLAG1 into register *r2*, then reads the semaphore into *r0* and writes the 1 value into the semaphore. The code then tests whether the semaphore fetched from memory is zero; if it was, the semaphore was not busy and we can enter the critical region that begins with the HASFLAG label. If the flag was nonzero, we loop back to try to get the flag once again.

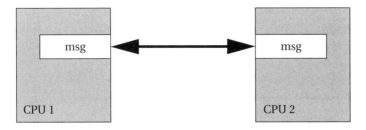

FIGURE 6.15

Message passing communication.

6.4.2 Message Passing

Message passing communication complements the shared memory model. As shown in Figure 6.15, each communicating entity has its own message send/receive unit. The message is not stored on the communications link, but rather at the senders/receivers at the end points. In contrast, shared memory communication can be seen as a memory block used as a communication device, in which all the data are stored in the communication link/memory.

Applications in which units operate relatively autonomously are natural candidates for message passing communication. For example, a home control system has one microcontroller per household device—lamp, thermostat, faucet, appliance, and so on. The devices must communicate relatively infrequently; furthermore, their physical separation is large enough that we would not naturally think of them as sharing a central pool of memory. Passing communication packets among the devices is a natural way to describe coordination between these devices. Message passing is the natural implementation of communication in many 8-bit microcontrollers that do not normally operate with external memory.

6.4.3 Signals

Another form of interprocess communication commonly used in Unix is the **_signal_**. A signal is simple because it does not pass data beyond the existence of the signal itself. A signal is analogous to an interrupt, but it is entirely a software creation. A signal is generated by a process and transmitted to another process by the operating system.

A UML signal is actually a generalization of the Unix signal. While a Unix signal carries no parameters other than a condition code, a UML signal is an object. As such, it can carry parameters as object attributes. Figure 6.16 shows the use of a signal in UML. The _sigbehavior_() behavior of the class is responsible for throwing the signal, as indicated by ≪ _send_ ≫. The signal object is indicated by the ≪ _signal_ ≫ stereotype.

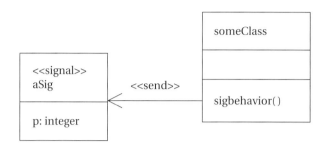

FIGURE 6.16

Use of a UML signal.

6.5 EVALUATING OPERATING SYSTEM PERFORMANCE

The scheduling policy does not tell us all that we would like to know about the performance of a real system running processes. Our analysis of scheduling policies makes some simplifying assumptions:

- We have assumed that context switches require zero time. Although it is often reasonable to neglect context switch time when it is much smaller than the process execution time, context switching can add significant delay in some cases.

- We have assumed that we know the execution time of the processes. In fact, we learned in Section 5.6 that program time is not a single number, but can be bounded by worst-case and best-case execution times.

- We probably determined worst-case or best-case times for the processes in isolation. But, in fact, they interact with each other in the cache. Cache conflicts among processes can drastically degrade process execution time.

The zero-time context switch assumption used in the analysis of RMS is not correct—we must execute instructions to save and restore context, and we must execute additional instructions to implement the scheduling policy. On the other hand, context switching can be implemented efficiently—context switching need not kill performance. The effects of nonzero context switching time must be carefully analyzed in the context of a particular implementation to be sure that the predictions of an ideal scheduling policy are sufficiently accurate.

Example 6.8 shows that context switching can, in fact, cause a system to miss a deadline.

Example 6.8

Scheduling and context switching overhead

Appearing below is a set of processes and their characteristics.

Process	Execution time	Deadline
P1	3	5
P2	3	10

First, let us try to find a schedule assuming that context switching time is zero. Following is a feasible schedule for a sequence of data arrivals that meets all the deadlines:

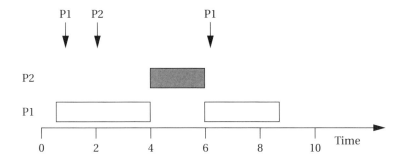

Now let us assume that the total time to initiate a process, including context switching and scheduling policy evaluation, is one time unit. It is easy to see that there is no feasible schedule for the above release time sequence, since we require a total of $2T_{P1} + T_{P2} = 2 \times (1 + 3) + (1 + 3) = 11$ time units to execute one period of P2 and two periods of P1.

In Example 6.8, overhead was a large fraction of the process execution time and of the periods. In most real-time operating systems, a context switch requires only a few hundred instructions, with only slightly more overhead for a simple real-time scheduler like RMS. When the overhead time is very small relative to the task periods, then the zero-time context switch assumption is often a reasonable approximation. Problems are most likely to manifest themselves in the highest-rate processes, which are often the most critical in any case. Completely checking that all deadlines will be met with nonzero context switching time requires checking all possible schedules for processes and including the context switch time at each preemption or process initiation. However, assuming an average number of context switches per process and computing CPU utilization can provide at least an estimate of how close the system is to CPU capacity.

Another important assumption we have made thus far is that process execution time is constant. As seen in Section 5.6, this is definitely not the case—both data-dependent behavior and caching effects can cause large variations in run times. If we can determine worst-case execution time, then shorter run times for a process simply mean unused CPU time. If we cannot accurately bound WCET, then we will be left with a very conservative estimate of execution time that will leave even more CPU time unused.

We also assumed that processes don't interact, but the cache causes the execution of one program to influence the execution time of other programs. The techniques for bounding the cache-based performance of a single program do not work when multiple programs are in the same cache. Many real-time systems have been designed based on the assumption that there is no cache present, even though one actually exists. This grossly conservative assumption is made because the system architects lack tools that permit them to analyze the effect of caching. Since they do not know where caching will cause problems, they are forced to retreat to the simplifying assumption that there is no cache. The result is extremely overdesigned hardware, which has much more computational power than is necessary. However, just as experience tells us that a well-designed cache provides significant performance benefits for a single program, a properly sized cache can allow a microprocessor to run a set of processes much more quickly. By analyzing the effects of the cache, we can make much better use of the available hardware.

Li and Wolf [Li99] developed a model for estimating the performance of multiple processes sharing a cache. In the model, some processes can be given reservations in the cache, such that only a particular process can inhabit a reserved section of the cache; other processes are left to share the cache. We generally want to use cache partitions only for performance-critical processes since cache reservations are wasteful of limited cache space. Performance is estimated by constructing a schedule, taking into account not just execution time of the processes but also the state of the cache. Each process in the shared section of the cache is modeled by a binary variable: 1 if present in the cache and 0 if not. Each process is also characterized by three total execution times: assuming no caching, with typical caching, and with all code always resident in the cache. The always-resident time is unrealistically optimistic, but it can be used to find a lower bound on the required schedule time. During construction of the schedule, we can look at the current cache state to see whether the no-cache or typical-caching execution time should be used at this point in the schedule. We can also update the cache state if the cache is needed for another process. Although this model is simple, it provides much more realistic performance estimates than assuming the cache either is nonexistent or is perfect. Example 6.9 shows how cache management can improve CPU utilization.

Example 6.9

Effects of scheduling on the cache

Consider a system containing the following three processes:

Process	Worst-case CPU time	Average-case CPU time
P1	8	6
P2	4	3
P3	4	3

Each process uses half the cache, so only two processes can be in the cache at the same time.

Appearing below is a first schedule that uses a least-recently-used cache replacement policy on a process-by-process basis.

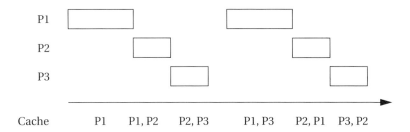

In the first iteration, we must fill up the cache, but even in subsequent iterations, competition among all three processes ensures that a process is never in the cache when it starts to execute. As a result, we must always use the worst-case execution time.

Another schedule in which we have reserved half the cache for P1 is shown below. This leaves P2 and P3 to fight over the other half of the cache.

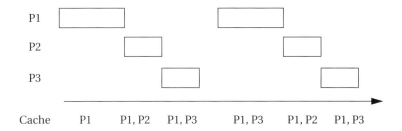

In this case, P2 and P3 still compete, but P1 is always ready. After the first iteration, we can use the average-case execution time for P1, which gives us some spare CPU time that could be used for additional operations.

6.6 POWER MANAGEMENT AND OPTIMIZATION FOR PROCESSES

We learned in Section 3.6 about the features that CPUs provide to manage power consumption. The RTOS and system architecture can use static and dynamic power management mechanisms to help manage the system's power consumption. A *power management policy* [Ben00] is a strategy for determining when to

perform certain power management operations. A power management policy in general examines the state of the system to determine when to take actions. However, the overall strategy embodied in the policy should be designed based on the characteristics of the static and dynamic power management mechanisms.

Going into a low-power mode takes time; generally, the more that is shut off, the longer the delay incurred during restart. Because power-down and power-up are not free, modes should be changed carefully. Determining when to switch into and out of a power-up mode requires an analysis of the overall system activity.

- Avoiding a power-down mode can cost unnecessary power.

- Powering down too soon can cause severe performance penalties.

Re-entering run mode typically costs a considerable amount of time.

A straightforward method is to power up the system when a request is received. This works as long as the delay in handling the request is acceptable. A more sophisticated technique is **predictive shutdown**. The goal is to predict when the next request will be made and to start the system just before that time, saving the requestor the start-up time. In general, predictive shutdown techniques are probabilistic—they make guesses about activity patterns based on a probabilistic model of expected behavior. Because they rely on statistics, they may not always correctly guess the time of the next activity. This can cause two types of problems:

- The requestor may have to wait for an activity period. In the worst case, the requestor may not make a deadline due to the delay incurred by system start-up.

- The system may restart itself when no activity is imminent. As a result, the system will waste power.

Clearly, the choice of a good probabilistic model of service requests is important. The policy mechanism should also not be too complex, since the power it consumes to make decisions is part of the total system power budget.

Several predictive techniques are possible. A very simple technique is to use fixed times. For instance, if the system does not receive inputs during an interval of length T_{on}, it shuts down; a powered-down system waits for a period T_{off} before returning to the power-on mode. The choice of T_{off} and T_{on} must be determined by experimentation. Srivastava and Eustace [Sri94] found one useful rule for graphics terminals. They plotted the observed idle time (T_{off}) of a graphics terminal versus the immediately preceding active time (T_{on}). The result was an L-shaped distribution as illustrated in Figure 6.17. In this distribution, the idle period after a long active period is usually very short, and the length of the idle period after a short active period is uniformly distributed. Based on this distribution, they proposed a shut down threshold that depended on the length of the last active period—they shut

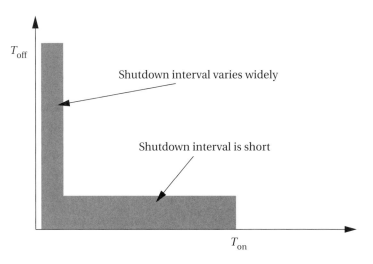

FIGURE 6.17

An L-shaped usage distribution.

down when the active period length was below a threshold, putting the system in the vertical portion of the L distribution.

The *Advanced Configuration and Power Interface (ACPI)* is an open industry standard for power management services. It is designed to be compatible with a wide variety of OSs. It was targeted initially to PCs. The role of ACPI in the system is illustrated in Figure 6.18. ACPI provides some basic power management facilities and abstracts the hardware layer, the OS has its own power management module that determines the policy, and the OS then uses ACPI to send the required controls to the hardware and to observe the hardware's state as input to the power manager.

ACPI supports the following five basic global power states:

- G3, the mechanical off state, in which the system consumes no power.

- G2, the soft off state, which requires a full OS reboot to restore the machine to working condition. This state has four substates:

 —S1, a low wake-up latency state with no loss of system context;

 —S2, a low wake-up latency state with a loss of CPU and system cache state;

 —S3, a low wake-up latency state in which all system state except for main memory is lost; and

 —S4, the lowest-power sleeping state, in which all devices are turned off.

- G1, the sleeping state, in which the system appears to be off and the time required to return to working condition is inversely proportional to power consumption.

FIGURE 6.18

The advanced configuration and power interface and its relationship to a complete system.

- G0, the working state, in which the system is fully usable.

- The legacy state, in which the system does not comply with ACPI.

The power manager typically includes an observer, which receives messages through the ACPI interface that describe the system behavior. It also includes a decision module that determines power management actions based on those observations.

<div style="display:flex"><div>*Design Example*</div></div>

6.7 TELEPHONE ANSWERING MACHINE

In this section we design a digital telephone answering machine. The system will store messages in digital form rather than on an analog tape. To make life more interesting, we use a simple algorithm to compress the voice data so that we can make more efficient use of the limited amount of available memory.

6.7.1 Theory of Operation and Requirements

In addition to studying the compression algorithm, we also need to learn a little about the operation of telephone systems.

The compression scheme we will use is known as ***adaptive differential pulse code modulation (ADPCM)***. Despite the long name, the technique is relatively simple but can yield 2 × compression ratios on voice data.

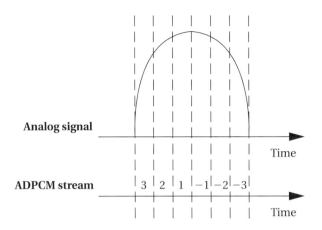

FIGURE 6.19

The ADPCM coding scheme.

The ADPCM coding scheme is illustrated in Figure 6.19. Unlike traditional sampling, in which each sample shows the magnitude of the signal at a particular time, ADPCM encodes changes in the signal. The samples are expressed in a *coding alphabet*, whose values are in a relative range that spans both negative and positive values. In this case, the value range is $\{-3, -2, -1, 1, 2, 3\}$. Each sample is used to predict the value of the signal at the current instant from the previous value. At each point in time, the sample is chosen such that the error between the predicted value and the actual signal value is minimized.

An ADPCM compression system, including an encoder and decoder, is shown in Figure 6.20. The encoder is more complex, but both the encoder and decoder use an integrator to reconstruct the waveform from the samples. The integrator simply computes a running sum of the history of the samples; because the samples are differential, integration reconstructs the original signal. The encoder compares the incoming waveform to the predicted waveform (the waveform that will be generated in the decoder). The quantizer encodes this difference as the best predictor of the next waveform value. The inverse quantizer allows us to map bit-level symbols onto real numerical values; for example, the eight possible codes in a 3-bit code can be mapped onto floating-point numbers. The decoder simply uses an inverse quantizer and integrator to turn the differential samples into the waveform.

The answering machine will ultimately be connected to a telephone *subscriber line* (although for testing purposes we will construct a simulated line). At the other end of the subscriber line is the *central office*. All information is carried on the phone line in analog form over a pair of wires. In addition to analog/digital and digital/analog converters to send and receive voice data, we need to sense two other characteristics of the line.

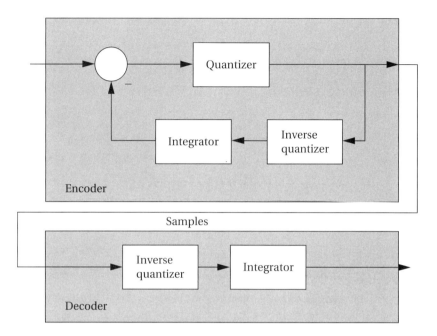

FIGURE 6.20

An ADPCM compression system.

- *Ringing:* The central office sends a ringing signal to the telephone when a call is waiting. The ringing signal is in fact a 90 V RMS sinusoid, but we can use analog circuitry to produce 0 for no ringing and 1 for ringing.

- *Off-hook:* The telephone industry term for answering a call is going *off-hook*; the technical term for hanging up is going *on-hook*. (This creates some initial confusion since *off-hook* means the telephone is active and *on-hook* means it is not in use, but the terminology starts to make sense after a few uses.) Our interface will send a digital signal to take the phone line off-hook, which will cause analog circuitry to make the necessary connection so that voice data can be sent and received during the call.

We can now write the requirements for the answering machine. We will assume that the interface is not to the actual phone line but to some circuitry that provides voice samples, off-hook commands, and so on. Such circuitry will let us test our system with a telephone line simulator and then build the analog circuitry necessary to connect to a real phone line. We will use the term ***outgoing message (OGM)*** to refer to the message recorded by the owner of the machine and played at the start of every phone call.

Name	Digital telephone answering machine
Purpose	Telephone answering machine with digital memory, using speech compression.
Inputs	*Telephone:* voice samples, ring indicator. *User interface:* microphone, play messages button, record OGM button.
Outputs	*Telephone:* voice samples, on-hook/off-hook command. *User interface:* speaker, # messages indicator, message light.
Functions	*Default mode:* When machine receives ring indicator, it signals off-hook, plays the OGM, and then records the incoming message. Maximum recording length for incoming message is 30 s, at which time the machine hangs up. If the machine runs out of memory, the OGM is played and the machine then hangs up without recording. *Playback mode:* When the play button is depressed, the machine plays all messages. If the play button is depressed again within five seconds, the messages are played again. Messages are erased after playback. *OGM editing mode:* When the user hits the record OGM button, the machine records an OGM of up to 10 s. When the user holds down the record OGM button and hits the play button, the OGM is played back.
Performance	Should be able to record about 30 min of total voice, including incoming and OGMs. Voice data are sampled at the standard telephone rate of 8 kHz.
Manufacturing cost	Consumer product range: approximately $50.
Power	Powered by AC through a standard power supply.
Physical size and weight	Comparable in size and weight to a desk telephone.

We have made a few arbitrary decisions about the user interface in these requirements. The amount of voice data that can be saved by the machine should in fact be determined by two factors: the price per unit of DRAM at the time at which the device goes into manufacturing (since the cost will almost certainly drop from the start of design to manufacture) and the projected retail price at which the machine must sell. The protocol when the memory is full is also arbitrary—it would make at least as much sense to throw out old messages and replace them with new ones, and ideally the user could select which protocol to use. Extra features such as an indicator showing the number of messages or a save messages feature would also be nice to have in a real consumer product.

6.7.2 Specification

Figure 6.21 shows the class diagram for the answering machine. In addition to the classes that perform the major functions, we also use classes to describe the incoming and OGMs. As seen below, these classes are related.

The definitions of the physical interface classes are shown in Figure 6.22. The buttons and lights simply provide attributes for their input and output values. The phone line, microphone, and speaker are given behaviors that let us sample their current values.

The message classes are defined in Figure 6.23. Since incoming and OGM types share many characteristics, we derive both from a more fundamental message type.

The major operational classes—*Controls*, *Record*, and *Playback*—are defined in Figure 6.24. The *Controls* class provides an *operate()* behavior that oversees the user-level operations. The *Record* and *Playback* classes provide behaviors that handle writing and reading sample sequences.

The state diagram for the *Controls activate* behavior is shown in Figure 6.25. Most of the user activities are relatively straightforward. The most complex is answering an incoming call. As with the software modem of Section 5.11, we want to be sure that a single depression of a button causes the required action to be taken exactly once; this requires edge detection on the button signal.

State diagrams for *record-msg* and *playback-msg* are shown in Figure 6.26. We have parameterized the specification for *record-msg* so that it can be used either from the phone line or from the microphone. This requires parameterizing the source itself and the termination condition.

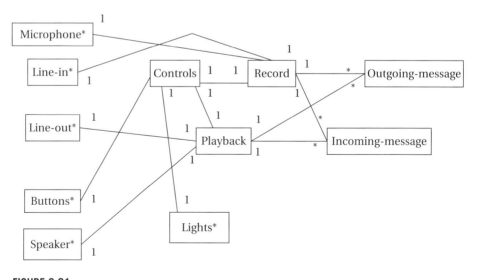

FIGURE 6.21

Class diagram for the answering machine.

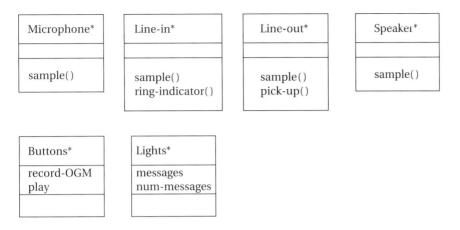

FIGURE 6.22

Physical class interfaces for the answering machine.

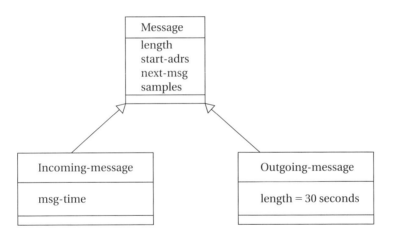

FIGURE 6.23

The message classes for the answering machine.

FIGURE 6.24

Operational classes for the answering machine.

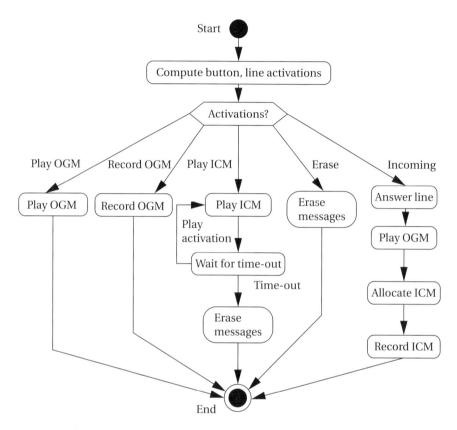

FIGURE 6.25

State diagram for the controls activate behavior.

6.7.3 System Architecture

The machine consists of two major subsystems from the user's point of view: the user interface and the telephone interface. The user and telephone interfaces both appear internally as I/O devices on the CPU bus with the main memory serving as the storage for the messages.

The software splits into the following seven major pieces:

- The *front panel module* handles the buttons and lights.

- The *speaker module* handles sending data to the user's speaker.

- The *telephone line module* handles off-hook detection and on-hook commands.

- The *telephone input and output modules* handle receiving samples from and sending samples to the telephone line.

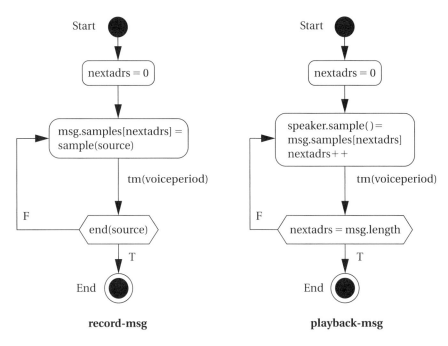

FIGURE 6.26

State diagrams for the record-msg and playback-msg behaviors.

■ The *compression module* compresses data and stores it in memory.

■ The *decompression module* uncompresses data and sends it to the speaker module.

We can determine the execution model for these modules based on the rates at which they must work and the ways in which they communicate.

■ The front panel and telephone line modules must regularly test the buttons and phone line, but this can be done at a fairly low rate. As seen below, they can therefore run as polled processes in the software's main loop.

```
while (TRUE) {
    check_phone_line();
    run_front_panel();
}
```

■ The speaker and phone input and output modules must run at higher, regular rates and are natural candidates for interrupt processing. These modules don't run all the time and so can be disabled by the front panel and telephone line modules when they are not needed.

■ The compression and decompression modules run at the same rate as the speaker and telephone I/O modules, but they are not directly connected to devices. We will therefore call them as subroutines to the interrupt modules.

One subtlety is that we must construct a very simple file system for messages, since we have a variable number of messages of variable lengths. Since messages vary in length, we must record the length of each one. In this simple specification, because we always play back the messages in the order in which they were recorded, we don't have to keep a full-fledged directory. If we allowed users to selectively delete messages and save others, we would have to build some sort of directory structure for the messages.

The hardware architecture is straightforward and illustrated in Figure 6.27. The speaker and telephone I/O devices appear as standard A/D and D/A converters. The telephone line appears as a one-bit input device (ring detect) and a one-bit output device (off-hook/on-hook). The compressed data are kept in main memory.

6.7.4 Component Design and Testing

Performance analysis is important in this case because we want to ensure that we don't spend so much time compressing that we miss voice samples. In a real consumer product, we would carefully design the code so that we could use the slowest, cheapest possible CPU that would still perform the required processing in the available time between samples. In this case, we will choose the microprocessor in advance for simplicity and simply ensure that all the deadlines are met.

An important class of problems that should be adequately tested is memory overflow. The system can run out of memory at any time, not just between messages. The modules should be tested to ensure that they do reasonable things when all the available memory is used up.

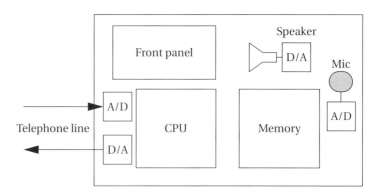

FIGURE 6.27

The hardware structure of the answering machine.

6.7.5 System Integration and Testing

We can test partial integrations of the software on our host platform. Final testing with real voice data must wait until the application is moved to the target platform.

Testing your system by connecting it directly to the phone line is not a very good idea. In the United States, the Federal Communications Commission regulates equipment connected to phone lines. Beyond legal problems, a bad circuit can damage the phone line and incur the wrath of your service provider. The required analog circuitry also requires some amount of tuning, and you need a second telephone line to generate phone calls for tests. You can build a telephone line simulator to test the hardware independently of a real telephone line. The phone line simulator consists of A/D and D/A converters plus a speaker and microphone for voice data, an LED for off-hook/on-hook indication, and a button for ring generation. The telephone line interface can easily be adapted to connect to these components, and for purposes of testing the answering machine the simulator behaves identically to the real phone line.

SUMMARY

The process abstraction is forced on us by the need to satisfy complex timing requirements, particularly for multirate systems. Writing a single program that simultaneously satisfies deadlines at multiple rates is too difficult because the control structure of the program becomes unintelligible. The process encapsulates the state of a computation, allowing us to easily switch among different computations.

The operating system encapsulates the complex control to coordinate the process. The scheme used to determine the transfer of control among processes is known as a scheduling policy. A good scheduling policy is useful across many different applications while also providing efficient utilization of the available CPU cycles.

It is difficult, however, to achieve 100% utilization of the CPU for complex applications. Because of variations in data arrivals and computation times, reserving some cycles to meet worst-case conditions is often necessary. Some scheduling policies achieve higher utilizations than others, but often at the cost of unpredictability—they may not guarantee that all deadlines are met. Knowledge of the characteristics of an application can be used to increase CPU utilization while also complying with deadlines.

What We Learned

- A process is a single thread of execution.

- Pre-emption is the act of changing the CPU's execution from one process to another.

- A scheduling policy is a set of rules that determines the process to run.

- Rate-monotonic scheduling (RMS) is a simple but powerful scheduling policy.

- Interprocess communication mechanisms allow data to be passed reliably between processes.

- Scheduling analysis often ignores certain real-world effects. Cache interactions between processes are the most important effects to consider when designing a system.

FURTHER READING

Gallmeister [Gal95] provides a thorough and very readable introduction to POSIX in general and its real-time aspects in particular. Liu and Layland [Liu73] introduce rate-monotonic scheduling; this paper became the foundation for real-time systems analysis and design. The book by Liu [Liu00] provides a detailed analysis of real-time scheduling. Benini et al. [Ben00] provide a good survey of system-level power management techniques. Falik and Intrater [Fal92] describe a custom chip designed to perform answering machine operations.

QUESTIONS

Q6-1 Identify activities that operate at different rates in

 a. a PDA;
 b. a laser printer; and
 c. an airplane.

Q6-2 Name an embedded system that requires both periodic and aperiodic computation.

Q6-3 An audio system processes samples at a rate of 44.1 kHz. At what rate could we sample the system's front panel to both simplify analysis of the system schedule and provide adequate response to the user's front panel requests?

Q6-4 Draw a UML class diagram for a process in an operating system. The process class should include the necessary attributes and behaviors required of a typical process.

Q6-5 What factors provide a lower bound on the period at which the system timer interrupts for preemptive context switching?

Q6-6 What factors provide an upper bound on the period at which the system timer interrupts for preemptive context switching?

Q6-7 You are given these periodic tasks:

Task	Period	Execution time
P1	5 ms	2 ms
P2	10 ms	3 ms
P3	10 ms	3 ms
P4	15 ms	6 ms

Compute the utilization of this set of tasks.

Q6-8 You are given these periodic tasks:

Task	Period	Execution time
P1	5 ms	1 ms
P2	10 ms	2 ms
P3	10 ms	2 ms
P4	15 ms	3 ms

a. Show a cyclostatic schedule for the tasks.

b. Compute the CPU utilization for the system.

Q6-9 For the task set of question Q6-8, show a round robin schedule assuming that P1 does not execute during its first period and P3 does not execute during its second period.

Q6-10 What is the distinction between the ready and waiting states of process scheduling?

Q6-11 Provide examples of

a. blocking interprocess communication, and

b. nonblocking interprocess communication.

Q6-12 Assuming that you have a routine called swap(int *a,int *b) that atomically swaps the values of the memory locations pointed to a and b, write C code for:

a. P(); and

b. V().

Q6-13 Draw UML sequence diagrams of two versions of P(): one that incorrectly uses a nonatomic operation to test and set the semaphore location and another that correctly uses an atomic test-and-set.

Q6-14 For the following periodic processes, what is the shortest interval we must examine to see all combinations of deadlines?

a.

Process	Deadline
P1	3
P2	5
P3	15

b.

Process	Deadline
P1	2
P2	3
P3	6
P4	10

c.

Process	Deadline
P1	3
P2	4
P3	5
P4	6
P5	10

Q6-15 Consider the following system of periodic processes executing on a single CPU:

Process	CPU time	Deadline
P1	4	200
P2	1	10
P3	2	40
P4	6	50

Can we add another instance of P1 to the system and still meet all the deadlines using RMS?

Q6-16 Given the following set of periodic processes running on a single CPU, what is the maximum execution time for P5 for which all the processes will be schedulable using RMS?

Process	CPU time	Deadline
P1	1	10
P2	18	100
P3	2	20
P4	5	50
P5	x	25

Q6-17 A set of periodic processes is scheduled using RMS. For the process execution times and periods shown below, show the state of the processes at the critical instant for each of these processes.

 a. P1

 b. P2

 c. P3

Process	CPU time	Deadline
P1	1	4
P2	2	5
P3	1	20

Q6-18 For the given periodic process execution times and periods, show how much CPU time of higher-priority processes will be required during one period of each of the following processes:

 a. P1

 b. P2

 c. P3

 d. P4

Process	CPU time	Deadline
P1	1	5
P2	2	10
P3	3	25
P4	4	50

Q6-19 For the periodic processes shown below:

 a. Schedule the processes using an RMS policy.

 b. Schedule the processes using an EDF policy.

In each case, compute the schedule for the hyperperiod of the processes. Time starts at $t = 0$.

Process	CPU time	Deadline
P1	1	3
P2	1	4
P3	1	12

Q6-20 For the periodic processes shown below:

 a. Schedule the processes using an RMS policy.

 b. Schedule the processes using an EDF policy.

In each case, compute the schedule for an interval equal to the hyperperiod of the processes. Time starts at $t = 0$.

Process	CPU time	Deadline
P1	1	3
P2	1	4
P3	2	8

Q6-21 For the given set of periodic processes, all of which share the same deadline of 12:

 a. Schedule the processes for the given arrival times using standard rate-monotonic scheduling (no data dependencies).

 b. Schedule the processes taking advantage of the data dependencies. By how much is the CPU utilization reduced?

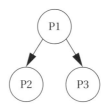

Process	CPU time
P1	2
P2	1
P3	2

Q6-22 For the periodic processes given below, find a valid schedule

 a. using standard RMS, and

 b. adding one unit of overhead for each context switch.

Process	CPU time	Deadline
P1	2	30
P2	4	40
P3	7	120
P4	5	60
P5	1	15

Q6-23 For the periodic processes and deadlines given below:

 a. Schedule the processes using RMS.

 b. Schedule using EDF and compare the number of context switches required for EDF and RMS.

Process	CPU time	Deadline
P1	1	5
P2	1	10
P3	2	20
P4	9	50
P5	7	100

Q6-24 In each circumstance below, would shared memory or message passing communication be better? Explain.

 a. A cascaded set of digital filters.

 b. A digital video decoder and a process that overlays user menus on the display.

 c. A software modem process and a printing process in a fax machine.

Q6-25 If you wanted to reduce the cache conflicts between the most computationally intensive parts of two processes, what are two ways that you could control the locations of the processes' cache footprints?

Q6-26 Draw a state diagram for the predictive shutdown mechanism of a cell phone. The cell phone wakes itself up once every five minutes for 0.01 second to listen for its address. It goes back to sleep if it does not hear its address or after it has received its message.

Q6-27 How would you use the ADPCM method to encode an unvarying (DC) signal with the coding alphabet $\{-3, -2, -1, 1, 2, 3\}$?

LAB EXERCISES

L6-1 Using your favorite operating system, write code to spawn a process that writes "Hello, world" to the screen or flashes an LED, depending on your available output devices.

L6-2 Build a small serial port device that lights LEDs based on the last character written to the serial port. Create a process that will light LEDs based on keyboard input.

L6-3 Write a driver for an I/O device.

L6-4 Write context switch code for your favorite CPU.

L6-5 Measure context switching overhead on an operating system.

L6-6 Using a CPU that runs an operating system that uses RMS, try to get the CPU utilization up to 100%. Vary the data arrival times to test the robustness of the system.

L6-7 Using a CPU that runs an operating system that uses EDF, try to get the CPU utilization as close to 100% as possible without failing. Try a variety of data arrival times to determine how sensitive your process set is to environmental variations.

Multiprocessors

- Why we design and use multiprocessors.
- Accelerators and hardware/software co-design.
- Performance analysis.
- Architectural templates.
- Architecture design: scheduling and allocation.
- Multiprocessor performance analysis.
- A video accelerator design.

INTRODUCTION

Multiprocessing—using computers that have more than one processor—has a long history in embedded computing. A surprising number of embedded systems are built on multiprocessor platforms. In fact, many of the least expensive embedded systems are built on sophisticated multiprocessors. Battery-powered devices that must deliver high performance at very low energy rates generally rely on multiprocessor platforms; this description fits a large part of the consumer electronics space.

The next section discusses why multiprocessors make sense for embedded systems. Section 7.2 introduces accelerators, a particular type of unit used in embedded multiprocessor systems and surveys the design process for accelerated and multiprocessors systems. Section 7.3 considers performance analysis of accelerators and multiprocessors. The next five sections discuss examples of real-world embedded multiprocessors in consumer electronics: Section 7.4 discusses some general properties of the architecture of consumer electronics devices; Section 7.5 describes cell phones; Section 7.6 discusses CD players; Section 7.7 describes audio players; and Section 7.8 describes digital still cameras. Section 7.9 designs a video accelerator as an example of an accelerated embedded system.

7.1 WHY MULTIPROCESSORS?

Programming a single CPU is hard enough. Why make life more difficult by adding more processors? A ***multiprocessor*** is, in general, any computer system with

two or more processors coupled together. Multiprocessors used for scientific or business applications tend to have regular architectures: several identical processors that can access a uniform memory space. We use the term ***processing element (PE)*** to mean any unit responsible for computation, whether it is programmable or not.

Embedded system designers must take a more general view of the nature of multiprocessors. As we will see, embedded computing systems are built on top of an astonishing array of different multiprocessor architectures.

Why is there no single multiprocessor architecture for all types of embedded computing applications? And why do we need embedded multiprocessors at all? The reasons for multiprocessors are the same reasons that drive all of embedded system design: real-time performance, power consumption, and cost.

The first reason for using an embedded multiprocessor is that they offer significantly better cost/performance—that is, performance and functionality per dollar spent on the system—than would be had by spending the same amount of money on a uniprocessor system. The basic reason for this is that processing element purchase price is a *nonlinear* function of performance [Wol08]. The cost of a microprocessor increases greatly as the clock speed increases. We would expect this trend as a normal consequence of VLSI fabrication and market economics. Clock speeds are normally distributed by normal variations in VLSI processes; because the fastest chips are rare, they naturally command a high price in the marketplace.

Because the fastest processors are very costly, splitting the application so that it can be performed on several smaller processors is usually much cheaper. Even with the added costs of assembling those components, the total system comes out to be less expensive. Of course, splitting the application across multiple processors does entail higher engineering costs and lead times, which must be factored into the project.

In addition to reducing costs, using multiple processors can also help with real-time performance. We can often meet deadlines and be responsive to interaction much more easily when we put those time-critical processes on separate processors. Given that scheduling multiple processes on a single CPU incurs overhead in most realistic scheduling models, as discussed in Chapter 6, putting the time-critical processes on PEs that have little or no time-sharing reduces scheduling overhead. Because we pay for that overhead at the nonlinear rate for the processor, as illustrated in Figure 7.1, the savings by segregating time-critical processes can be large—it may take an extremely large and powerful CPU to provide the same responsiveness that can be had from a distributed system.

Many of the technology trends that encourage us to use multiprocessors for performance also lead us to multiprocessing for low power embedded computing. Several processors running at slower clock rates consume less power than a single large processor: performance scales linearly with power supply voltage but power scales with V^2.

Austin *et al.* [Aus04] showed that general-purpose computing platforms are not keeping up with the strict energy budgets of battery-powered embedded

FIGURE 7.1

Scheduling overhead is paid for at a nonlinear rate.

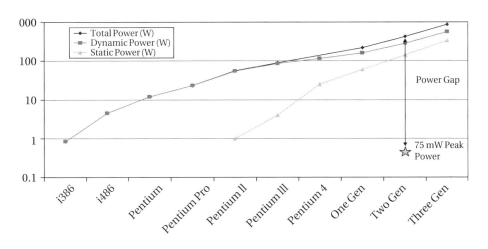

FIGURE 7.2

Power consumption trends for desktop processors [Aus04]. © 2004 IEEE Computer Society.

computing. Figure 7.2 compares the performance of power requirements of desktop processors with available battery power. Batteries can provide only about 75 mW of power. Desktop processors require close to 1000 times that amount of power to run. That huge gap cannot be solved by tweaking processor architectures or software. Multiprocessors provide a way to break through this power barrier and build substantially more efficient embedded computing platforms.

7.2 CPUs AND ACCELERATORS

One important category of PE for embedded multiprocessor is the ***accelerator***. An accelerator is attached to CPU buses to quickly execute certain key functions. Accelerators can provide large performance increases for applications with ***computational kernels*** that spend a great deal of time in a small section of code. Accelerators can also provide critical speedups for low-latency I/O functions.

The design of accelerated systems is one example of ***hardware/software co-design***—the simultaneous design of hardware and software to meet system objectives. Thus far, we have taken the computing platform as a given; by adding accelerators, we can customize the embedded platform to better meet our application's demands.

As illustrated in Figure 7.3, a CPU accelerator is attached to the CPU bus. The CPU is often called the ***host***. The CPU talks to the accelerator through data and control registers in the accelerator. These registers allow the CPU to monitor the accelerator's operation and to give the accelerator commands.

The CPU and accelerator may also communicate via shared memory. If the accelerator needs to operate on a large volume of data, it is usually more efficient to leave the data in memory and have the accelerator read and write memory directly rather than to have the CPU shuttle data from memory to accelerator registers and back. The CPU and accelerator use synchronization mechanisms like those described in Section 6.5 to ensure that they do not destroy each other's data.

FIGURE 7.3

CPU accelerators in a system.

An accelerator is not a co-processor. A co-processor is connected to the internals of the CPU and processes instructions as defined by opcodes. An accelerator interacts with the CPU through the programming model interface; it does not execute instructions. Its interface is functionally equivalent to an I/O device, although it usually does not perform input or output.

Both CPUs and accelerators perform computations required by the specification; at some level we do not care whether the work is done on a programmable CPU or on a hardwired unit.

The first task in designing an accelerator is determining that our system actually needs one. We have to make sure that the function we want to accelerate will run more quickly on our accelerator than it will by executing as software on a CPU. If our system CPU is a small microcontroller, the race may be easily won, but competing against a high-performance CPU is a challenge. We also have to make sure that the accelerated function will speed up the system. If some other operation is in fact the bottleneck, or if moving data into and out of the accelerator is too slow, then adding the accelerator may not be a net gain.

Once we have analyzed the system, we need to design the accelerator itself. In order to have identified our need for an accelerator, we must have a good understanding of the algorithm to be accelerated, which is often in the form of a high-level language program. We must translate the algorithm description into a hardware design, a considerable task in itself. We must also design the interface between the accelerator core and the CPU bus. The interface includes more than bus handshaking logic. For example, we have to determine how the application software on the CPU will communicate with the accelerator and provide the required registers; we may have to implement shared memory synchronization operations; and we may have to add address generation logic to read and write large amounts of data from system memory.

Finally, we will have to design the CPU-side interface to the accelerator. The application software will have to talk to the accelerator, providing it data and telling it what to do. We have to somehow synchronize the operation of the accelerator with the rest of the application so that the accelerator knows when it has the required data and the CPU knows when it has received the desired results.

7.2.1 System Architecture Framework

The complete architectural design of the accelerated system depends on the application being implemented. However, it is helpful to think of an ***architectural framework*** into which our accelerator fits. Because the same basic techniques for connecting the CPU and accelerator can be applied to many different problems, understanding the framework helps us quickly identify what is unique about our application.

An accelerator can be considered from two angles: its core functionality and its interface to the CPU bus. We often start with the accelerator's basic functionality and work our way out to the bus interface, but in some cases the bus interface and

the internal logic are closely intertwined in order to provide high-performance data access.

The accelerator core typically operates off internal registers. How many registers are required is an important design decision. Main memory accesses will probably take multiple clock cycles, slowing down the accelerator. If the algorithm to be accelerated can predict which data values it will use, the data can be prefetched from main memory and stored in internal registers.

The accelerator will almost certainly use registers for basic control. Status registers like those of I/O devices are a good way for the CPU to test the accelerator's state and to perform basic operations such as starting, stopping, and resetting the accelerator.

Large-volume data transfers may be performed by special-purpose read/write logic. Figure 7.4 illustrates an accelerator with read/write units that can supply higher volumes of data without CPU intervention. A register file in the accelerator acts as a buffer between main memory and the accelerator core. The read unit can read ahead of the accelerator's requirements and load the registers with the next required data; similarly, the write unit can send recently completed values to main memory while the core works with other values. In order to avoid tying up the CPU, the data transfers can be performed in DMA mode, which means that the accelerator must have the required logic to become a bus master and perform DMA operations.

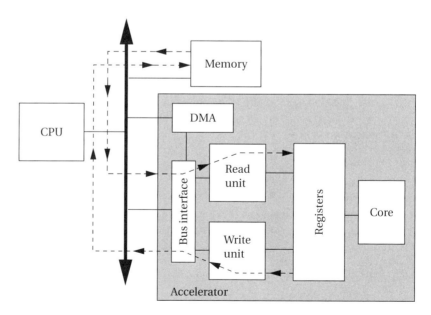

FIGURE 7.4

Read/write units in an accelerator.

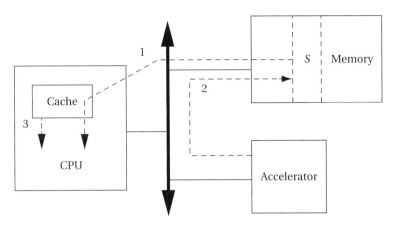

FIGURE 7.5

A cache updating problem in an accelerated system.

The CPU cache can cause problems for accelerators. Consider the following sequence of operations as illustrated in Figure 7.5:

1. The *CPU* reads location *S*.

2. The *accelerator* writes *S*.

3. The *CPU* again reads *S*.

If the *CPU* has cached location *S*, the program will not see the value of *S* written by the *accelerator*. It will instead get the old value of *S* stored in the *cache*. To avoid this problem, the CPU's cache must be updated to reflect the fact that this cache entry is invalid. Your CPU may provide cache invalidation instructions; you can also remove the location from the cache by reading another location that is mapped to the same cache line (or, in the case of set-associative caches, enough such locations to replace all the cache sets). Some CPUs are designed to support multiprocessing. The bus interface of such machines provides mechanisms for other processors to tell the CPU of required cache changes. This mechanism can be used by the accelerator to update the cache.

If the CPU and accelerator operate concurrently and communicate via shared memory, it is possible that similar problems will occur in main memory, not just in the cache. If one PE reads a value and then updates it, the other PE may change the value, causing the first PE's update to be invalid. In some cases, it may be possible to use a very simple synchronization scheme for communication: the CPU writes data into a memory buffer, starts the accelerator, waits for the accelerator to finish, and then reads the shared memory area. This amounts to using the accelerator's status registers as a simple semaphore system. If the CPU and accelerator both want access to the same block of memory at the same time, then the accelerator will need to implement a test-and-set operation in order to implement semaphores. Many CPU

buses implement test-and-set atomic operations that the accelerator can use for the semaphore operation.

7.2.2 System Integration and Debugging

Design of an accelerated system requires both designing your own components and interfacing them to a hardware platform. It is usually a good policy to separately debug the basic interface between the accelerator and the rest of the system before integrating the full accelerator into the platform.

Hardware/software co-simulation can be very useful in accelerator design. Because the co-simulator allows you to run software relatively efficiently alongside a hardware simulation, it allows you to exercise the accelerator in a realistic but simulated environment. It is especially difficult to exercise the interface between the accelerator core and the host CPU without running the CPU's accelerator driver. It is much better to do so in a simulator before fabricating the accelerator, rather than to have to modify the hardware prototype of the accelerator.

7.3 MULTIPROCESSOR PERFORMANCE ANALYSIS

Analyzing the performance of a system with multiple processors is not easy. We saw a glimpse of some of the difficulties in Section 4.7 when we studied the performance of a simple system with a CPU, an I/O device, and a bus. That basic uniprocessor architecture still shows some opportunity for parallelism. In this section we will consider multiprocessor performance in more detail. We will start by analyzing accelerators, then move on to more general instances of multiprocessors.

7.3.1 Accelerators and Speedup

The most basic question that we can ask about our accelerator is *speedup*: how much faster is the system with the accelerator than the system without it? We may, of course, be concerned with other metrics such as power consumption and manufacturing cost. However, if the accelerator does not provide an attractive speedup, questions of cost and power will be moot.

The speedup factor depends in part on whether the system is *single threaded* or *multithreaded*, that is, whether the CPU sits idle while the accelerator runs in the single-threaded case or the CPU can do useful work in parallel with the accelerator in the multithreaded case. Another equivalent description is *blocking* vs. *nonblocking*. Does the CPU's scheduler block other operations and wait for the accelerator call to complete, or does the CPU allow some other process to run in parallel with the accelerator? The possibilities are shown in Figure 7.6. Data dependencies allow P2 and P3 to run independently on the CPU, but P2 relies on the results of the A1 process that is implemented by the accelerator. However, in the single-threaded case, the CPU blocks to wait for the accelerator to return the results of its computation. As a result, it does not matter whether P2 or P3 runs next on the CPU. In the multithreaded case, the CPU continues to do useful work while

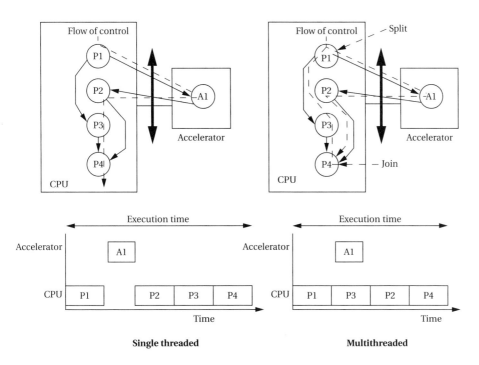

FIGURE 7.6

Single-threaded versus multithreaded control of an accelerator.

the accelerator runs, so the CPU can start P3 just after starting the accelerator and finish the task earlier.

The first task is to analyze the performance of the accelerator. As illustrated in Figure 7.7, the execution time for the accelerator depends on more than just the time required to execute the accelerator's function. It also depends on the time required to get the data into the accelerator and back out of it. Since the CPU's registers are probably not addressable by the accelerator, the data probably reside in main memory.

A simple accelerator will read all its input data, perform the required computation, and then write all its results. In this case, the total execution time may be written as

$$t_{accel} = t_{in} + t_x + t_{out} \tag{7.1}$$

where t_x is the execution time of the accelerator assuming all data are available, and t_{in} and t_{out} are the times required for reading and writing the required variables, respectively. The values for t_{in} and t_{out} must reflect the time required for the bus transactions, including the following factors:

- the time required to flush any register or cache values to main memory, if those values are needed in main memory to communicate with the accelerator; and

- the time required for transfer of control between the CPU and accelerator.

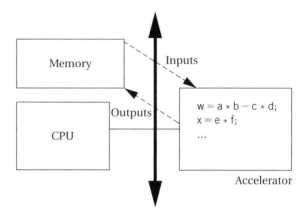

FIGURE 7.7

Components of execution time for an accelerator.

Transferring data into and out of the accelerator may require the accelerator to become a bus master. Since the CPU may delay bus mastership requests, some worst-case value for bus mastership acquisition must be determined based on the CPU characteristics.

A more sophisticated accelerator could try to overlap input and output with computation. For example, it could read a few variables and start computing on those values while reading other values in parallel. In this case, the t_{in} and t_{out} terms would represent the nonoverlapped read/write times rather than the complete input and output times. One important example of overlapped I/O and computation is streaming data applications such as digital filtering. As illustrated in Figure 7.8, an accelerator may take in one or more streams of data and output a stream. Latency requirements generally require that outputs be produced on the fly rather than storing up all the data and then computing; furthermore, it may be impractical to store long streams at all. In this case, the t_{in} and t_{out} terms are determined by the amount of data read in before starting computation and the length of time between the last computation and the last data output. We discussed the performance of bus-based systems with overlapped communication and computation in Section 4.7.

We are most interested in the speedup obtained by replacing the software implementation with the accelerator. The total speedup S for a kernel can be written as [Hen94]:

$$S = n(t_{CPU} - t_{accel})$$
$$= n[t_{CPU} - (t_{in} + t_x + t_{out})] \tag{7.2}$$

where t_{CPU} is the execution time of the equivalent function in software on the CPU and n is the number of times the function will be executed. We can use the techniques of Chapter 5 to determine the value of t_{CPU}. Clearly, the more times the function is evaluated, the more valuable the speedup provided by the accelerator becomes.

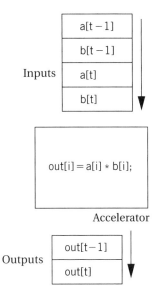

Inputs

| a[t − 1] |
| b[t − 1] |
| a[t] |
| b[t] |

out[i] = a[i] * b[i];

Accelerator

Outputs

| out[t−1] |
| out[t] |

FIGURE 7.8

Streaming data in and out of an accelerator.

Ultimately, we don't care so much about the accelerator's speedup as the speedup for the complete system—that is, how much faster the entire application completes execution. In a single-threaded system, the evaluation of the accelerator's speedup to the total system speedup is simple: The system execution time is reduced by S. The reason is illustrated in Figure 7.9—the single thread of control gives us a single path whose length we can measure to determine the new execution speed.

Evaluating system speedup in a multithreaded environment requires more subtlety. As shown in Figure 7.10, there is now more than one execution path. The total system execution time depends on the ***longest path*** from the beginning of execution to the end of execution. In this case, the system execution time depends on the relative speeds of P3 and P2 plus A1. If P2 and A1 together take the most time, P3 will not play a role in determining system execution time. If P3 takes longer, then P2 and A1 will not be a factor. To determine system execution time, we must label each node in the graph with its execution time.

In simple cases we can enumerate the paths, measure the length of each, and select the longest one as the system execution time. Efficient graph algorithms can also be used to compute the longest path.

This analysis shows the importance of selecting the proper functions to be moved to the accelerator. Clearly, if the function selected for speedup isn't a big portion of system execution time, taking the number of times it is executed into account, you won't see much system speedup. We also learned from Equation 7.1 that if too

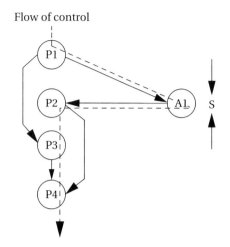

FIGURE 7.9

Evaluating system speedup in a single-threaded implementation.

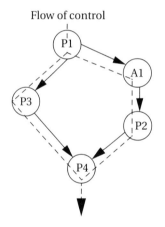

FIGURE 7.10

Evaluating system speedup in a multithreaded implementation.

much overhead is incurred getting data into and out of the accelerator, we won't see much speedup.

7.3.2 Performance Effects of Scheduling and Allocation

When we design a multiprocessor system, we must allocate tasks to PEs; we must also schedule both the computations on the PEs and schedule the communication between the processes on the buses in the system. The next example considers the interaction between scheduling and allocation in a two-processor system.

Example 7.1

Performance effects of scheduling and allocation

We want to execute a simple task graph:

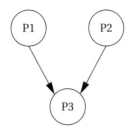

We want to execute it on a platform that has two processors connected by a bus:

One obvious way to allocate the tasks to the processors would be by precedence: put P1 and P2 onto M1; put the task that receives their outputs, namely P3, onto M2. When we look at the schedule for this system, we see that M2 sits idle for quite some time:

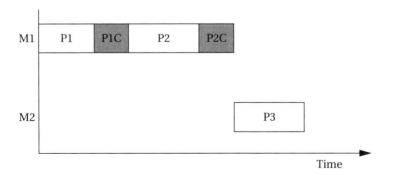

In this timing graph, P1C is the time required to communicate P1's output to P3 and P2C is the communication time for P2 to P3. M2 sits idle as P3 waits for its inputs.

Let's change the allocation so that P1 runs on M1 while P2 and P3 run on M2. This gives us a new schedule:

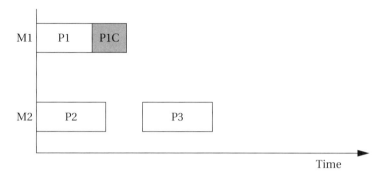

Eliminating P2C gives us some benefit, but the biggest benefit comes from the fact that P1 and P2 run concurrently.

If we can change the code for our tasks, then we can extract even more opportunities for parallelism. The next example looks at how to split computations into smaller pieces to expose more parallelism opportunities.

Example 7.2

Overlapping computation and communication

In some cases, we can redesign our computations to increase the available parallelism. Assume we want to implement the following task graph:

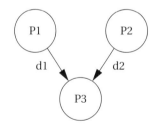

Assume also that we want to implement the task graph on this network:

We will allocate P1 to M1, P2 to M2, and P3 to M3. P1 and P2 run for three time units while P3 runs for four time units. A complete transmission of either d1 or d2 takes four time units.

The task graph shows that P3 cannot start until it receives its data from both P1 and P2 over the bus network.

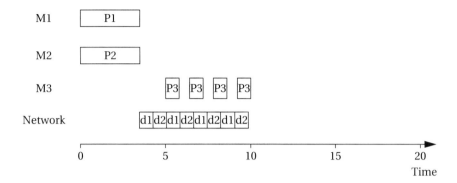

The simplest implementation transmits all the required data in one large message, which is four packets long in this case. Appearing below is a schedule based on that message structure.

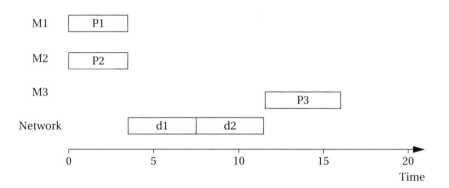

P3 does not start until time 11, when the transmission of the second message has been completed. The total schedule length is 15.

Let's redesign P3 so that it does not require all of both messages to begin. We modify the program so that it reads one packet of data each from d1 and d2 and start computing on that. If it finishes what it can do on that data before the next packets from d1 and d2 arrive, it waits; otherwise, it picks up the packets and keeps computing. This organization allows us to take advantage of concurrency between the M3 processing element (PE) and the network as shown by the schedule below.

Reorganizing the messages so that they can be sent concurrently with P3's execution reduces the schedule length from 15 to 12, even with P3 stopping to wait for more data from P1 and P2.

7.3.3 Buffering and Performance

Moving data in a multiprocessor can incur significant and sometimes unpredictable costs. When we move data in a uniprocessor, we are copying from one part of memory to another, we are doing so within the same memory system. When we move data in a multiprocessor, we may exercise several different parts of the system, and we have to be careful to understand the costs of those transfers.

Consider, as an example, copying an array. If the source and destination are in different memories, then the data transfer rate will be limited by the slowest element along the path: the source memory, the bus, or the destination memory. The energy required to copy the data will be the sum of the energy costs of all those components.

The schedule that we use for the transfers also affects latency, as illustrated by the next example.

Example 7.3

Buffers and latency

Our system needs to process data in three stages:

The data arrives in blocks of *n* data elements, so we use buffers in between the stages. Since the data arrives in blocks and not one item at a time, we have some flexibility in the order in which we process the blocks. Perhaps the easiest schedule for data processing does all the A operations, then all the Bs, then all the Cs:

```
A[0]
A[1]
. . .
a[n-1]
B[0]
B[1]
. . .
C[0]
C[1]
. . .
```

Note that no output is generated until after all of the A and B operations have finished—the C[0] output is the first to be generated after $2n + 1$ operations have been performed. It then produces all of the outputs on successive cycles (assuming, for simplicity, that the operations each take one clock cycle).

But it is not necessary to wait so long for some data. Consider this schedule:

```
A[0]
B[0]
C[0]
A[1]
B[1]
C[1]
. . .
```

This schedule generates the first output after three cycles and generates new outputs every three cycles thereafter.

Equally important, as we include more components in the transfer, we introduce more opportunities for interruptions and variations in execution time. Any resource that is shared may be subject to delays caused by other processes that use the resource. Buses may handle other transfers; memories may also be shared among several processors.

7.4 CONSUMER ELECTRONICS ARCHITECTURE

Although some predict the complete convergence of all consumer electronic functions into a single device, much as the personal computer now relies on a common platform, we still have a variety of devices with different functions. However, consumer electronics devices have converged over the past decade around a set of common features that are supported by common architectural features. Not all devices have all features, depending on the way the device is to be used, but most devices select features from a common menu. Similarly, there is no single platform for consumer electronics devices, but the architectures in use are organized around some common themes.

This convergence is possible because these devices implement a few basic types of functions in various combinations: multimedia, communications, and data storage and management. The style of multimedia or communications may vary, and different devices may use different formats, but this causes variations in hardware and software components within the basic architectural templates. In this section we will look at general features of consumer electronics devices; in the following sections we will study a few devices in more detail.

7.4.1 Use Cases and Requirements

Consumer electronics devices provide several types of services in different combinations:

- *Multimedia:* The media may be audio, still images, or video (which includes both motion pictures and audio). These multimedia objects are generally

stored in compressed form and must be uncompressed to be played (audio playback, video viewing, etc.). A large and growing number of standards have been developed for multimedia compression: MP3, Dolby Digital(TM), etc. for audio; JPEG for still images; MPEG-2, MPEG-4, H.264, etc. for video.

- *Data storage and management:* Because people want to select what multimedia objects they save or play, data storage goes hand-in-hand with multimedia capture and display. Many devices provide PC-compatible file systems so that data can be shared more easily.

- *Communications:* Communications may be relatively simple, such as a USB interface to a host computer. The communications link may also be more sophisticated, such as an Ethernet port or a cellular telephone link.

Consumer electronics devices must meet several types of strict nonfunctional requirements as well. Many devices are battery-operated, which means that they must operate under strict energy budgets. A typical battery for a portable device provides only about 75 mW, which must support not only the processors and digital electronics but also the display, radio, etc. Consumer electronics must also be very inexpensive. A typical primary processing chip must sell in the neighborhood of $10. These devices must also provide very high performance—sophisticated networking and multimedia compression require huge amounts of computation.

Let's consider some basic use cases of some basic operations. Figure 7.11 shows a use case for selecting and playing a multimedia object (an audio clip, a picture, etc.). Selecting an object makes use of both the user interface and the file system. Playing also makes use of the file system as well as the decoding subsystem and I/O subsystem.

Figure 7.12 shows a use case for connecting to a client. The connection may be either over a local connection like USB or over the Internet. While some operations

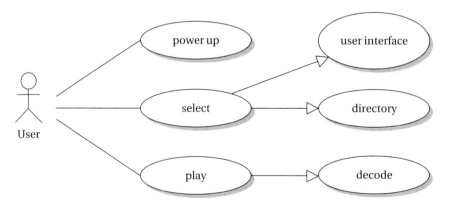

FIGURE 7.11

Use case for playing multimedia.

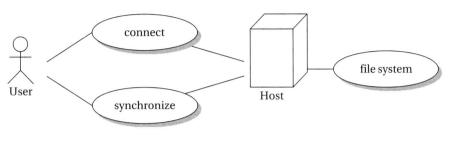

FIGURE 7.12

Use case of synchronizing with a host system.

FIGURE 7.13

Functional architecture of a generic consumer electronics device.

may be performed locally on the client device, most of the work is done on the host system while the connection is established.

7.4.2 Platforms and Operating Systems

Given these types of usage scenarios, we can deduce a few basic characteristics of the underlying architecture of these devices. Figure 7.13 shows a functional block diagram of a typical device. The storage system provides bulk, permanent storage. The network interface may provide a simple USB connection or a full-blown Internet connection.

Multiprocessor architectures are common in many consumer multimedia devices. Figure 7.13 shows a two-processor architecture; if more computation is required, more DSPs and CPUs may be added. The RISC CPU runs the operating system, runs the user interface, maintains the file system, etc. The DSP performs signal processing. The DSP may be programmable in some systems; in other cases, it may be one or more hardwired accelerators.

The operating system that runs on the CPU must maintain processes and the file system. Processes are necessary to provide concurrency—for example, the user wants to be able to push a button while the device is playing back audio. Depending on the complexity of the device, the operating system may not need to create tasks dynamically. If all tasks can be created using initialization code, the operating system can be made smaller and simpler.

7.4.3 Flash File Systems

Many consumer electronics devices use flash memory for mass storage. Flash memory is a type of semiconductor memory that, unlike DRAM or SRAM, provides permanent storage. Values are stored in the flash memory cell as electric charge using a specialized capacitor that can store the charge for years. The flash memory cell does not require an external power supply to maintain its value. Furthermore, the memory can be written electrically and, unlike previous generations of electrically-erasable semiconductor memory, can be written using standard power supply voltages and so does not need to be disconnected during programming.

Disk drives, which use rotating magnetic platters, are the most common form of mass storage in PCs. Disk drives have some advantages: they are much cheaper than flash memory (at this writing, disk storage costs $0.50 per gigabyte, while flash memory is slightly less than $50/gigabyte) and they have much greater capacity. But disk drives also consume more power than flash storage. When devices need a moderate amount of storage, they often use flash memory.

The file system of a device is typically shared with a PC. In many cases the memory device is read directly by the PC through a flash card reader or a USB port. The device must therefore maintain a PC-compatible file system, using the same directory structure, file names, etc. as are used on a PC.

However, flash memory has one important limitation that must be taken into account. Writing a flash memory cell causes mechanical stress that eventually wears out the cell. Today's flash memories can reliably be written a million times but at some point they will fail. While a million write cycles may sound like enough to ensure that the memory will never wear out, creating a single file may require many write operations, particularly to the part of the memory that stores the directory information.

A wear-leveling flash file system [Ban95] manages the use of flash memory locations to equalize wear while maintaining compatibility with existing file systems. A simple model of a standard file system has two layers: the bottom layer handles physical reads and writes on the storage device; the top layer provides a logical view of the file system. A flash file system imposes an intermediate layer that allows the logical-to-physical mapping of files to be changed. This layer keeps track of how frequently different sections of the flash memory have been written and allocates data to equalize wear. It may also move the location of the directory structure while the file system is operating. Because the directory system receives the most

wear, keeping it in one place may cause part of the memory to wear out before the rest, unnecessarily reducing the useful life of the memory device. Several flash file systems have been developed, such as Yet Another Flash Filing System (YAFFS) [Ale05].

7.5 CELL PHONES

Design Example

The cell phone is the most popular consumer electronics device in history. The Motorola DynaTAC portable cell phone was introduced in 1973. Today, about one billion cell phones are sold each year. The cell phone is part of a larger cellular telephony network, but even as a standalone device the cell phone is a sophisticated instrument.

As shown in Figure 7.14, cell phone networks are built from a system of base stations. Each base station has a coverage area known as a *cell*. A **handset** belonging to a user establishes a connection to a base station within its range. If the cell phone moves out of range, the base stations arrange to hand off the handset to another base station. The *handoff* is made seamlessly without losing service.

A cell phone performs several very different functions:

- It transmits and receives digital data over a radio and may provide analog voice service as well.

- It executes a protocol that manages its relationship to the cellular network.

- It provides a basic user interface to the cell phone.

- It performs some functions of a PC, such as contact management, multimedia capture and playback, etc.

Let's understand these functions one at a time.

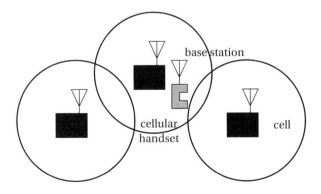

FIGURE 7.14

Cells in a cellular telephone network.

Early cell phones transmitted voice using analog methods. Today, analog voice is used only in low-cost cell phones, primarily in the developing world; the voice signal in most systems is transmitted digitally. A wireless data link must perform two basic functions: it must modulate or demodulate the data during transmission or reception; and it must correct errors using error correcting codes.

Today's cell phones generally use traditional radios that use analog and digital circuits to modulate and demodulate the signal and decode the bits during reception. A processor in the cell phone sets various radio parameters, such as power level and frequency. However, the processor does not process the radio frequency signal itself.

As low power, high performance processors become available, we will see more cell phones perform at least some of the radio frequency processing in programmable processors. This technique is often called *software radio* or *software-defined radio (SDR)*. SDR helps the cell phone support multiple standards and a wider variety of signal processing parameters.

Error correction algorithms detect and correct errors in the raw data stream. Radio channels are sufficiently noisy that powerful error correction algorithms are necessary to provide reasonable service. Error correction algorithms, such as Viterbi coding or turbo coding, require huge amounts of computation. Many handset platforms provide specialized hardware to implement error correction.

Many cell phone standards transmit compressed audio. The audio compression algorithms have been optimized to provide adequate speech quality. The handset must compress the audio stream before sending it to the radio and must decompress the audio stream during reception.

The network protocol that manages the communication between the cell phone and the network performs several tasks: it sets up and tears down calls; it manages the hand-off when a handset moves from one base station to another; it manages the power at which the cell phone transmits, etc.

The protocol's events are generated at a fairly low rate. These events can be handled by a CPU. The protocol itself is implemented in software that is handed from project to project. Since the network protocols change very slowly, this software is a prime candidate for reuse.

The cell phone may also be used as a data connection for a computer. In this case, the handset must perform a separate protocol to manage the data flow to and from the PC.

The basic user interface for a cell phone is straightforward: a few buttons and a simple display. Early cell phones used microcontrollers to implement their user interface.

However, modern cell phones do much more than make phone calls. Cell phones have taken over many of the functions of the PDA, such as contact lists and calendars. Even mid-range cell phones not only play audio and image or video files, they can also capture still images and video using built-in cameras. They provide these functions using a graphical user interface.

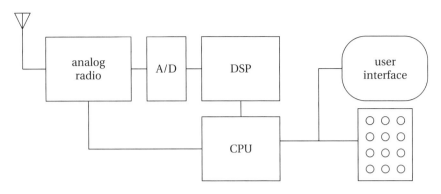

FIGURE 7.15

Baseband processing in cell phones.

Figure 7.15 shows a sketch of the architecture of a typical high-end cell phone. The radio frequency processing is performed in analog circuits. The baseband processing is handled by a combination of a RISC-style CPU and a DSP. The CPU runs the host operating system and handles the user interface, controlling the radio, and a variety of other control functions. The DSP performs signal processing: audio compression and decompression, multimedia operations, etc. The DSP can perform the signal processing functions at lower power consumption levels than can the RISC processor. The CPU acts as the master, sending requests to the DSP.

7.6 COMPACT DISCs AND DVDs

Design Example

Compact DiscTM was introduced in 1980 to provide a mass storage medium for digital audio. It has since become widely used for general purpose data storage and to record MP3 files for playback. Compact discs use optical storage—the data is read off the disc using a laser. The design of the CD system is a triumph of signal processing over mechanics—CD players perform a great deal of signal processing to compensate for the limitations of a cheap, inaccurate player mechanism. The DVDTM and more recently, Blu-RayTM provide higher density optical storage. However, the basic principles governing their operation are the same as those for CD. In this section we will concentrate on the CD as an example of optical disc technology.

As shown in Figure 7.16, data is stored in pits on the bottom of a compact disc. A laser beam is reflected or not reflected by the absence or presence of a pit. The pits are very closely spaced: pits range from 0.8 to 3 μm long and 0.5 μm wide. The pits are arranged in tracks with 1.6 μm between adjacent tracks.

Unlike magnetic disks, which arrange data in concentric circles, CD data is stored in a spiral as shown in Figure 7.17. The spiral organization makes sense if the data is

FIGURE 7.16

Data stored on a compact disc.

FIGURE 7.17

Spiral data organization of a compact disc.

to be played from beginning to end. But as we will see, the spiral complicates some aspect of CD operation.

The data on a CD is divided into sectors. Each sector has an address so that the drive can determine its location on the CD. Sectors also contain several bits of control: P is 1 during music or lead-in and 0 at the start of a selection; Q contains track number, time, etc.

The compact disc mechanism is shown in Figure 7.18. A sled moves radially across the CD to be positioned at different points in the spiral data. The sled carries a laser, optics, and a photo detector. The laser illuminates the CD through the optics. The same optics capture the reflected light and pass it onto the photo detector.

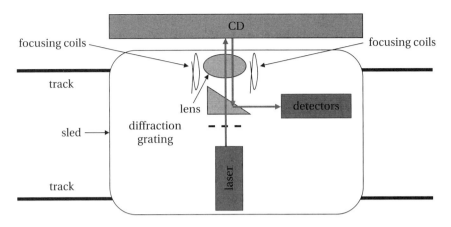

FIGURE 7.18

A compact disc mechanism.

Out-of-focus In-focus Out-of-focus

FIGURE 7.19

Laser focusing in a CD.

The optics can be focused using some simple electric coils. Laser focus adjusts for variations in the distance to the CD. As shown in Figure 7.19, an in-focus beam produces a circular spot, while an out-of-focus beam produces an elliptical spot with the beam's major axis indicating the direction of focus. The focus can change relatively quickly depending on how the CD is seated on the spindle, so the focus needs to be continuously adjusted.

As shown in Figure 7.20, the laser pickup is divided into six regions, named A, B, C, D, E, and F. The basic four regions—A, B, C, and D—are used to determine whether the laser is focused. The focus error signal is $(A + C) - (B + D)$. The magnitude of the signal gives the amount of focus error and the sign determines the orientation of the elliptical spot's major axis. The sum of the four basic regions, $A + B + C + D$, gives the laser level to determine whether a pit is being illuminated. Two additional detectors, E and F, are used to determine when the laser has gone far off the track. Tracking error is given by $E - F$.

The sled, focus system, and detector form a servo system. Several different systems must be controlled: laser focus and tracking must each be controlled at a sample

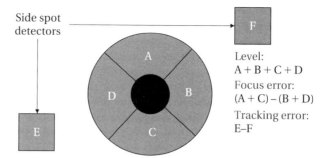

FIGURE 7.20

CD laser pickup regions.

rate of 245 kHz; the sled is controlled at 800 Hz. Control algorithms monitor the level and error signals and determine how to adjust focus, tracking, and sled signals. These control algorithms are very sophisticated. Each control may require digital filters with 30 or more coefficients. Several control modes must be programmed, such as seeking vs. playback. The development of the control algorithms usually requires several person-years of effort.

The servo control algorithms are generally performed on a programmable DSP. Although a CD is a very low power device which could benefit from the lower energy consumption of hardwired servo control, the complexity of the servo algorithms requires programmability. Not only are the algorithms complex, but different CD mechanisms may require different control algorithms.

The complete control system for the drive requires more than simple closed-loop control of the data. For example, when a CD is bumped, the system must reacquire the proper position on the track. Because the track is arranged in a spiral, and because the sled mechanism is inaccurate, positioning the read head is harder than in a magnetic disk. The sled must be positioned to a point before the data's location; the system must start reading data and watch for the proper sector to appear, then start reading again.

The bits on the CD are not encoded directly. To help with tracking, the data stream must be organized to produce 0–1 transitions at some minimum interval. An **eight-to-fourteen (EFM) encoding** is used to ensure a minimum transition rate. For example, the 8 bits of user data 00000011 is mapped to the 14-bit code 00100100000000. The data are reconstructed from the EFM code using tables.

CD use powerful error correction codes to compensate for inexpensive CD manufacturing processes and problems during readback. A CD contains 6.99 GB of raw bits but provides only about 700 MB of formatted data. CDs use a form of Reed–Solomon coding; the codes are also block interleaved to reduce the effects of scratches and other bursty errors. Reed–Solomon decoding determines data and erasure bits. The time required to complete Reed–Solomon coding depends greatly

on the number of erasure bits. As a result, the system may declare an entire block to be bad if decoding takes too long. Error correction is typically performed by hardwired units.

CD players are very vulnerable to shaking. Early players could be disrupted by walking on the floor near the player. Clearly, portable or automotive players would need even stronger protection against mechanical disturbance. Memory is much cheaper today than it was when CD players were introduced. A **jog memory** is used to buffer data to maintain playing during a jog to the drive. The player reads ahead and puts data into the jog memory. During a jog, the audio output system reads data stored in the jog memory while the drive tries to find the proper point on the CD to continue reading.

Jog control memories also help reduce power consumption. The drive can read ahead, put a large block of data into the jog memory, then turn the drive off and play from jog memory. Because the drive motors consume a considerable amount of power, this strategy saves battery life. When reading compressed music from data discs, a large part of a song can be put into jog memory.

The result of error correction is the sector data. This can be easily parsed to determine the audio samples and control information. In the case of an audio disc, the samples may be directly provided to the audio output subsystem; some players use digital filters to perform part of the anti-aliasing filtering. In the case of a data disc, the sector data may be sent to the output registers.

Figure 7.21 shows the hardware architecture of a CD player. The player includes several processors: servo processor, error correction unit, and audio unit. These processors operate in parallel to process the stream of data coming from the read mechanism.

FIGURE 7.21

Hardware architecture of a CD player.

Writable CDs provide a pilot track that allows the laser and servo to position the head. The CD system must compute Reed–Solomon codes and EFM codes to feed the DVD. Data must be provided to the write system continuously, so the host system must properly buffer data to ensure that it can be delivered on time.

Several CD formats have been defined. Each standard is published in a separate document: the *Red Book* defines the CD digital audio standard; the *Yellow Book* defines CD-ROM; the *Orange Book* defines CD-RW.

Design Example

7.7 AUDIO PLAYERS

Audio players are often called ***MP3 players*** after the popular audio data format. The earliest portable MP3 players were based on compact disc mechanisms. Modern MP3 players use either flash memory or disk drives to store music.

An MP3 player performs three basic functions: audio storage, audio decompression, and user interface. Although audio compression is computationally intensive, audio decompression is relatively lightweight. The incoming bit stream has been encoded using a Huffman-style code, which must be decoded. The audio data itself is applied to a reconstruction filter, along with a few other parameters. MP3 decoding can, for example, be executed using only 10% of an ARM7 CPU.

The user interface of an MP3 player is usually kept simple to minimize both the physical size and power consumption of the device. Many players provide only a simple display and a few buttons.

FIGURE 7.22

Architecture of a Cirrus audio processor for CD/MP3 players.

The file system of the player generally must be compatible with PCs. CD/MP3 players used compact discs that had been created on PCs. Today's players can be plugged into USB ports and treated as disk drives on the host processor.

The Cirrus CS7410 [Cir04B] is an audio controller designed for CD/MP3 players. The audio controller includes two processors. The 32-bit RISC processor is used to perform system control and audio decoding. The 16-bit DSP is used to perform audio effects such as equalization. The memory controller can be interfaced to several different types of memory: flash memory can be used for data or code storage; DRAM can be used as a buffer to handle temporary disruptions of the CD data stream. The audio interface unit puts out audio in formats that can be used by A/D converters. General-purpose I/O pins can be used to decode buttons, run displays, etc. Cirrus provides a reference design for a CD/MP3 player [Cir04A].

7.8 DIGITAL STILL CAMERAS

Design Example

The digital still camera bears some resemblance to the film camera but is fundamentally different in many respects. The digital still camera not only captures images, it also performs a substantial amount of image processing that formerly was done by photofinishers.

Digital image processing allows us to fundamentally rethink the camera. A simple example is digital zoom, which is used to extend or replace optical zoom. Many cell phones include digital cameras, creating a hybrid imaging/communication device.

Digital still cameras must perform many functions:

- It must determine the proper exposure for the photo.

- It must display a preview of the picture for framing.

- It must capture the image from the image sensor.

- It must transform the image into usable form.

- It must convert the image into a usable format, such as JPEG, and store the image in a file system.

A typical hardware architecture for a digital still camera is shown in Figure 7.23. Most cameras use two processors. The controller sequences operations on the camera and performs operations like file system management. The DSP concentrates on image processing. The DSP may be either a programmable processor or a set of hardwired accelerators. Accelerators are often used to minimize power consumption.

The picture taking process can be divided into three main phases: composition, capture, and storage. We can better understand the variety of functions that must be performed by the camera through a sequence diagram. Figure 7.24 shows a

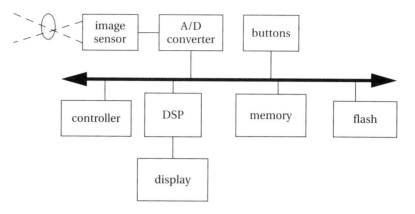

FIGURE 7.23

Architecture of a digital still camera.

sequence diagram for taking a picture using a point-and-shoot digital still camera. As we walk through this sequence diagram, we can introduce some concepts in digital photography.

When the camera is turned on, it must start to display the image on the camera's screen. That imagery comes from the camera's image sensor. To provide a reasonable image, it must adjust the image exposure. The camera mechanism provides two basic exposure controls: shutter speed and aperture. The camera also displays what is seen through the lens on the camera's display. In general, the display has fewer pixels than does the image sensor; the image processor must generate a smaller version of the image.

When the user depresses the shutter button, a number of steps occur. Before the image is captured, the final exposure must be determined. Exposure is computed by analyzing the image characteristics; histograms of the distribution of pixel brightness are often used to determine focus. The camera must also determine **white balance**. Different sources of light, such as sunlight and incandescent lamps, provide light of different colors. The eye naturally compensates for the color of incident light; the camera must perform comparable processing to avoid giving the picture a color cast. White balance algorithms generally use color histograms to determine the range of colors and re-weigh colors to reduce casts.

The image captured from the image sensor is not directly usable, even after exposure and white balance. Virtually all still cameras use a single image sensor to capture a color image. Color is captured using microscopic color filters, each the size of a pixel, over the image sensor. Since each pixel can capture only one color, the color filters must be arranged in a pattern across the image sensor. A commonly used pattern is the Bayer pattern [Bay75] shown in Figure 7.25. This pattern uses two greens for every red and blue pixel since the human eye is most sensitive to green. The camera must interpolate colors so that every pixel has red, green, and blue values.

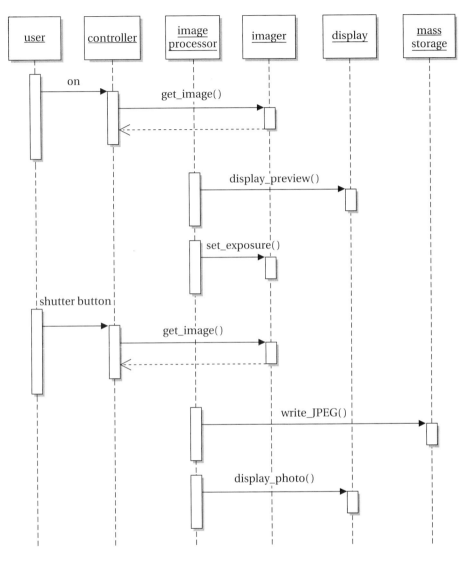

FIGURE 7.24

Sequence diagram for taking a picture with a digital still camera.

After this image processing is complete, the image must be compressed and saved. Images are often compressed in JPEG format, but other formats, such as GIF, may also be used. The EXIF standard (http://www.exif.org) defines a file format for data interchange. Standard compressed image formats such as JPEG are components of an EXIF image file; the EXIF file may also contain a thumbnail image for preview, metadata about the picture such as when it was taken, etc.

FIGURE 7.25

The Bayer pattern for color image pixels.

Image compression need not be performed strictly in real time. However, many cameras allow users to take a burst of images, in which case the images must be compressed quickly to make room in the image processing pipeline for the next image.

Buffering is very important in digital still cameras. Image processing often takes longer than capturing an image. Users often want to take a burst of several pictures, for example during sports events. A buffer memory is used to capture the image from the sensor and store it until it can be processed by the DSP [Sas91].

The display is often connected to the DSP rather than the system bus. Because the display is of lower resolution than the image sensor, the images from the image sensor must be reduced in resolution. Many still cameras use displays originally designed for camcorders, so the DSP may also need to clip the image to accommodate the differing aspect ratios of the display and image sensor.

Design Example

7.9 VIDEO ACCELERATOR

In this section we use a video accelerator as an example of an accelerated embedded system. Digital video is still a computationally intensive task, so it is well suited to acceleration. Motion estimation engines are used in real-time search engines; we may want to have one attached to our personal computer to experiment with video processing techniques.

7.9.1 Algorithm and Requirements

We could build an accelerator for any number of digital video algorithms. We will choose **block motion estimation** as our example here because it is very computation and memory intensive but it is relatively easy to explain.

Block motion estimation is used in digital video compression algorithms so that one frame in the video can be described in terms of the differences between it and another frame. Because objects in the frame often move relatively little, describing one frame in terms of another greatly reduces the number of bits required to describe the video.

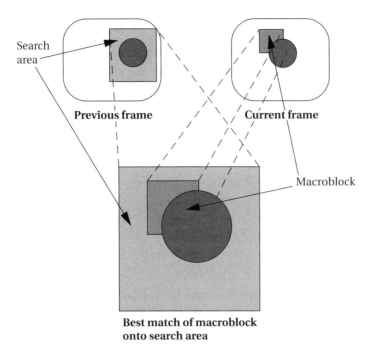

Search
area

Previous frame **Current frame**

Macroblock

**Best match of macroblock
onto search area**

FIGURE 7.26

Block motion estimation.

The concept of block motion estimation is illustrated in Figure 7.26. The goal is to perform a two-dimensional correlation to find the best match between regions in the two frames. We divide the current frame into **macroblocks** (typically, 16×16). For every macroblock in the frame, we want to find the region in the previous frame that most closely matches the macroblock. Searching over the entire previous frame would be too expensive, so we usually limit the search to a given area, centered around the macroblock and larger than the macroblock. We try the macroblock at various offsets in the search area. We measure similarity using the following sum-of-differences measure:

$$\sum_{1 \leq i,\ i \leq n} \left| M(i,j) - S(i - o_x, j - o_y) \right|, \tag{7.3}$$

where $M(i,j)$ is the intensity of the macroblock at pixel i, j, $S(i,j)$ is the intensity of the search region, n is the size of the macroblock in one dimension, and $\langle o_x, o_y \rangle$ is the offset between the macroblock and search region. Intensity is measured as an 8-bit luminance that represents a monochrome pixel—color information is not used in motion estimation. We choose the macroblock position relative to the search area that gives us the smallest value for this metric. The offset at this chosen position describes a vector from the search area center to the macroblock's center that is called the **motion vector**.

For simplicity, we will build an engine for a full search, which compares the macroblock and search area at every possible point. Because this is an expensive operation, a number of methods have been proposed for conducting a sparser search of the search area. These methods introduce extra control that would cloud our discussion, but these algorithms may provide good examples.

A good way to describe the algorithm is in C. Some basic parameters of the algorithm are illustrated in Figure 7.27. Appearing below is the C code for a single search, which assumes that the search region does not extend past the boundary of the frame.

```
bestx = 0; besty = 0; /* initialize best location-none yet */
bestsad = MAXSAD;      /* best sum-of-difference thus far */
for (ox = -SEARCHSIZE; ox < SEARCHSIZE; ox++) {
    /* x search ordinate */
    for (oy = -SEARCHSIZE; oy < SEARCHSIZE; oy++) {
        /* y search ordinate */
        int result = 0;
        for (i = 0; i < MBSIZE; i++) {
            for (j = 0; j < MBSIZE; j++) {
            result = result + iabs(mb[i][j] - search[i - ox
                            + XCENTER][j - oy + YCENTER]);
            }
        }
    if (result <= bestsad) { /* found better match */
        bestsad = result;
        bestx = ox; besty = oy;
    }
}
```

The arithmetic on each pixel is simple, but we have to process a lot of pixels. If MBSIZE is 16 and SEARCHSIZE is 8, and remembering that the search distance in each dimension is $8 + 1 + 8$, then we must perform

$$n_{\mathrm{ops}} = (16 \times 16) \times (17 \times 17) = 73984 \tag{7.4}$$

different operations to find the motion vector for a single macroblock, which requires looking at twice as many pixels, one from the search area and one from the macroblock. (We can now see the interest in algorithms that do not require a full search.) To process video, we will have to perform this computation on every macroblock of every frame. Adjacent blocks have overlapping search areas, so we will try to avoid reloading pixels we already have.

One relatively low-resolution standard video format, common intermediate format, has a frame size of 352×288, which gives an array of 22×18 macroblocks. If we want to encode video, we would have to perform motion estimation on

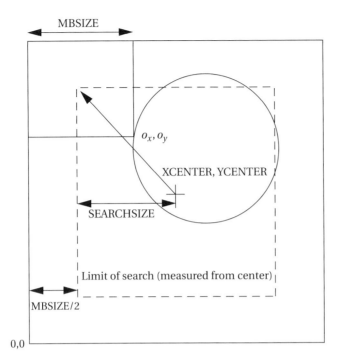

MBSIZE

o_x, o_y

XCENTER, YCENTER

SEARCHSIZE

Limit of search (measured from center)

MBSIZE/2

0,0

FIGURE 7.27

Block motion search parameters.

every macroblock of most frames (some frames are sent without using motion compensation).

We will build the system using an FPGA connected to the PCI bus of a personal computer. We clearly need a high-bandwidth connection such as the PCI between the accelerator and the CPU. We can use the accelerator to experiment with video processing, among other things. Appearing below are the requirements for the system.

Name	Block motion estimator
Purpose	Perform block motion estimation within a PC system
Inputs	Macroblocks and search areas
Outputs	Motion vectors
Functions	Compute motion vectors using full search algorithms
Performance	As fast as we can get
Manufacturing cost	Hundreds of dollars
Power	Powered by PC power supply
Physical size and weight	Packaged as PCI card for PC

7.9.2 Specification

The specification for the system is relatively straightforward because the algorithm is simple. Figure 7.28 defines some classes that describe basic data types in the system: the motion vector, the macroblock, and the search area. These definitions are straightforward. Because the behavior is simple, we need to define only two classes to describe it: the accelerator itself and the PC. These classes are shown in Figure 7.29. The PC makes its memory accessible to the accelerator. The accelerator provides a behavior compute-mv() that performs the block motion estimation algorithm. Figure 7.30 shows a sequence diagram that describes the operation of compute-mv(). After initiating the behavior, the accelerator reads the search area and macroblock from the PC; after computing the motion vector, it returns it to the PC.

7.9.3 Architecture

The accelerator will be implemented in an FPGA on a card connected to a PC's PCI slot. Such accelerators can be purchased or they can be designed from scratch. If you design such a card from scratch, you have to decide early on whether the card will be used only for this video accelerator or if it should be made general enough to support other applications as well.

The architecture for the accelerator requires some thought because of the large amount of data required by the algorithm. The macroblock has $16 \times 16 = 256$; the

FIGURE 7.28

Classes describing basic data types in the video accelerator.

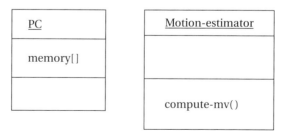

FIGURE 7.29

Basic classes for the video accelerator.

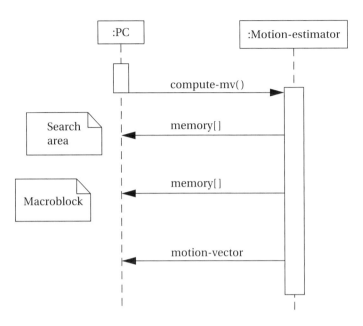

FIGURE 7.30

Sequence diagram for the video accelerator.

search area has $(8 + 8 + 1 + 8 + 8)^2 = 1,089$ pixels. The FPGA probably will not have enough memory to hold 1,089 8-bit values. We have to use a memory external to the FPGA but on the accelerator board to hold the pixels.

There are many possible architectures for the motion estimator. One is shown in Figure 7.31. The machine has two memories, one for the macroblock and another for the search memories. It has 16 PEs that perform the difference calculation on a pair of pixels; the comparator sums them up and selects the best value to find the motion vector. This architecture can be used to implement algorithms other than a full search by changing the address generation and control. Depending on the number of different motion estimation algorithms that you want to execute on the machine, the networks connecting the memories to the PEs may also be simplified.

Figure 7.32 shows how we can schedule the transfer of pixels from the memories to the PEs in order to efficiently compute a full search on this architecture. The schedule fetches one pixel from the macroblock memory and (in steady state) two pixels from the search area memory per clock cycle. The pixels are distributed to the PEs in a regular pattern as shown by the schedule. This schedule computes 16 correlations between the macroblock and search area simultaneously. The computations for each correlation are distributed among the PEs; the comparator is responsible for collecting the results, finding the best match value, and remembering the corresponding motion vector.

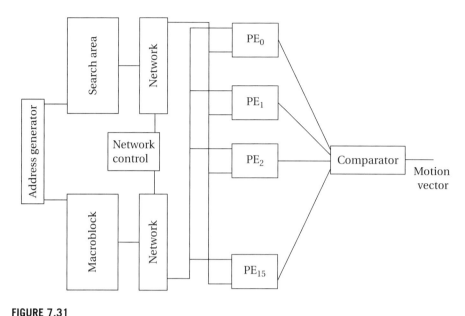

FIGURE 7.31

An architecture for the motion estimation accelerator [Dut96].

Based on our understanding of efficient architectures for accelerating motion estimation, we can derive a more detailed definition of the architecture in UML, which is shown in Figure 7.33. The system includes the two memories for pixels, one a single-port memory and the other dual ported. A bus interface module is responsible for communicating with the PCI bus and the rest of the system. The estimation engine reads pixels from the M and S memories, and it takes commands from the bus interface and returns the motion vector to the bus interface.

7.9.4 Component Design

If we want to use a standard FPGA accelerator board to implement the accelerator, we must first make sure that it provides the proper memory required for M and S. Once we have verified that the accelerator board has the required structure, we can concentrate on designing the FPGA logic. Designing an FPGA is, for the most part, a straightforward exercise in logic design. Because the logic for the accelerator is very regular, we can improve the FPGA's clock rate by properly placing the logic in the FPGA to reduce wire lengths.

If we are designing our own accelerator board, we have to design both the video accelerator design proper and the interface to the PCI bus. We can create and exercise the video accelerator architecture in a hardware description language like VHDL or Verilog and simulate its operation. Designing the PCI interface requires somewhat different techniques since we may not have a simulation model for a PCI

t	M	S	S9	PE$_0$	PE$_1$	PE$_2$
0	M(0,0)	S(0,0)		\|M(0,0) – S(0,0)\|		
1	M(0,1)	S(0,1)		\|M(0,1) – S(0,1)\|	\|M(0,0) – S(0,1)\|	
2	M(0,2)	S(0,2)		\|M(0,2) – S(0,2)\|	\|M(0,1) – S(0,2)\|	\|M(0,0) – S(0,2)\|
3	M(0,3)	S(0,3)		\|M(0,3) – S(0,3)\|	\|M(0,2) – S(0,3)\|	\|M(0,1) – S(0,3)\|
4	M(0,4)	S(0,4)		\|M(0,4) – S(0,4)\|	\|M(0,3) – S(0,4)\|	\|M(0,2) – S(0,4)\|
5	M(0,5)	S(0,5)		\|M(0,5) – S(0,5)\|	\|M(0,4) – S(0,5)\|	\|M(0,3) – S(0,5)\|
6	M(0,6)	S(0,6)		\|M(0,6) – S(0,6)\|	\|M(0,5) – S(0,6)\|	\|M(0,4) – S(0,6)\|
7	M(0,7)	S(0,7)		\|M(0,7) – S(0,7)\|	\|M(0,6) – S(0,7)\|	\|M(0,5) – S(0,7)\|
8	M(0,8)	S(0,8)		\|M(0,8) – S(0,8)\|	\|M(0,7) – S(0,8)\|	\|M(0,6) – S(0,8)\|
9	M(0,9)	S(0,9)		\|M(0,9) – S(0,9)\|	\|M(0,8) – S(0,9)\|	\|M(0,7) – S(0,9)\|
10	M(0,10)	S(0,10)		\|M(0,10) – S(0,10)\|	\|M(0,9) – S(0,10)\|	\|M(0,8) – S(0,10)\|
11	M(0,11)	S(0,11)		\|M(0,11) – S(0,11)\|	\|M(0,10) – S(0,11)\|	\|M(0,9) – S(0,11)\|
12	M(0,12)	S(0,12)		\|M(0,12) – S(0,12)\|	\|M(0,11) – S(0,12)\|	\|M(0,10) – S(0,12)\|
13	M(0,13)	S(0,13)		\|M(0,13) – S(0,13)\|	\|M(0,12) – S(0,13)\|	\|M(0,11) – S(0,13)\|
14	M(0,14)	S(0,14)		\|M(0,14) – S(0,14)\|	\|M(0,13) – S(0,14)\|	\|M(0,12) – S(0,14)\|
15	M(0,15)	S(0,15)		\|M(0,15) – S(0,15)\|	\|M(0,14) – S(0,15)\|	\|M(0,13) – S(0,15)\|
16	M(1,0)	S(1,0)	S(0,16)	\|M(1,0) – S(1,0)\|	\|M(0,15) – S(0,16)\|	\|M(0,14) – S(0,16)\|
17	M(1,1)	S(1,1)	S(0,17)	\|M(1,1) – S(1,1)\|	\|M(1,0) – S(1,1)\|	\|M(0,15) – S(0,17)\|

FIGURE 7.32

A schedule of pixel fetches for a full search [Yan89].

FIGURE 7.33

Object diagram for the video accelerator.

bus. We may want to verify the operation of the basic PCI interface before we finish implementing the video accelerator logic.

The host PC will probably deal with the accelerator as an I/O device. The accelerator board will have its own driver that is responsible for talking to the board. Since most of the data transfers are performed directly by the board using DMA, the driver can be relatively simple.

7.9.5 System Testing

Testing video algorithms requires a large amount of data. Luckily, the data represents images and video, which are plentiful. Because we are designing only a motion estimation accelerator and not a complete video compressor, it is probably easiest to use images, not video, for test data. You can use standard video tools to extract a few frames from a digitized video and store them in JPEG format. Open source for JPEG encoders and decoders is available. These programs can be modified to read JPEG images and put out pixels in the format required by your accelerator. With a little more cleverness, the resulting motion vector can be written back onto the image for a visual confirmation of the result. If you want to be adventurous and try motion estimation on video, open source MPEG encoders and decoders are also available.

SUMMARY

Although the design of an accelerator itself is a hardware design task, the design of an accelerated system requires that we go to a higher level of abstraction. Interactions between the accelerator and the host system, particularly if the host and accelerator execute in parallel, make performance analysis a challenge. Based on the results of performance analysis, we can determine which operations need to go into the accelerator and how to coordinate the actions of the host CPU and the accelerator. Many general-purpose computer systems use accelerators of various types, particularly to support I/O. Adding an accelerator to an embedded system can be an effective way of meeting design requirements.

What We Learned

- Multiprocessors are common in embedded systems because they provide higher performance and lower power consumption at lower cost.

- An accelerated system is an example of a custom multiprocessor.

- Performance analysis of a multiprocessor is challenging. We must consider the performance of several implementations of an algorithm (CPU, accelerator) as well as communication costs for various configurations.

- We must partition the behavior, schedule operations in time, and allocate operations to processing elements in order to design the system.

- Consumer electronics devices share many characteristics under the hood. Multiprocessors are commonly used in consumer electronics devices to provide real-time performance at low energy consumption levels.

FURTHER READING

Staunstrup and Wolf's edited volume [Sta97B] surveys hardware/software co-design, including techniques for accelerated systems like those described in this chapter. The volume edited by De Micheli *et al.* [DeM01] includes a number of basic papers on hardware/software co-design. Callahan *et al.* [Cal00] describe an on-chip reconfigurable co-processor connected to a CPU. Some information on the history of cell phones can be found at www.motorola.com. The book *DVD Demystified* [Tay06] gives a thorough introduction to the DVD; technical information is also available at the "DVD Technical Guide" section of www.pioneerelectronics.com. The Blu-Ray Association Web site is www.blu-raydisc.com.

QUESTIONS

Q7-1 You are designing an embedded system using an Intel Xeon as a host. Does it make sense to add an accelerator to implement the function $z = ax + by + c$? Explain.

Q7-2 You are designing an embedded system using an embedded processor with no floating-point support as host. Does it make sense to add an accelerator to implement the floating-point function $s = A \sin(2\pi f + \varphi)$? Explain.

Q7-3 You are designing an embedded system using a high-performance embedded processor with floating point as host. Does it make sense to add an accelerator to implement the floating-point function $s = A \sin(2\pi f + \varphi)$? Explain.

Q7-4 You are designing an accelerated system that performs the following function as its main task:

```
for (i = 0; i < M; i++)
    for (j = 0; j < N; j++)
        f[i][j] = (pix[i][j - 1] + pix[i - 1][j] +
                   pix[i][j] + pix[i + 1][j] +
                   pix[i][j + 1])/(5*MAXVAL);
```

Assume that the accelerator has the entire pix and *f* arrays in its internal memory during the entire computation—pix is read into the accelerator before the operations begin and *f* is written out after all computations have been completed.

 a. Show a system schedule for the host, accelerator, and bus assuming that the accelerator is inactive during all data transfers. (All data are sent to the accelerator before it starts and data are read from the accelerator after the computations are finished.)

 b. Show a system schedule for the host, accelerator, and bus assuming that the accelerator has enough memory for two pix and *f* arrays and that the host can transfer data for one set of computations while another set is being performed.

Q7-5 Find the longest path through the graph below, using the computation times on the nodes and the communication times on the edges.

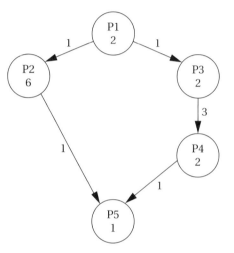

Q7-6 Each of these task graphs will be run on a two-PE multiprocessor; the two processing elements are identical. For each of the task graphs, including the process execution times and communication times, determine the allocation of processes to PEs that minimizes total execution time.

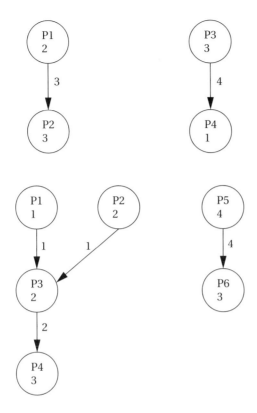

Q7-7 Write pseudocode for an algorithm to determine the longest path through a system execution graph. The longest path is to be measured from one designated entry point to one exit point. Each node in the graph is labeled with a number giving the execution time of the process represented by that node.

Q7-8 Write pseudocode that describes the schedules shown in Example 7.3:

 a. The schedule that performs all As and Bs before any Cs.

 b. The schedule that performs A, B, and C on one data element at a time.

Q7-9 Assuming that you can control when the data inputs arrive, which schedule in Example 7.3 requires the least amount of total buffer space? Justify your answer.

LAB EXERCISES

L7-1 Determine how much logic in an FPGA must be devoted to a PCI bus interface and how much would be left for an accelerator core.

L7-2 Develop a debugging scheme for an accelerator. Determine how you would easily enter data into the accelerator and easily observe its behavior. You will need to verify the system thoroughly, starting with basic communication and going through algorithmic verification.

L7-3 Develop a generic streaming interface for an accelerator. The interface should allow streaming data to be read by the accelerator from the host's memory. It should also allow streaming data to be written from the accelerator back to memory. The interface should include a host-side mechanism for filling and draining the streaming data buffers.

Networks

- Why we build networked embedded systems.

- General network architectures and the ISO network layers.

- Several networks: I^2C, CAN, and Ethernet.

- Internet-enabled embedded systems.

- Sensor networks.

- Elevator controller design example.

INTRODUCTION

In this chapter we study **networks** that can be used to build **distributed embedded systems**. In a distributed embedded system, several **processing elements (PEs)** (either microprocessors or ASICs) are connected by a network that allows them to communicate. The application is distributed over the PEs, and some of the work is done at each node in the network.

There are several reasons to build network-based embedded systems. When the processing tasks are physically distributed, it may be necessary to put some of the computing power near where the events occur. Consider, for example, an automobile: the short time delays required for tasks such as engine control generally mean that at least parts of the task are done physically close to the engine. Data reduction is another important reason for distributed processing. It may be possible to perform some initial signal processing on captured data to reduce its volume—for example, detecting a certain type of event in a sampled data stream. Reducing the data on a separate processor may significantly reduce the load on the processor that makes use of that data. Modularity is another motivation for network-based design. For instance, when a large system is assembled out of existing components, those components may use a network port as a clean interface that does not interfere with the internal operation of the component in ways that using the microprocessor bus would. A distributed system can also be easier to debug—the microprocessors in one part of the network can be used to probe components in another part of the network. Finally, in some cases, networks are used to build fault tolerance into systems. Distributed embedded system design is another example of hardware/software co-design, since we

must design the network topology as well as the software running on the network nodes.

Of course, the microprocessor bus is a simple type of network. However, we use the term *network* to mean an interconnection scheme that does not provide shared memory communication. In the next section, we develop the basic principles of hardware and software architectures for networks. Section 8.2 examines several different networking systems. Section 8.3 considers techniques for the design of distributed embedded systems. Section 8.4 focuses on how embedded systems can be designed to talk to the Internet. Section 8.5 looks at the networked electronics in automobiles and airplanes. Section 8.6 introduces some basic principles of wireless sensor networks. Section 8.7 presents an elevator system as an example of network-based design.

8.1 DISTRIBUTED EMBEDDED ARCHITECTURES

A distributed embedded system can be organized in many different ways, but its basic units are the PE and the network as illustrated in Figure 8.1. A PE may be an instruction set processor such as a *DSP*, *CPU*, or *microcontroller*, as well as a nonprogrammable unit such as the *ASICs* used to implement *PE 4*. An I/O device such as *PE 1* (which we call here a **sensor** or **actuator**, depending on whether it provides input or output) may also be a PE, so long as it can speak the network protocol to communicate with other PEs. The network in this case is a bus, but other network topologies are also possible. It is also possible that the system can use more than one network, such as when relatively independent functions require relatively little communication among them. We often refer to the connection between PEs provided by the network as a **communication link**.

FIGURE 8.1

An example of a distributed embedded system.

The system of PEs and networks forms the ***hardware platform*** on which the application runs.

However, unlike the system bus of Chapter 4, the distributed embedded system does not have memory on the bus (unless a memory unit is organized as an I/O device that speaks the network protocol). In particular, PEs do not fetch instructions over the network as they do on the microprocessor bus. We take advantage of this fact when analyzing network performance—the speed at which PEs can communicate over the bus would be difficult if not impossible to predict if we allowed arbitrary instruction and data fetches as we do on microprocessor buses.

8.1.1 Why Distributed?

Building an embedded system with several PEs talking over a network is definitely more complicated than using a single large microprocessor to perform the same tasks. So why would anyone build a distributed embedded system? All the reasons for designing accelerator systems also apply to distributed embedded systems, and several more reasons are unique to distributed systems.

In some cases, distributed systems are necessary because the devices that the PEs communicate with are physically separated. If the deadlines for processing the data are short, it may be more cost-effective to put the PEs where the data are located rather than build a higher-speed network to carry the data to a distant, fast PE.

An important advantage of a distributed system with several CPUs is that one part of the system can be used to help diagnose problems in another part. Whether you are debugging a prototype or diagnosing a problem in the field, isolating the error to one part of the system can be difficult when everything is done on a single CPU. If you have several CPUs in the system, you can use one to generate inputs for another and to watch its output.

8.1.2 Network Abstractions

Networks are complex systems. Ideally, they provide high-level services while hiding many of the details of data transmission from the other components in the system. In order to help understand (and design) networks, the International Standards Organization has developed a seven-layer model for networks known as Open Systems Interconnection (OSI) models [Sta97A]. Understanding the OSI layers will help us to understand the details of real networks.

The seven layers of the **OSI model**, shown in Figure 8.2, are intended to cover a broad spectrum of networks and their uses. Some networks may not need the services of one or more layers because the higher layers may be totally missing or an intermediate layer may not be necessary. However, any data network should fit into the OSI model. The OSI layers from lowest to highest level of abstraction are described below.

- *Physical:* The physical layer defines the basic properties of the interface between systems, including the physical connections (plugs and wires),

Application	End-use interface
Presentation	Data format
Session	Application dialog control
Transport	Connections
Network	End-to-end service
Data link	Reliable data transport
Physical	Mechanical, electrical

FIGURE 8.2

The OSI model layers.

electrical properties, basic functions of the electrical and physical components, and the basic procedures for exchanging bits.

- *Data link:* The primary purpose of this layer is error detection and control across a single link. However, if the network requires multiple hops over several data links, the data link layer does not define the mechanism for data integrity between hops, but only within a single hop.

- *Network:* This layer defines the basic end-to-end data transmission service. The network layer is particularly important in multihop networks.

- *Transport:* The transport layer defines connection-oriented services that ensure that data are delivered in the proper order and without errors across multiple links. This layer may also try to optimize network resource utilization.

- *Session:* A session provides mechanisms for controlling the interaction of end-user services across a network, such as data grouping and checkpointing.

- *Presentation:* This layer defines data exchange formats and provides transformation utilities to application programs.

- *Application:* The application layer provides the application interface between the network and end-user programs.

Although it may seem that embedded systems would be too simple to require use of the OSI model, the model is in fact quite useful. Even relatively simple embedded networks provide physical, data link, and network services. An increasing number of embedded systems provide Internet service that requires implementing the full range of functions in the OSI model.

8.1.3 **Hardware and Software Architectures**

Distributed embedded systems can be organized in many different ways depending upon the needs of the application and cost constraints. One good way to understand possible architectures is to consider the different types of interconnection networks that can be used.

A ***point-to-point*** link establishes a connection between exactly two PEs. Point-to-point links are simple to design precisely because they deal with only two components. We do not have to worry about other PEs interfering with communication on the link.

Figure 8.3 shows a simple example of a distributed embedded system built from point-to-point links. The input signal is sampled by the input device and passed to the first digital filter, $F1$, over a point-to-point link. The results of that filter are sent through a second point-to-point link to filter $F2$. The results in turn are sent to the output device over a third point-to-point link. A digital filtering system requires that its outputs arrive at strict intervals, which means that the filters must process their inputs in a timely fashion. Using point-to-point connections allows both $F1$ and $F2$ to receive a new sample and send a new output at the same time without worrying about collisions on the communications network.

It is possible to build a ***full-duplex***, point-to-point connection that can be used for simultaneous communication in both directions between the two PEs. (A half-duplex connection allows for only one-way communication.)

A bus is a more general form of network since it allows multiple devices to be connected to it. Like a microprocessor bus, PEs connected to the bus have addresses. Communications on the bus generally take the form of ***packets*** as illustrated in Figure 8.4. A packet contains an address for the destination and the

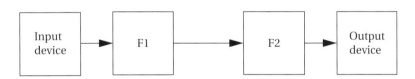

FIGURE 8.3

A signal processing system built from print-to-point links.

FIGURE 8.4

Format of a typical message on a bus.

data to be delivered. It frequently includes error detection/correction information such as parity. It also may include bits that serve to signal to other PEs that the bus is in use, such as the header shown in the figure. The data to be transmitted from one PE to another may not fit exactly into the size of the **data payload** on the packet. It is the responsibility of the transmitting PE to divide its data into packets; the receiving PE must of course reassemble the complete data message from the packets.

Distributed system buses must be arbitrated to control simultaneous access, just as with microprocessor buses. Arbitration scheme types are summarized below.

- *Fixed-priority arbitration* always gives priority to competing devices in the same way. If a high-priority and a low-priority device both have long data transmissions ready at the same time, it is quite possible that the low-priority device will not be able to transmit anything until the high-priority device has sent all its data packets.

- *Fair arbitration* schemes make sure that no device is starved. **Round-robin arbitration** is the most commonly used of the fair arbitration schemes. The PCI bus requires that the arbitration scheme used on the bus must be fair, although it does not specify a particular arbitration scheme. Most implementations of PCI use round-robin arbitration.

A bus has limited available bandwidth. Since all devices connect to the bus, communications can interfere with each other. Other network topologies can be used to reduce communication conflicts. At the opposite end of the generality spectrum from the bus is the **crossbar** network shown in Figure 8.5. A crossbar not only allows any input to be connected to any output, it also allows all combinations of input/output connections to be made. Thus, for example, we can simultaneously

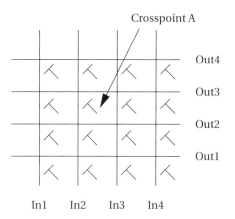

FIGURE 8.5

A crossbar network.

connect *in1* to *out4*, *in2* to *out3*, *in3* to *out2*, and *in4* to *out1* or any other combinations of inputs. (Multicast connections can also be made from one input to several outputs.) A **crosspoint** is a switch that connects an input to an output. To connect an input to an output, we activate the crosspoint at the intersection between the corresponding input and output lines in the crossbar. For example, to connect *in2* and *out3* in the figure, we would activate crossbar *A* as shown. The major drawback of the crossbar network is expense: The size of the network grows as the square of the number of inputs (assuming the numbers of inputs and outputs are equal).

Many other networks have been designed that provide varying amounts of parallel communication at varying hardware costs. Figure 8.6 shows an example **multistage network**. The crossbar of Figure 8.5 is a **direct network** in which messages go from source to destination without going through any memory element. Multistage networks have intermediate routing nodes to guide the data packets.

Most networks are **blocking**, meaning that there are some combinations of sources and destinations for which messages cannot be delivered simultaneously. A bus is a maximally blocking network since any message on the bus blocks messages from any other node. A crossbar is non-blocking.

In general, networks differ from microprocessor buses in how they implement communication protocols. Both need handshaking to ensure that PEs do not interfere with each other. But in most networks, most of the protocol is performed in software. Microprocessors rely on bus hardware for fast transfers of instructions and data to and from the CPU. Most embedded network ports on microprocessors implement the basic communication functions (such as driving the communications medium) in hardware and implement many other operations in software.

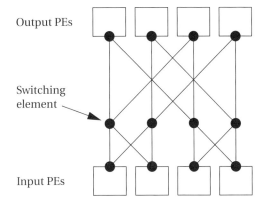

Output PEs

Switching
element

Input PEs

FIGURE 8.6

A multistage network.

An alternative to a non-bus network is to use multiple networks. As with PEs, it may be cheaper to use two slow, inexpensive networks than a single high-performance, expensive network. If we can segregate critical and noncritical communications onto separate networks, it may be possible to use simpler topologies such as buses. Many systems use serial links for low-speed communication and CPU buses for higher speed and volume data transfers.

8.1.4 Message Passing Programming

Distributed embedded systems do not have shared memory, so they must communicate by passing *messages*. We will refer to a message as the natural communication unit of an algorithm; in general, a message must be broken up into packets to be sent on the network. A procedural interface for sending a packet might look like the following:

```
send_packet(address,data);
```

The routine should return a value to indicate whether the message was sent successfully if the network includes a handshaking protocol. If the message to be sent is longer than a packet, it must be broken up into packet-size data segments as follows:

```
for (i = 0; i < message.length; i = i + PACKET_SIZE)
    send_packet(address,&message.data[i]);
```

The above code uses a loop to break up an arbitrary-length message into packet-size chunks. However, clever system design may be able to recast the message to take advantage of the packet format. For example, clever encoding may reduce the length of the message enough so that it fits into a single packet. On the other hand, if the message is shorter than a packet or not an even multiple of the packet data size, some extra information may be packed into the remaining bits of a packet.

Reception of a packet will probably be implemented with interrupts. The simplest procedural interface will simply check to see whether a received message is waiting in a buffer. In a more complex RTOS-based system, reception of a packet may enable a process for execution.

As seen in Section 6.4, communication may be blocking or non-blocking. Of course, the simplest implementation of message passing is blocking, with the routine not returning until it has transmitted or received. A non-blocking network interface requires a queue of data to be sent, with the network driver sending packets off the head of the queue and placing received packets on the tail of the queue. A non-blocking communication mechanism makes sense only when concurrency is available between computing and data transfer.

Network protocols may encourage a **data-push** design style for the system built around the network. In a single-CPU environment, a program typically initiates a read whenever it wants data. In many networked systems, nodes send values out

without any request from the intended user of the system. Data-push programming makes sense for periodic data—if the data will always be used at regular intervals, we can reduce data traffic on the network by automatically sending it when it is needed. Example 8.1 shows an application that can make good use of the data-push architecture.

Example 8.1

Data-push network architectures

Consider the following automobile in which distributed sensors and actuators talk to a central controller:

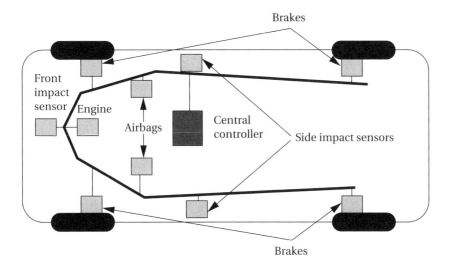

The sensors generally need to be sampled periodically. In such a system, it makes sense for sensors to transmit their data automatically rather than waiting for the controller to request it.

8.2 NETWORKS FOR EMBEDDED SYSTEMS

Networks for embedded computing span a broad range of requirements; many of those requirements are very different from those for general-purpose networks. Some networks are used in safety-critical applications, such as automotive control. Some networks, such as those used in consumer electronics systems, must be very inexpensive. Other networks, such as industrial control networks, must be extremely rugged and reliable.

Several interconnect networks have been developed especially for distributed embedded computing:

- The I^2C bus is used in microcontroller-based systems.

- The Controller Area Network (CAN) bus was developed for automotive electronics. It provides megabit rates and can handle large numbers of devices.

- Ethernet and variations of standard Ethernet are used for a variety of control applications.

In addition, many networks designed for general-purpose computing have been put to use in embedded applications as well.

In this section, we study some commonly used embedded networks, including the I^2C bus and Ethernet; we will also briefly discuss networks for industrial applications.

8.2.1 The I^2C Bus

The *I^2C bus* [Phi92] is a well-known bus commonly used to link microcontrollers into systems. It has even been used for the command interface in an MPEG-2 video chip [van97]; while a separate bus was used for high-speed video data, setup information was transmitted to the on-chip controller through an I^2C bus interface.

I^2C is designed to be low cost, easy to implement, and of moderate speed (up to 100 KB/s for the standard bus and up to 400 KB/s for the extended bus). As a result, it uses only two lines: the *serial data line (SDL)* for data and the *serial clock line (SCL)*, which indicates when valid data are on the data line. Figure 8.7 shows the structure of a typical I^2C bus system. Every node in the network is connected to both SCL and SDL. Some nodes may be able to act as bus masters and the bus

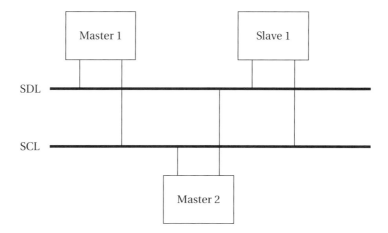

FIGURE 8.7

Structure of an I^2C bus system.

may have more than one master. Other nodes may act as slaves that only respond to requests from masters.

The basic electrical interface to the bus is shown in Figure 8.8. The bus does not define particular voltages to be used for high or low so that either bipolar or MOS circuits can be connected to the bus. Both bus signals use open collector/open drain circuits.[1] A pull-up resistor keeps the default state of the signal high, and transistors are used in each bus device to pull down the signal when a 0 is to be transmitted. Open collector/open drain signaling allows several devices to simultaneously write the bus without causing electrical damage.

The open collector/open drain circuitry allows a slave device to stretch a clock signal during a read from a slave. The master is responsible for generating the SCL clock, but the slave can stretch the low period of the clock (but not the high period) if necessary.

The I^2C bus is designed as a multimaster bus—any one of several different devices may act as the master at various times. As a result, there is no global master to generate the clock signal on *SCL*. Instead, a master drives both *SCL* and *SDL* when it is sending data. When the bus is idle, both *SCL* and *SDL* remain high. When two devices try to drive either *SCL* or *SDL* to different values, the open collector/open drain circuitry prevents errors, but each master device must listen to the bus while transmitting to be sure that it is not interfering with another message—if the device receives a different value than it is trying to transmit, then it knows that it is interfering with another message.

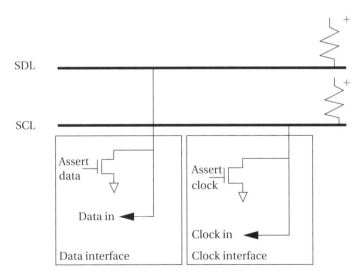

FIGURE 8.8

Electrical interface to the I^2C bus.

[1]An open collector uses a bipolar transistor, while an open drain circuit uses an MOS transistor.

Every I^2C device has an address. The addresses of the devices are determined by the system designer, usually as part of the program for the I^2C driver. The addresses must of course be chosen so that no two devices in the system have the same address. A device address is 7 bits in the standard I^2C definition (the extended I^2C allows 10-bit addresses). The address 0000000 is used to signal a **general call** or bus broadcast, which can be used to signal all devices simultaneously. The address 11110XX is reserved for the extended 10-bit addressing scheme; there are several other reserved addresses as well.

A **bus transaction** comprised a series of 1-byte **transmissions** and an address followed by one or more data bytes. I^2C encourages a data-push programming style. When a master wants to write a slave, it transmits the slave's address followed by the data. Since a slave cannot initiate a transfer, the master must send a read request with the slave's address and let the slave transmit the data. Therefore, an address transmission includes the 7-bit address and 1 bit for data direction: 0 for writing from the master to the slave and 1 for reading from the slave to the master. (This explains the 7-bit addresses on the bus.) The format of an address transmission is shown in Figure 8.9.

A bus transaction is initiated by a start signal and completed with an end signal as follows:

- A start is signaled by leaving the SCL high and sending a 1 to 0 transition on SDL.

- A stop is signaled by setting the SCL high and sending a 0 to 1 transition on SDL.

However, starts and stops must be paired. A master can write and then read (or read and then write) by sending a start after the data transmission, followed by another address transmission and then more data. The basic state transition graph for the master's actions in a bus transaction is shown in Figure 8.10.

The formats of some typical complete bus transactions are shown in Figure 8.11. In the first example, the master writes 2 bytes to the addressed slave. In the second, the master requests a read from a slave. In the third, the master writes 1 byte to the slave, and then sends another start to initiate a read from the slave.

Figure 8.12 shows how a data byte is transmitted on the bus, including start and stop events. The transmission starts when SDL is pulled low while SCL remains high.

FIGURE 8.9

Format of an I^2C address transmission.

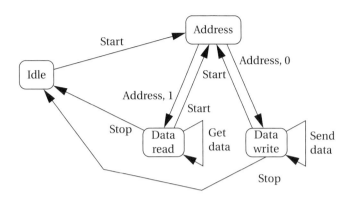

FIGURE 8.10

State transition graph for an I²C bus master.

FIGURE 8.11

Typical bus transactions on the I²C bus.

FIGURE 8.12

Transmitting a byte on the I²C bus.

After this start condition, the clock line is pulled low to initiate the data transfer. At each bit, the clock line goes high while the data line assumes its proper value of 0 or 1. An acknowledgment is sent at the end of every 8-bit transmission, whether it is an address or data. For acknowledgment, the transmitter does not pull down the SDL, allowing the receiver to set the SDL to 0 if it properly received the byte. After acknowledgment, the SDL goes from low to high while the SCL is high, signaling the stop condition.

The bus uses this feature to arbitrate on each message. When sending, devices listen to the bus as well. If a device is trying to send a logic 1 but hears a logic 0, it immediately stops transmitting and gives the other sender priority. (The devices should be designed so that they can stop transmitting in time to allow a valid bit to be sent.) In many cases, arbitration will be completed during the address portion of a transmission, but arbitration may continue into the data portion. If two devices are trying to send identical data to the same address, then of course they never interfere and both succeed in sending their message.

The I^2C interface on a microcontroller can be implemented with varying percentages of the functionality in software and hardware [Phi89]. As illustrated in Figure 8.13, a typical system has a 1-bit hardware interface with routines for byte-level functions. The I^2C device takes care of generating the clock and data. The application code calls routines to send an address, send a data byte, and so on, which then generates the SCL and SDL, acknowledges, and so forth. One of the microcontroller's timers is typically used to control the length of bits on the bus. Interrupts may be used to recognize bits. However, when used in master mode, polled I/O may be acceptable if no other pending tasks can be performed, since masters initiate their own transfers.

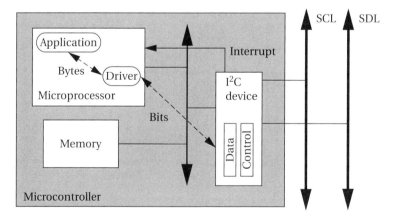

FIGURE 8.13

An I^2C interface in a microcontroller.

8.2.2 **Ethernet**

Ethernet is very widely used as a local area network for general-purpose computing. Because of its ubiquity and the low cost of Ethernet interfaces, it has seen significant use as a network for embedded computing. Ethernet is particularly useful when PCs are used as platforms, making it possible to use standard components, and when the network does not have to meet rigorous real-time requirements.

The physical organization of an Ethernet is very simple, as shown in Figure 8.14. The network is a bus with a single signal path; the Ethernet standard allows for several different implementations such as twisted pair and coaxial cable.

Unlike the I^2C bus, nodes on the Ethernet are not synchronized—they can send their bits at any time. I^2C relies on the fact that a collision can be detected and quashed within a single bit time thanks to synchronization. But since Ethernet nodes are not synchronized, if two nodes decide to transmit at the same time, the message will be ruined. The Ethernet arbitration scheme is known as **Carrier Sense Multiple Access** with **Collision Detection (CSMA/CD)**. The algorithm is outlined in Figure 8.15. A node that has a message waits for the bus to become silent and then starts transmitting. It simultaneously listens, and if it hears another transmission that interferes with its transmission, it stops transmitting and waits to retransmit. The waiting time is random, but weighted by an exponential function of the number of times the message has been aborted. Figure 8.16 shows the exponential backoff function both before and after it is modulated by the random wait time. Since a message may be interfered with several times before it is successfully transmitted, the **exponential backoff** technique helps to ensure that the network does not become overloaded at high demand factors. The random factor in the wait time minimizes the chance that two messages will repeatedly interfere with each other.

The maximum length of an Ethernet is determined by the nodes' ability to detect collisions. The worst case occurs when two nodes at opposite ends of the bus are transmitting simultaneously. For the collision to be detected by both nodes, each node's signal must be able to travel to the opposite end of the bus so that it can be heard by the other node. In practice, Ethernets can run up to several hundred meters.

FIGURE 8.14

Ethernet organization.

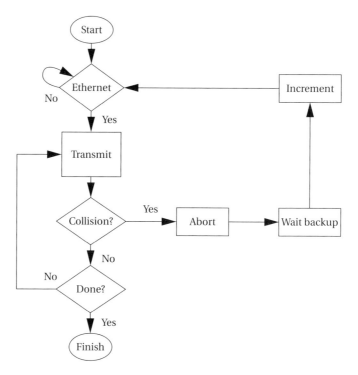

FIGURE 8.15

The Ethernet CSMA/CD algorithm.

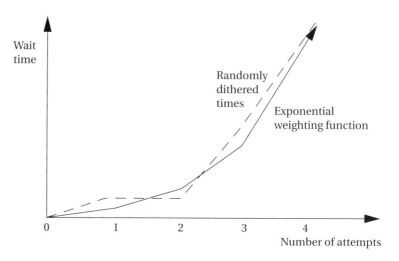

FIGURE 8.16

Exponential backoff times.

Preamble	Start frame	Destination address	Source address	Length	Data	Padding	CRC

FIGURE 8.17

Ethernet packet format.

Figure 8.17 shows the basic format of an Ethernet packet. It provides addresses of both the destination and the source. It also provides for a variable-length data payload.

The fact that it may take several attempts to successfully transmit a message and that the waiting time includes a random factor makes Ethernet performance difficult to analyze. It is possible to perform data streaming and other real-time activities on Ethernets, particularly when the total network load is kept to a reasonable level, but care must be taken in designing such systems.

Ethernet was not designed to support real-time operations; the exponential backoff scheme cannot guarantee delivery time of any data. Because so much Ethernet hardware and software is available, many different approaches have been developed to extend Ethernet to real-time operation; some of these are compatible with the standard while others are not. As Decotignie points out [Dec05], there are three ways to reduce the variance in Ethernet's packet delivery time: suppress collisions on the network, reduce the number of collisions, or resolve collisions deterministically. Felser [Fel05] describes several real-time Ethernet architectures.

8.2.3 Fieldbus

Manufacturing systems require networked sensors and actuators. Fieldbus (http://www.fieldbus.org) is a set of standards for industrial control and instrumentation systems.

The H1 standard uses a twisted-pair physical layer that runs at 31.25 MB/s. It is designed for device integration and process control.

The High Speed Ethernet standard is used for backbone networks in industrial plants. It is based on the 100 MB/s Ethernet standard. It can integrate devices and subsystems.

8.3 NETWORK-BASED DESIGN

Designing a distributed embedded system around a network involves some of the same design tasks we faced in accelerated systems. We must schedule computations in time and allocate them to PEs. Scheduling and allocation of communication are important additional design tasks required for many distributed networks. Many embedded networks are designed for low cost and therefore do not provide excessively high communication speed. If we are not careful, the network can become

the bottleneck in system design. In this section we concentrate on design tasks unique to network-based distributed embedded systems.

We know how to analyze the execution time of programs and systems of processes on single CPUs, but to analyze the performance of networks we must know how to determine the delay incurred by transmitting messages. Let us assume for the moment that messages are sent reliably—we do not have to retransmit a message. The **message delay** for a single message with no contention (as would be the case in a point-to-point connection) can be modeled as

$$t_m = t_x + t_n + t_r \qquad (8.1)$$

where t_x is the transmitter-side overhead, t_n is the network transmission time, and t_r is the receiver-side overhead. In I^2C, t_x and t_r are negligible relative to t_n, as illustrated by Example 8.2.

Example 8.2

Simple message delay for an I^2C message

Let's assume that our I^2C bus runs at the rate of 100 KB/s and that we need to send one 8-bit byte. Based on the message format shown in Figure 8.9, we can compute the number of bits in the complete packet:

$$n_{packet} = \text{startbit} + \text{address} + \text{data} + \text{stopbit}$$
$$= 1 + 8 + 8 + 1 = 18 \, \text{bits}$$

The time required, then, to transmit the packet is

$$t_n = n_{packet} \times t_{bit} = 1.8 \times 10^{-4} \, \text{s}.$$

Some of the instructions in the transmitter and receiver drivers—namely, the loops that send bytes to and receive bytes from the network interface—will run concurrently with the message transmission. If we assume that 20 instructions outside of these loops are executed by the transmitter and receiver, overheads on an 8 MHz microcontroller would be as follows:

$$t_x = t_r = 20 \times 0.125 \times 10^{-6} = 2.5 \times 10^{-6}.$$

The total message delay is:

$$t_m = 2.5 \times 10^{-6} + 1.8 \times 10^{-4} + 2.5 \times 10^{-6} = 1.85 \times 10^{-4}.$$

Overhead is <3% of the total message time in this case.

If messages can interfere with each other in the network, analyzing communication delay becomes difficult. In general, because we must wait for the network to become available and then transmit the message, we can write the **message delay** as

$$t_y = t_d + t_m \qquad (8.2)$$

where t_d is the **network availability delay** incurred waiting for the network to become available. The main problem, therefore, is calculating t_d. That value depends on the type of arbitration used in the network.

- If the network uses fixed-priority arbitration, the network availability delay is unbounded for all but the highest-priority device. Since the highest-priority device always gets the network first, unless there is an application-specific limit on how long it will transmit before relinquishing the network, it can keep blocking the other devices indefinitely.

- If the network uses fair arbitration, the network availability delay is bounded. In the case of round-robin arbitration, if there are N devices, then the worst-case network availability delay is $N(t_x + t_{arb})$, where t_{arb} is the delay incurred for arbitration. t_{arb} is usually small compared to transmission time.

Even when round-robin arbitration is used to bound the network availability delay, the waiting time can be very long. If we add acknowledgment and data corruption into the analysis, figuring network delay is more difficult. Assuming that errors are random, we cannot predict a worst-case delay since every packet may contain an error. We can, however, compute the probability that a packet will be delayed for more than a given amount of time. However, such analysis is beyond the scope of this book.

Arbitration on networks is a form of prioritization. Therefore, we can use the techniques we learned for process scheduling in Chapter 6 to help us schedule communications. In a rate-monotonic communication scheme, the task with the shortest deadline should be assigned the highest priority in the network.

Our process scheduling model assumed that we could interrupt processes at any point. But network communications are organized into packets. In most networks we cannot interrupt a packet transmission to take over the network for a higher-priority packet. As a result, networks exhibit priority inversion like that introduced in Chapter 6. When a low-priority message is on the network, the network is effectively allocated to that low-priority message, allowing it to block higher-priority messages. This cannot cause deadlock since each message has a bounded length, but it can slow down critical communications. The only solution is to analyze network behavior to determine whether priority inversion causes some messages to be delayed for too long.

Of course, a round-robin arbitrated network puts all communications at the same priority. This does not eliminate the priority inversion problem because processes still have priorities.

Thus far we have assumed a *single-hop network:* A message is received at its intended destination directly from the source, without going through any other network node. It is possible to build *multihop networks* in which messages are routed through network nodes to get to their destinations. (Using a multistage network does not necessarily mean using a multihop network—the stages in a multistage network are generally much smaller than the network PEs.) Figure 8.18 shows an example of a multihop communication. The hardware platform has two separate networks (perhaps so that communications between subsets of the PEs do not interfere), but there is no direct path from $M1$ to $M5$. The message is therefore routed through $M3$, which reads it from one network and sends it on to the other one. Analyzing delays

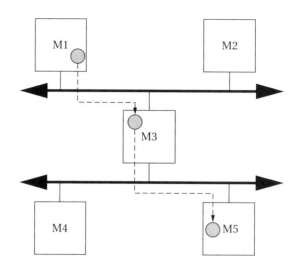

FIGURE 8.18

A multihop communication.

through multihop systems is very difficult. For example, the time that the message is held at $M3$ depends on both the computational load of $M3$ and the other messages that it must handle.

If there is more than one network, we must allocate communications to the networks. We may establish multiple networks so that lower-priority communications can be handled separately without interfering with high-priority communications on the primary network.

Scheduling and allocation of computations and communications are clearly interrelated. If we change the allocation of computations, we change not only the scheduling of processes on those PEs but also potentially the schedules of PEs with which they communicate. For example, if we move a computation to a slower PE, its results will be available later, which may mean rescheduling both the process that uses the value and the communication that sends the value to its destination.

8.4 INTERNET-ENABLED SYSTEMS

Some very different types of distributed embedded system are rapidly emerging—the **Internet-enabled embedded system** and **Internet appliances**. The Internet is not well suited to the real-time tasks that are the bread and butter of embedded computing, but it does provide a rich environment for non–real-time interaction. In this section we will discuss the Internet and how it can be used by embedded computing systems.

8.4.1 Internet

The **Internet Protocol (IP)** [Los97, Sta97A] is the fundamental protocol on the *Internet*. It provides connectionless, packet-based communication. Industrial automation has long been a good application area for Internet-based embedded systems. Information appliances that use the Internet are rapidly becoming another use of IP in embedded computing.

Internet protocol is not defined over a particular physical implementation—it is an *internetworking* standard. Internet packets are assumed to be carried by some other network, such as an Ethernet. In general, an Internet packet will travel over several different networks from source to destination. The IP allows data to flow seamlessly through these networks from one end user to another. The relationship between IP and individual networks is illustrated in Figure 8.19. IP works at the network layer. When node A wants to send data to node B, the application's data pass through several layers of the protocol stack to send to the IP. IP creates packets for routing to the destination, which are then sent to the *data link* and *physical* layers. A node that transmits data among different types of networks is known as a *router*. The router's functionality must go up to the IP layer, but since it is not running applications, it does not need to go to higher levels of the OSI model. In general, a packet may go through several routers to get to its destination. At the destination, the IP layer provides data to the transport layer and ultimately the receiving application. As the data pass through several layers of the protocol stack, the IP packet data are encapsulated in packet formats appropriate to each layer.

The basic format of an IP packet is shown in Figure 8.20. The header and data payload are both of variable length. The maximum total length of the header and data payload is 65,535 bytes.

An Internet address is a number (32 bits in early versions of IP, 128 bits in IPv6). The IP address is typically written in the form xxx.xx.xx.xx. The names by which users and applications typically refer to Internet nodes, such as foo.baz.com,

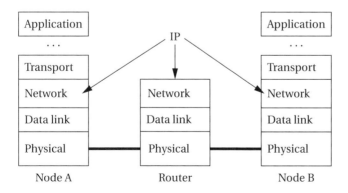

FIGURE 8.19

Protocol utilization in Internet communication.

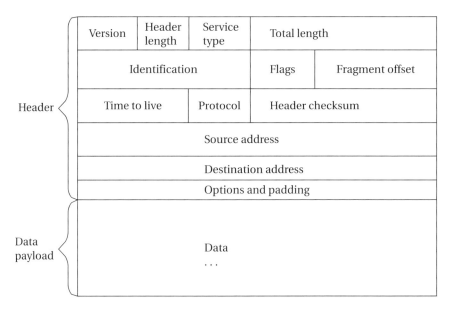

FIGURE 8.20

IP packet structure.

are translated into IP addresses via calls to a ***Domain Name Server***, one of the higher-level services built on top of IP.

The fact that IP works at the network layer tells us that it does not guarantee that a packet is delivered to its destination. Furthermore, packets that do arrive may come out of order. This is referred to as ***best-effort routing***. Since routes for data may change quickly with subsequent packets being routed along very different paths with different delays, real-time performance of IP can be hard to predict. When a small network is contained totally within the embedded system, performance can be evaluated through simulation or other methods because the possible inputs are limited. Since the performance of the Internet may depend on worldwide usage patterns, its real-time performance is inherently harder to predict.

The Internet also provides higher-level services built on top of IP. The ***Transmission Control Protocol (TCP)*** is one such example. It provides a connection-oriented service that ensures that data arrive in the appropriate order, and it uses an acknowledgment protocol to ensure that packets arrive. Because many higher-level services are built on top of TCP, the basic protocol is often referred to as TCP/IP.

Figure 8.21 shows the relationships between IP and higher-level Internet services. Using IP as the foundation, TCP is used to provide ***File Transport Protocol*** for batch file transfers, ***Hypertext Transport Protocol (HTTP)*** for World Wide Web service, ***Simple Mail Transfer Protocol*** for email, and Telnet for virtual terminals. A separate transport protocol, ***User Datagram Protocol***, is used as

FTP	HTTP	SMTP	Telnet		SNMP
	TCP				UDP
		IP			

FIGURE 8.21

The Internet service stack.

the basis for the network management services provided by the *Simple Network Management Protocol*.

8.4.2 Internet Applications

The Internet provides a standard way for an embedded system to act in concert with other devices and with users, such as:

- One of the earliest Internet-enabled embedded systems was the laser printer. High-end laser printers often use IP to receive print jobs from host machines.

- Portable Internet devices can display Web pages, read email, and synchronize calendar information with remote computers.

- A home control system allows the homeowner to remotely monitor and control home cameras, lights, and so on.

Although there are higher-level services that provide more time-sensitive delivery mechanisms for the Internet, the basic incarnation of the Internet is not well suited to hard real-time operations. However, IP is a very good way to let the embedded system talk to other systems. IP provides a way for both special-purpose and standard programs (such as Web browsers) to talk to the embedded system. This non–real-time interaction can be used to monitor the system, set its configuration, and interact with it.

As seen in Section 8.4.1, the Internet provides a wide range of services built on top of IP. Since code size is an important issue in many embedded systems, one architectural decision that must be made is to determine which Internet services will be needed by the system. This choice depends on the type of data service required, such as connectionless versus connection oriented, streaming vs. non-streaming, and so on. It also depends on the application code and its services: does the system look to the rest of the Internet like a terminal, a Web server, or something else?

Application Example 8.1 describes an Internet appliance that runs Java to provide useful services.

Application Example 8.1

An Internet video camera

Javacam [McD98] is an Internet-accessible video camera that was designed as a demonstration of a Java Nanokernel. The Java Nanokernel is designed to require very little memory. As a result, Javacam can provide an Internet interface with a National Semiconductor NS486SXF microprocessor and 1.5 MB of memory.

Javacam was built from a Connectix QuickCam, a widely available, low-cost video camera for PCs that can send and receive data on a standard PC parallel port.

The illustration below shows how QuickCam operates as a Java applet.

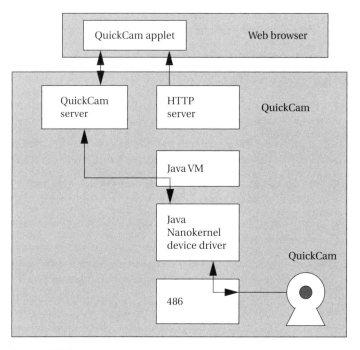

From [McD98].

The HTTP server returns a page containing a piece of Java code that acts as an applet to talk to the device. That Java code running on the Web browser requests an image from the QuickCam server on the QuickCam. The QuickCam server, which executes on top of the Java virtual machine and Java Nanokernel, grabs an image from the QuickCam, performs required transformations, and returns the data to the applet running on the Web browser.

The QuickCam driver communicates with the camera over a parallel port. It provides three basic functions: qc_initialize(); qc_send_command(), which sends commands to the camera; and qc_take_picture(), which returns a picture. Those functions are implemented in C.

8.4.3 Internet Security

Connecting an embedded system to the Internet opens up the system to the same sorts of attacks that are made on PCs and servers every day. However, attacks on embedded systems can destroy not only information but also the physical devices connected to the embedded processor. Dzung et al. [Dzu05] listed several example attacks that caused significant damage:

- A work infected the computer network of the CSX railway, causing all trains in the Washington, DC area to be shut down for a half day.

- A worm disabled the computer-based safety monitoring system at the Davis-Besse nuclear power plant in Ohio.

- A former consultant to a waste water plant in Australia used its computers to release one million liters of sewage into the area waterways.

They point out that security can be enforced at all levels of the network stack. General network security principles can be applied to Internet-enabled embedded systems; various industrial standards also deal with measures specific to industrial networks.

8.5 VEHICLES AS NETWORKS

Modern cars and planes rely on electronics to operate. About one-third of the total cost of an airplane or car comes from its electronics. Electronic systems are used in all aspects of the vehicle—safety-critical control, navigation and systems monitoring, and passenger comfort. These electronic devices are connected using data networks.

Networks are used for a variety of purposes in vehicles, with varying requirements on reliability and performance:

- Vehicle control (steering and brakes in cars, flight control surfaces in airplanes) is the most critical operation in the vehicle since it determines vehicle stability.

- Instruments for the driver or pilot must be reliable but often operate at higher data rates than do the vehicle control systems.

- Crew information systems may provide intercom functions, etc.

- Passenger systems provide entertainment, Internet access, etc.

Early vehicle networks assigned a separate processor to each physical device. Today, network designers tend to combine several functions onto one processor. In cars, the engine controller is the prime candidate for system compute server. This trend plays out more slowly in automobiles, but modern systems assign multiple tasks to a CPU in order to reduce the number of processors and their associated support hardware.

Automotive and aviation electronics *(avionics)* are similar in many respects but have some important differences. We will start with automotive electronics and then go on to discuss avionics.

8.5.1 Automotive Networks

The **CAN bus** [Bos07] was designed for automotive electronics and was first used in production cars in 1991. CAN is very widely used in cars as well as in other applications.

The CAN bus uses bit-serial transmission. CAN runs at rates of 1 MB/s over a twisted pair connection of 40 m. An optical link can also be used. The bus protocol supports multiple masters on the bus. Many of the details of the CAN and I^2C buses are similar, but there are also significant differences.

As shown in Figure 8.22, each node in the CAN bus has its own electrical drivers and receivers that connect the node to the bus in wired-AND fashion. In CAN terminology, a logical 1 on the bus is called *recessive* and a logical 0 is *dominant*. The driving circuits on the bus cause the bus to be pulled down to 0 if any node on the bus pulls the bus down (making 0 dominant over 1). When all nodes are transmitting 1s, the bus is said to be in the recessive state; when a node transmits a 0, the bus is in the dominant state. Data are sent on the network in packets known as *data frames*.

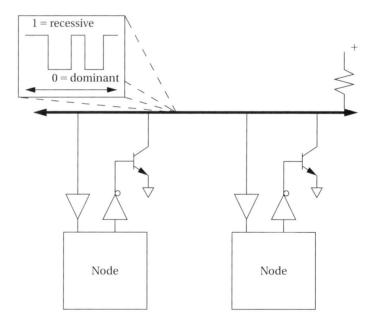

FIGURE 8.22

Physical and electrical organization of a CAN bus.

CAN is a synchronous bus—all transmitters must send at the same time for bus arbitration to work. Nodes synchronize themselves to the bus by listening to the bit transitions on the bus. The first bit of a data frame provides the first synchronization opportunity in a frame. The nodes must also continue to synchronize themselves against later transitions in each frame.

The format of a CAN data frame is shown in Figure 8.23. A data frame starts with a 1 and ends with a string of seven zeroes. (There are at least three bit fields between data frames.) The first field in the packet contains the packet's destination address and is known as the arbitration field. The destination identifier is 11 bits long. The trailing remote transmission request (RTR) bit is set to 0 if the data frame is used to request data from the device specified by the identifier. When RTR = 1, the packet is used to write data to the destination identifier. The control field provides an identifier extension and a 4-bit length for the data field with a 1 in between. The data field is from 0 to 64 bytes, depending on the value given in the control field. A cyclic redundancy check (CRC) is sent after the data field for error detection. The acknowledge field is used to let the identifier signal whether the frame was correctly received: The sender puts a recessive bit (1) in the ACK slot of the acknowledge field; if the receiver detected an error, it forces the value to a dominant (0) value. If the sender sees a 0 on the bus in the ACK slot, it knows that it must retransmit. The ACK slot is followed by a single bit delimiter followed by the end-of-frame field.

Control of the CAN bus is arbitrated using a technique known as Carrier Sense Multiple Access with Arbitration on Message Priority (CSMA/AMP). (As seen in

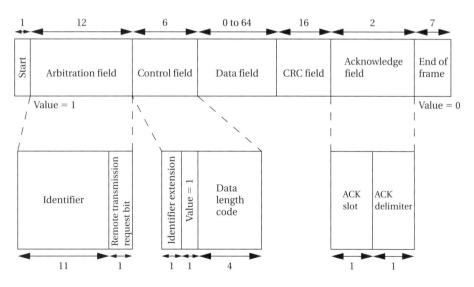

FIGURE 8.23

The CAN data frame format.

Section 8.2.2, Ethernet uses CSMA without AMP.) This method is similar to the I^2C bus's arbitration method; like I^2C, CAN encourages a data-push programming style. Network nodes transmit synchronously, so they all start sending their identifier fields at the same time. When a node hears a dominant bit in the identifier when it tries to send a recessive bit, it stops transmitting. By the end of the arbitration field, only one transmitter will be left. The identifier field acts as a priority identifier, with the all-0 identifier having the highest priority.

A remote frame is used to request data from another node. The requestor sets the RTR bit to 0 to specify a remote frame; it also specifies zero data bits. The node specified in the identifier field will respond with a data frame that has the requested value. Note that there is no way to send parameters in a remote frame—for example, you cannot use an identifier to specify a device and provide a parameter to say which data value you want from that device. Instead, each possible data request must have its own identifier.

An error frame can be generated by any node that detects an error on the bus. Upon detecting an error, a node interrupts the current transmission with an error frame, which consists of an error flag field followed by an error delimiter field of 8 recessive bits. The error delimiter field allows the bus to return to the quiescent state so that data frame transmission can resume. The bus also supports an overload frame, which is a special error frame sent during the interframe quiescent period. An overload frame signals that a node is overloaded and will not be able to handle the next message. The node can delay the transmission of the next frame with up to two overload frames in a row, hopefully giving it enough time to recover from its overload. The CRC field can be used to check a message's data field for correctness.

If a transmitting node does not receive an acknowledgment for a data frame, it should retransmit the data frame until the frame is acknowledged. This action corresponds to the data link layer in the OSI model.

Figure 8.24 shows the basic architecture of a typical CAN controller. The controller implements the physical and data link layers; since CAN is a bus, it does not need network layer services to establish end-to-end connections. The protocol control block is responsible for determining when to send messages, when a message must be resent due to arbitration losses, and when a message should be received.

The FlexRay network has been designed as the next generation of system buses for cars. FlexRay provides high data rates—up to 10 MB/s—with deterministic communication. It is also designed to be fault-tolerant.

The Local Interconnect Network (LIN) bus [Bos07] was created to connect components in a small area, such as a single door. The physical medium is a single wire that provides data rates of up to 20 KB/s for up to 16 bus subscribers. All transactions are initiated by the master and responded to by a frame. The software for the network is often generated from a LIN description file that describes the network subscribers, the signals to be generated, and the frames.

Several buses have come into use for passenger entertainment. Bluetooth is becoming the standard mechanism for cars to interact with consumer electronics devices such as audio players or phones. The Media Oriented Systems Transport

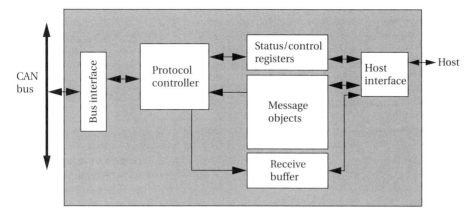

FIGURE 8.24

Architecture of a CAN controller.

(MOST) bus [Bos07] was designed for entertainment and multimedia information. The basic MOST bus runs at 24.8 MB/s and is known as MOST 25; 50 and 150 MB/s versions have also been developed. MOST can support up to 64 devices. The network is organized as a ring.

Data transmission is divided into channels. A control channel transfers control and system management data. Synchronous channels are used to transmit multimedia data; MOST 25 provides up to 15 audio channels. An asynchronous channel provides high data rates but without the quality-of-service guarantees of the synchronous channels.

8.5.2 Avionics

The most fundamental difference between avionics and automotive electronics is **certification**. Anything that is permanently attached to the aircraft must be certified. The certification process for production aircraft is twofold: first, the design is certified in a process known as **type certification**; then, the manufacture of each aircraft is certified during production.

The certification process is a prime reason why avionics architectures are more conservative than automotive electronics systems. The traditional architecture [Hel04] for an avionics system has a separate unit for each function: artificial horizon, engine control, flight surfaces, etc. These units are known as **line replaceable units** and are designed to be easily plugged and unplugged into the aircraft during maintenance.

A more sophisticated system is bus-based. The Boeing 777 avionics [Mor07], for example, is built from a series of racks. Each rack is a set of core processor modules (CPMs), I/O modules, and power supplies. The CPMs may implement one or more functions. A bus known as SAFEbus connects the modules. Cabinets are connected together using serial bus known as ARINC 629.

A more distributed approach to avionics is the ***federated network***. In this architecture, a function or several functions have their own network. The networks share data as necessary for the interaction of these functions. A federated architecture is designed so that a failure in one network will not interfere with the operation of the other networks.

The Genesis Platform [Wal07] is a next-generation architecture for avionics and safety-critical systems; it is used on the Boeing 787. Unlike federated architectures, it does not require a one-to-one correspondence between application groups and network units. In contrast, Genesis defines a virtual system for the avionics applications that are then mapped onto a physical network that may have a different topology.

8.6 SENSOR NETWORKS

Sensor networks are large-scale embedded systems that may contain tens of thousands or millions of nodes. Sensors are used in a wide variety of applications: manufacturing plants, weather reporting, etc. Traditional sensor systems use custom wiring to bring data to centralized computers for analysis. Sensor networks use standardized platforms to transport data either for analysis at a server or for computing directly in the network.

Sensor networks generally rely on battery-operated nodes and wireless data communication. Eliminating wires for power and data allows sensors to be deployed in environments that are not feasible for traditional sensors.

However, this combination of components presents many challenges. Because batteries have only a limited energy capacity, that energy must be rigorously conserved. But wireless communication requires much more energy than does communication by wire. In addition to traditional energy conservation techniques, we must develop new networking methods that conserve energy in wireless environments.

The Internet is designed to be resilient, but it is still too structured for many sensor network applications. Sensor networks must be installed by non-computer scientists. The nodes in the network are physically distributed and nodes may fail or be introduced to the network over time. Because they do not use wires, the structure of the connections between nodes is not designed in advance.

An ***ad hoc network*** organizes itself without intervention of a network administrator. *Ad hoc* networks allow users to distribute a set of wireless sensor network nodes and let the nodes organize their communication links for themselves.

An *ad hoc* network must be able to do several things:

- Nodes must be able to declare themselves to be part of the network and to determine what other nodes are in the network. Admission control policies determine how nodes can be admitted.

- The network must determine how to route data. Sensor networks generally establish a network structure, such as a grid, based upon the relative locations of the network nodes. Data are then routed based upon these selected communication paths.

- When nodes enter or leave the network, the network must update its configuration and routing.

In order to support these network operations, nodes in the network must be fairly capable:

- The node must be able to turn its radio on and off quickly and efficiently. Power wasted during power-up and power-down is not available for operating the network.

- The radios in the nodes may need to operate at several different power levels to avoid interference and save battery power. They may also need to operate at several frequencies to avoid interference.

- The node must be able to buffer network traffic and make routing decisions.

Power management and networking are intimately related. The power profiles of sensor nodes help determine the characteristics of networking protocols.

The sensor node's radio consumes much more energy than its processor. Transmitting 1 bit of information takes roughly 100 times more energy than an arithmetic operation. As a result, nodes can save energy by spending computing cycles to determine when to turn their radios on and off.

Furthermore, sensor node radios spend more energy receiving than transmitting. In most radio applications, transmitting is assumed to take more energy than receiving. However, sensor nodes spend most of their time listening. Therefore, power management protocols must take into account the energy consumption of reception.

A basic sensor network moves data from sensors to servers for processing. This approach makes sense for low data rate applications. However, higher data rate applications like audio and video benefit from performing at least some of the data analysis in network nodes.

Because communication costs more energy than computation, in-network processing saves energy if it reduces the volume of data transmitted over the network. In many cases, we can generate abstractions of the raw data that can be transmitted at much less cost. However, processing may require trading data between nodes, so the net amount of communication must be carefully considered.

8.7 ELEVATOR CONTROLLER

Design Example

We willuse the principles of distributed system design by designing an elevator controller. The components are physically distributed among the elevators and

floors of the building, and the system must meet both hard (making sure the elevator stops at the right point) and soft (responding to requests for elevators) deadlines.

8.7.1 Theory of Operation and Requirements

We design a multiple elevator system to increase the challenge. The configuration of a bank of elevators is shown in Figure 8.25. The *elevator car* is the unit that runs up and down the *hoistway* (also known as the shaft) carrying passengers; we will use N to represent the number of hoistways. Each car runs in a hoistway and can stop at any of F floors. (For convenience we will number the floors 1 through F, although some of the elevator doors may in fact be in the basement.) Every elevator car has a *car control panel* that allows the passengers to select floors to stop at. Each floor has a single *floor control panel* that calls for an elevator. Each floor also has a set of displays to show the current state of the elevator systems.

The user interface consists of the *elevator control panels*, floor control panels, and *displays*. The car control panels have F buttons to request the floors plus an emergency stop button. Each floor control panel has an up button and a down button that request an elevator going in the chosen direction. There is one display

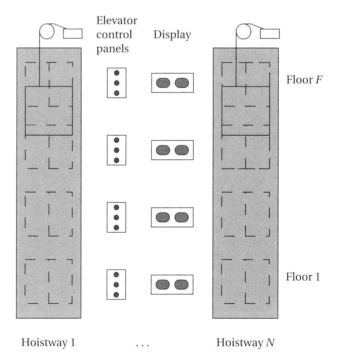

FIGURE 8.25

A bank of elevators.

per *hoistway* on each floor. Each *display* has an up light and a down light; if the elevator is idle, neither light is on. The displays for a *hoistway* always show the same state on all floors.

The elevator control system consists of two types of components. First, a single master controller governs the overall behavior of all elevators, and second, on each elevator a car controller runs everything that must be done within the car. The car controller must of course sense button presses on the car control panel, but it must also sense the current position of the elevator. As shown in Figure 8.26, the car controller reads two sets of indicators on the wall of the elevator hoistway to sense position. The *coarse indicators* run the entire length of the hoistway and a *sensor* determines when the elevator passes each one. *Fine indicators* are located only around the stopping point for each floor. There are $2S + 1$ fine indicators on each floor, one at the exact stopping point and S on each side of it. The sensor also reads fine indicators; it puts out separate signals for the coarse and fine indicators. The elevator system can stop at the proper position by counting coarse and fine indicators.

The elevator's movement is controlled by two motor control inputs: one for up and one for down. When both are disabled, the elevator does not move. The system should not enable both up and down on a single hoistway simultaneously.

The master controller has several tasks—it must read inputs from the floor control panels, send signals to the lights on the floor displays, read floor requests from the car controllers, and take inputs from the car sensors. Most importantly, it must tell the elevators when to move and when to stop. It must also schedule the elevators to efficiently answer passenger requests.

The basic requirements for the elevator system follow.

Name	Elevator system
Inputs	F floor control inputs, N position sensors, N car control panels, one master control panel
Outputs	F displays, N motor controllers
Functions	Responds to floor, car, and master control panels; operates cars safely
Performance	Control of elevators is time critical
Manufacturing cost	Cost of electronics is small compared to mechanical systems
Power	Not important
Physical size and weight	Cabling is the major concern

In this design, we are much more aware of the surrounding mechanical elements than we have been in previous examples. The electronics are clearly a small part of

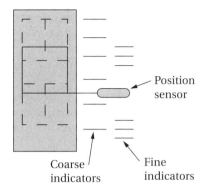

FIGURE 8.26

Sensing elevator position.

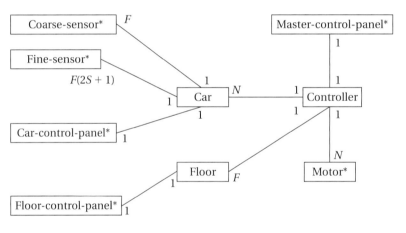

FIGURE 8.27

Basic class diagram for the elevator system.

the cost and bulk of the elevator system. But because the elevators are controlled by the computers, the proper operation of the embedded hardware and software is very important.

8.7.2 Specification

The basic class diagram for the elevator system is shown in Figure 8.27. This diagram concentrates on the relationships among the classes and the number of objects of each type that the system requires.

The physical interface classes are defined in more detail in Figure 8.28. We have used inheritance to define the sensors, even though these classes represent physical

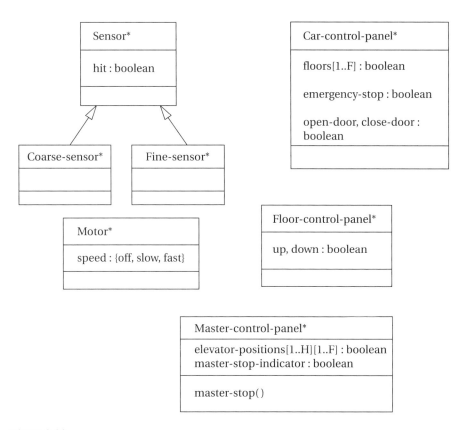

FIGURE 8.28

Physical interface classes for the elevator system.

objects. The only difference among the sensors to the elevator controller is whether they indicate coarse or fine positions; other physical distinctions among the sensors do not matter.

The *Car* and *Floor* classes, which describe the control panels on the floors and in the cars, are shown in Figure 8.29. These classes define the basic attributes of the car and floor control panels.

The *Controller* class is defined in Figure 8.30. This class defines attributes that describe the state of the system, including where each car is and whether the system has made an emergency stop. It also defines several behaviors, such as an *operate* behavior and behaviors to check the state of parts of the system.

8.7.3 Architecture

Computation and I/O occur at three major locations in this system: the floor control panels/displays, the elevator cabs, and the system controller. Let's consider the basic

Car
request lights[1..F] : integer current-floor : integer

Floor
up-light, down-light : boolean

FIGURE 8.29

The Car and Floor classes.

Controller
car-floor[1..H] : integer emergency-stop[1..H] : boolean
scan-cars() scan-floors() scan-master-panel() operate()

FIGURE 8.30

The Controller class for the elevator system.

operation of each of these subsystems one at a time and then go back and design the network that connects them.

The floor control panels and displays are relatively simple since they have no hard real-time requirements. Each one takes a set of inputs for the up/down indicators and lights the appropriate lights. Each also watches for button events and sends the results to the system controller. We can use a simple microcontroller for all these tasks.

The cab controller must read the cab's buttons and send events to the system controller. It must also read the sensor inputs and send them to the system controller. Reading the sensors is a hard real-time task—proper operation of the elevator requires that the cab controller not miss any of the indicators. We have to decide whether to use one or two PEs in the cab. A conservative design would use separate PEs for the button panel and the sensor. We could also use a single processor to handle both the buttons and the sensor.

The system controller must take inputs from all these units. Its control of the elevators has both hard and soft real-time aspects: It must constantly monitor all moving elevators to be sure they stop properly, as well as choose which elevator to dispatch to a request.

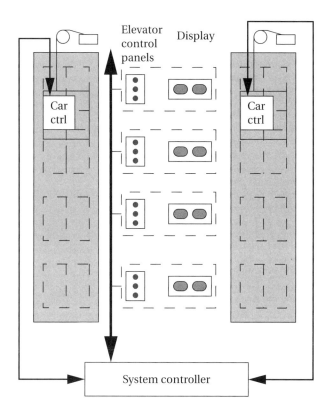

FIGURE 8.31

The networks in the elevator.

Figure 8.31 shows the set of networks we will use in the system. The floor control panels/displays are connected along a single bus network. Each elevator car has its own point-to-point link with the system controller.

8.7.4 Testing

The simplest way to test the controllers is to build an elevator simulator using an FPGA. We can easily program an FPGA to simulate several elevators by keeping registers for the current position of each elevator and using counters to control how often the elevators change state. Using an FPGA-based elevator simulator provides good motivation for this example because we can design the FPGA to indicate when an elevator has crashed through the floor or the ceiling of its shaft. Working with a real-time-oriented elevator simulator helps illustrate the challenges presented by real-time control. We can use a serial link from a PC to provide button inputs, or we can wire up panels of buttons and indicators ourselves.

SUMMARY

We often need or want to build an embedded system out of a network of connected processors. The uses of distributed embedded systems vary greatly, ranging from the real-time networks in an automobile to Internet-enabled information appliances. There are a great many networks that we can choose from to build embedded systems based on constraints of cost, overall throughput, and real-time behavior.

What We Learned

- Distributed embedded systems often make sense for cost, performance, and physical reasons.

- The OSI layer model breaks down the structure of a network into seven layers.

- A large number of networks, many with very different characteristics, are used in embedded systems.

- Performance analysis must take into account network delay.

- The Internet is not ideally suited to hard real-time operation, but it can be very useful in building a user interface and in simplifying the integration of systems with multiple nodes.

- Sensor networks use ad hoc networking techniques to simplify installation and operation.

FURTHER READING

Kopetz [Kop97] provides a thorough introduction to the design of distributed embedded systems. Stallings [Sta97A] provides a good introduction to data networking. A variety of manufacturers make components for interfacing to popular networks and microprocessors with built-in network interfaces. The book by Robert Bosch GmbH [Bos07] discusses automotive electronics in detail. The *Digital Aviation Handbook* [Spi07] describes the avionics systems of several aircraft. Wireless sensor networks are discussed in books by Karl and Willig [Kar06] and Zhao and Guibas [Zha04].

QUESTIONS

Q8-1 Describe an I^2C bus at the following OSI-compliant levels of detail:

 a. physical,

 b. data link,

 c. network, and

 d. transport.

Q8-2 Describe a 10Base-T Ethernet at the following OSI-compliant levels of detail:

 a. physical,

 b. data link,

 c. network, and

 d. transport.

Q8-3 Show the order in which requests would be answered in the timeline below, assuming that each takes one time unit to satisfy, under the following arbitration schemes:

 a. fixed: a highest, b middle, c lowest, and

 b. round robin.

$c\,(t=0)$	$a,b\,(t=5)$	$c\,(t=6)$	$a,c\,(t=15)$	$b\,(t=16)$	$a,b\,(t=25)$	$a,b,c\,(t=30)$

0	5	10	15	20	25	30	35

Q8-4 Answer question Q8-3, assuming that each request takes two time units to satisfy.

Q8-5 Answer question Q8-3, using the arrival times below and a request satisfaction time of two time units.

$a,b,c\,(t=0)$	$c\,(t=7)$	$a\,(t=8)$	$c(t=10)$	$a,b,c\,(t=15)$	$a\,(t=20)$	$b,c\,(t=22)$

0	5	10	15	20	25	30

Q8-6 Describe how an IP packet may be sent from a client on one Ethernet to a client on a second Ethernet. The two Ethernets are connected by a router.

Q8-7 What services would the Javacam of Application Example 8.1 require at the following levels of the OSI model:

 a. application,

 b. presentation,

 c. session, and

 d. transport.

Q8-8 Using the methodology of Example 8.2, plot both the transmission time for 1 byte as a function of the I^2C clock speed and the microcontroller overhead as a function of the number of instructions executed. Determine the values for bus clock speed and the number of instructions at which the transmission delay equals the overhead.

Q8-9 What is the longest time that a processing element may have to wait between two successive data transmissions on a round-robin arbitrated bus? Assume that each data transmission requires one time unit.

Q8-10 How can an automotive network ensure that safety-critical components are not starved of bus access—that they are guaranteed to be able to transmit within a certain amount of time?

Q8-11 Give examples of the component networks in a federated network for an automobile.

Q8-12 Give an example of a simple protocol that would allow sensor nodes in a sensor network to determine the other nodes with which they can communicate.

LAB EXERCISES

L8-1 Build an experimental setup that lets you monitor messages on an embedded network.

L8-2 Measure the effects of collisions on an Ethernet (doing so, of course, on a network where you will not disturb other users). Plot the amount of time required to successfully deliver a message as a function of network load.

System Design Techniques

- A deeper look into design methodologies, requirements, specification, and system analysis.

- Formal and informal methods for system specification.

- Quality assurance.

INTRODUCTION

In this chapter we consider the techniques required to create complex embedded systems. Thus far, our design examples have been small so that important concepts can be conveyed relatively simply. However, most real embedded system designs are inherently complex, given that their functional specifications are rich and they must obey multiple other requirements on cost, performance, and so on. We need methodologies to help guide our design decisions when designing large systems.

In the next section we look at design methodologies in more detail. Section 9.2 studies requirements analysis, which captures informal descriptions of what a system must do, while Section 9.3 considers techniques for more formally specifying system functionality. Section 9.4 focuses on details of system analysis methodologies. Section 9.5 discusses the topic of quality assurance (QA), which must be considered throughout the design process to ensure a high-quality design.

9.1 DESIGN METHODOLOGIES

This section considers the complete ***design methodology***—a ***design process***—for embedded computing systems. We will start with the rationale for design methodologies, then look at several different methodologies.

9.1.1 Why Design Methodologies?

Process is important because without it, we can't reliably deliver the products we want to create. Thinking about the sequence of steps necessary to build something may seem superfluous. But the fact is that everyone has their own design process, even if they don't articulate it. If you are designing embedded systems in your basement by yourself, having your own work habits is fine. But when

several people work together on a project, they need to agree on who will do things and how they will get done. Being explicit about process is important when people work together. Therefore, since many embedded computing systems are too complex to be designed and built by one person, we have to think about design processes.

The obvious goal of a design process is to create a product that does something useful. Typical specifications for a product will include functionality (e.g., cell phone), manufacturing cost (must have a retail price below $200), performance (must power up within 3 s), power consumption (must run for 12 h on two AA batteries), or other properties. Of course, a design process has several important goals beyond function, performance, and power. Three of these goals are summarized below.

- *Time-to-market:* Customers always want new features. The product that comes out first can win the market, even setting customer preferences for future generations of the product. The profitable market life for some products is 3–6 months—if you are 3 months late, you will never make money. In some categories, the competition is against the calendar, not just competitors. Calculators, for example, are disproportionately sold just before school starts in the fall. If you miss your market window, you have to wait a year for another sales season.

- *Design cost:* Many consumer products are very cost sensitive. Industrial buyers are also increasingly concerned about cost. The costs of designing the system are distinct from manufacturing cost—the cost of engineers' salaries, computers used in design, and so on must be spread across the units sold. In some cases, only one or a few copies of an embedded system may be built, so design costs can dominate manufacturing costs. Design costs can also be important for high-volume consumer devices when time-to-market pressures cause teams to swell in size.

- *Quality:* Customers not only want their products fast and cheap, they also want them to be right. A design methodology that cranks out shoddy products will soon be forced out of the marketplace. Correctness, reliability, and usability must be explicitly addressed from the beginning of the design job to obtain a high-quality product at the end.

Processes evolve over time. They change due to external and internal forces. Customers may change, requirements change, products change, and available components change. Internally, people learn how to do things better, people move on to other projects and others come in, and companies are bought and sold to merge and shape corporate cultures.

Software engineers have spent a great deal of time thinking about software design processes. Much of this thinking has been motivated by mainframe software such as databases. But embedded applications have also inspired some important thinking about software design processes.

A good methodology is critical to building systems that work properly. Delivering buggy systems to customers always causes dissatisfaction. But in some applications, such as medical and automotive systems, bugs create serious safety problems that can endanger the lives of users. We discuss quality in more detail in Section 9.5. As an introduction, Application Example 9.1 describes problems that led to the loss of an unmanned Martian space probe.

Application Example 9.1

Loss of the Mars Climate Observer

In September 1999, the Mars Climate Observer, an unmanned U.S. spacecraft designed to study Mars, was lost—it most likely exploded as it heated up in the atmosphere of Mars after approaching the planet too closely. The spacecraft came too close to Mars because of a series of problems, according to an analysis by *IEEE Spectrum* and contributing editor James Oberg [Obe99]. From an embedded systems perspective, the first problem is best classified as a requirements problem. The contractors who built the spacecraft at Lockheed Martin calculated values for use by the flight controllers at the Jet Propulsion Laboratory (JPL). JPL did not specify the physical units to be used, but they expected them to be in Newtons. The Lockheed Martin engineers returned values in units of pound force. This discrepancy resulted in trajectory adjustments being 4.45 times larger than they should have been. The error was not caught by a software configuration process nor was it caught by manual inspections. Although there were concerns about the spacecraft's trajectory, errors in the calculation of the spacecraft's position were not caught in time.

9.1.2 Design Flows

A ***design flow*** is a sequence of steps to be followed during a design. Some of the steps can be performed by tools, such as compilers or CAD systems; other steps can be performed by hand. In this section we look at the basic characteristics of design flows.

Figure 9.1 shows the ***waterfall model*** introduced by Royce [Dav90], the first model proposed for the software development process. The waterfall development model consists of five major phases: *requirements* analysis determines the basic characteristics of the system; *architecture* design decomposes the functionality into major components; *coding* implements the pieces and integrates them; *testing* uncovers bugs; and *maintenance* entails deployment in the field, bug fixes, and upgrades. The waterfall model gets its name from the largely one-way flow of work and information from higher levels of abstraction to more detailed design steps (with a limited amount of feedback to the next-higher level of abstraction). Although top–down design is ideal since it implies good foreknowledge of the implementation during early design phases, most designs are clearly not quite so top–down. Most design projects entail experimentation and changes that require bottom–up feedback. As a result, the waterfall model is today cited as an unrealistic

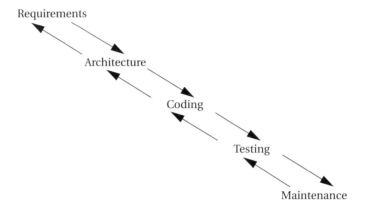

FIGURE 9.1

The waterfall model of software development.

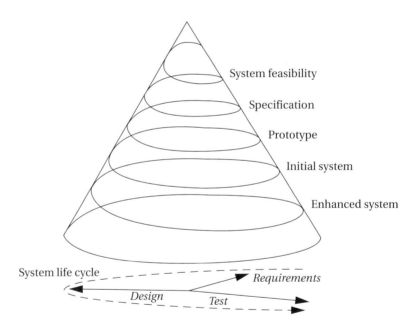

FIGURE 9.2

The spiral model of software design.

design process. However, it is important to know what the waterfall model is to be able to understand and how others are reacting against it.

Figure 9.2 illustrates an alternative model of software development called the **spiral model** [Boe87]. While the waterfall model assumes that the system is built once in its entirety, the spiral model assumes that several versions of the system

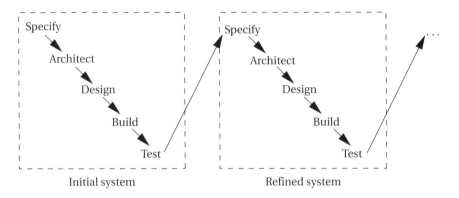

FIGURE 9.3

A successive refinement development model.

will be built. Early systems will be simple mock-ups constructed to aid designers' intuition and to build experience with the system. As design progresses, more complex systems will be constructed. At each level of design, the designers go through requirements, construction, and testing phases. At later stages when more complete versions of the system are constructed, each phase requires more work, widening the design spiral. This successive refinement approach helps the designers understand the system they are working on through a series of design cycles. The first cycles at the top of the spiral are very small and short, while the final cycles at the spiral's bottom add detail learned from the earlier cycles of the spiral. The spiral model is more realistic than the waterfall model because multiple iterations are often necessary to add enough detail to complete a design. However, a spiral methodology with too many spirals may take too long when design time is a major requirement.

Figure 9.3 shows a *successive refinement* design methodology. In this approach, the system is built several times. A first system is used as a rough prototype, and successive models of the system are further refined. This methodology makes sense when you are relatively unfamiliar with the application domain for which you are building the system. Refining the system by building several increasingly complex systems allows you to test out architecture and design techniques. The various iterations may also be only partially completed; for example, continuing an initial system only through the detailed design phase may teach you enough to help you avoid many mistakes in a second design iteration that is carried through to completion.

Embedded computing systems often involve the design of hardware as well as software. Even if you aren't designing a board, you may be selecting boards and plugging together multiple hardware components as well as writing code. Figure 9.4 shows a design methodology for a combined hardware/software project. Front-end activities such as specification and architecture simultaneously consider

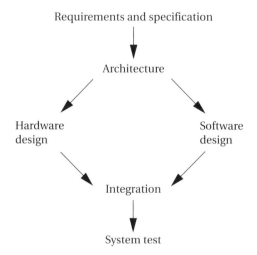

FIGURE 9.4

A simple hardware/software design methodology.

hardware and software aspects. Similarly, back-end integration and testing consider the entire system. In the middle, however, development of hardware and software components can go on relatively independently—while testing of one will require stubs of the other, most of the hardware and software work can proceed relatively independently.

In fact, many complex embedded systems are themselves built of smaller designs. The complete system may require the design of significant software components, custom logic, and so on, and these in turn may be built from smaller components that need to be designed. The design flow follows the levels of abstraction in the system, from complete system design flows at the most abstract to design flows for individual components. The design flow for these complex systems resembles the flow shown in Figure 9.5. The implementation phase of a flow is itself a complete flow from specification through testing. In such a large project, each flow will probably be handled by separate people or teams. The teams must rely on each other's results. The component teams take their requirements from the team handling the next higher level of abstraction, and the higher-level team relies on the quality of design and testing performed by the component team. Good communication is vital in such large projects.

When designing a large system along with many people, it is easy to lose track of the complete design flow and have each designer take a narrow view of his or her role in the design flow. *Concurrent engineering* attempts to take a broader approach and optimize the total flow. Reduced design time is an important goal for concurrent engineering, but it can help with any aspect of the design that cuts across the design flow, such as reliability, performance, power consumption, and so on. It tries to eliminate "over-the-wall" design steps, in which one designer

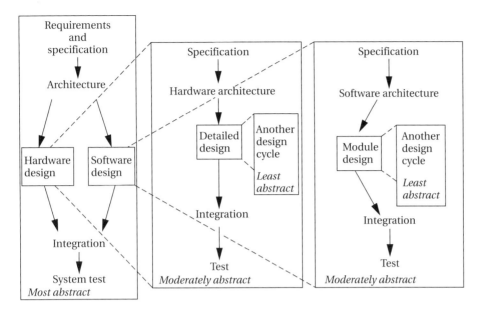

FIGURE 9.5

A hierarchical design flow for an embedded system.

performs an isolated task and then throws the result over the wall to the next designer, with little interaction between the two. In particular, reaping the most benefits from concurrent engineering usually requires eliminating the wall between design and manufacturing. Concurrent engineering efforts are comprised of several elements:

- *Cross-functional teams* include members from various disciplines involved in the process, including manufacturing, hardware and software design, marketing, and so forth.

- *Concurrent product realization* process activities are at the heart of concurrent engineering. Doing several things at once, such as designing various subsystems simultaneously, is critical to reducing design time.

- *Incremental information sharing* and use helps minimize the chance that concurrent product realization will lead to surprises. As soon as new information becomes available, it is shared and integrated into the design. Cross-functional teams are important to the effective sharing of information in a timely fashion.

- *Integrated project management* ensures that someone is responsible for the entire project, and that responsibility is not abdicated once one aspect of the work is done.

- *Early and continual supplier involvement* helps make the best use of suppliers' capabilities.

- *Early and continual customer focus* helps ensure that the product best meets customers' needs.

Example 9.1 describes the experiences of a telephone system design organization with concurrent engineering.

Example 9.1

Concurrent engineering applied to telephone systems

A group at AT&T applied concurrent engineering to the design of PBXs (telephone switching systems) [Gat94]. The company had a large existing organization and methodology for designing PBXs; their goal was to re-engineer their process to reduce design time and make other improvements to the end product. They used the seven-step process described below.

1. *Benchmarking:* They compared themselves to competitors and found that it took them 30% longer to introduce a new product than their best competitors. Based on this study, they decided to shoot for a 40% reduction in design time.

2. *Breakthrough improvement:* Next, they identified the factors that would influence their effort. Three major factors were identified: increased partnership between design and manufacturing; continued existence of the basic organization of design labs and manufacturing; and support of managers at least two levels above the working level. As a result, three groups were established to help manage the effort. A *steering committee* was formed by midlevel managers to provide feedback on the project. A *project office* was formed by an engineering manager and an operations analyst from the AT&T internal consulting organization. Finally, a *core team* of engineers and analysts was formed to make things happen.

3. *Characterization of the current process:* The core team built flowcharts and used other techniques to understand the current product development process. The existing design and manufacturing process resembled the figure below. The core team identified several root causes of delays that had to be remedied. First, too many design and manufacturing tasks were performed sequentially. Second, groups tended to focus on intermediate milestones related to their narrow job descriptions, rather than trying to take into account the effects of their decisions on other aspects of the development process. Third, too much time was spent waiting in queues—jobs were handed off from one person to another very frequently. In many cases, the recipient of a set of jobs didn't know how to best prioritize the incoming tasks. Fixing this problem was deemed to be fundamentally a managerial problem, not a technical one. Finally, the team found that too many groups had their own design databases, creating redundant data that had to be maintained and synchronized.

4. *Create the target process:* Based on its studies, the core team created a model for the new development process, which is reproduced below.

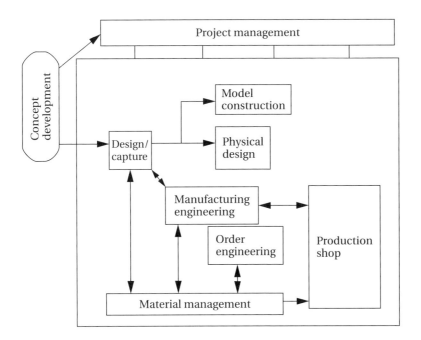

5. *Verify the new process:* The team undertook a pilot product development project to test the new process. The process was found to be basically sound. Some challenges were identified; for example, in the sequential project the design of circuit boards took longer than that of the mechanical enclosures, while in the new process the

enclosures ended up taking longer, pointing out the need to start designing them earlier.

6. *Implement across the product line:* After the pilot project, the new methodology was rolled out across the product lines. This activity required training of person-nel, documentation of the new standards and procedures, and improvements to information systems.

7. *Measure results and improve:* The performance of the new design flow was measured. The team found that product development time had been reduced from 18–30 months to 11 months.

9.2 REQUIREMENTS ANALYSIS

Before designing a system, we need to know what we are designing. The terms "requirements" and "specifications" are used in a variety of ways—some people use them as synonyms, while others use them as distinct phases. We use them to mean related but distinct steps in the design process. ***Requirements*** are informal descrip-tions of what the customer wants, while ***specifications*** are more detailed, precise, and consistent descriptions of the system that can be used to create the architec-ture. Both requirements and specifications are, however, directed to the outward behavior of the system, not its internal structure.

The overall goal of creating a requirements document is effective communication between the customers and the designers. The designers should know what they are expected to design for the customers; the customers, whether they are known in advance or represented by marketing, should understand what they will get.

We have two types of requirements: ***functional*** and ***nonfunctional***. A func-tional requirement states what the system must do, such as compute an FFT. A nonfunctional requirement can be any number of other attributes, including physical size, cost, power consumption, design time, reliability, and so on.

A good set of requirements should meet several tests [Dav90]:

- *Correctness:* The requirements should not mistakenly describe what the customer wants. Part of correctness is avoiding over-requiring—the require-ments should not add conditions that are not really necessary.

- *Unambiguousness:* The requirements document should be clear and have only one plain language interpretation.

- *Completeness:* All requirements should be included.

- *Verifiability:* There should be a cost-effective way to ensure that each require-ment is satisfied in the final product. For example, a requirement that the system package be "attractive" would be hard to verify without some agreed upon definition of attractiveness.

- *Consistency:* One requirement should not contradict another requirement.

- *Modifiability:* The requirements document should be structured so that it can be modified to meet changing requirements without losing consistency, verifiability, and so forth.

- *Traceability:* Each requirement should be traceable in the following ways:

 — We should be able to trace backward from the requirements to know why each requirement exists.

 — We should also be able to trace forward from documents created before the requirements (e.g., marketing memos) to understand how they relate to the final requirements.

 — We should be able to trace forward to understand how each requirement is satisfied in the implementation.

 — We should also be able to trace backward from the implementation to know which requirements they were intended to satisfy.

How do you determine requirements? If the product is a continuation of a series, then many of the requirements are well understood. But even in the most modest upgrade, talking to the customer is valuable. In a large company, marketing or sales departments may do most of the work of asking customers what they want, but a surprising number of companies have designers talk directly with customers. Direct customer contact gives the designer an unfiltered sample of what the customer says. It also helps build empathy with the customer, which often pays off in cleaner, easier-to-use customer interfaces. Talking to the customer may also include conducting surveys, organizing focus groups, or asking selected customers to test a mock-up or prototype.

9.3 SPECIFICATIONS

In this section we take a look at some advanced techniques for specification and how they can be used.

9.3.1 Control-Oriented Specification Languages

We have already seen how to use state machines to specify control in UML. An example of a widely used state machine specification language is the **SDL language** [Roc82], which was developed by the communications industry for specifying communication protocols, telephone systems, and so forth. As illustrated in Figure 9.6, SDL specifications include states, actions, and both conditional and unconditional transitions between states. SDL is an event-oriented state

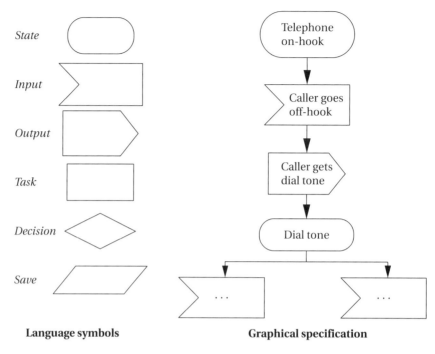

FIGURE 9.6

The SDL specification language.

machine model since transitions between states are caused by internal and external events.

Other techniques can be used to eliminate clutter and clarify the important structure of a state-based specification. The **Statechart** [Har87] is one well-known technique for state-based specification that introduced some important concepts. The Statechart notation uses an event-driven model. Statecharts allow states to be grouped together to show common functionality. There are two basic groupings: OR and AND. Figure 9.7 shows an example of an OR state by comparing a traditional state transition diagram with a Statechart described via an OR state. The state machine specifies that the machine goes to state $s4$ from any of $s1$, $s2$, or $s3$ when they receive the input $i2$. The Statechart denotes this commonality by drawing an OR state around $s1$, $s2$, and $s3$ (the name of the OR state is given in the small box at the top of the state). A single transition out of the OR state $s123$ specifies that the machine goes into state $s4$ when it receives the $i2$ input while in any state included in $s123$. The OR state still allows interesting transitions between its member states. There can be multiple ways to get into $s123$ (via $s1$ or $s2$), and there can be transitions between states within the OR state (such as from $s1$ to $s3$ or $s2$ to $s3$). The OR state is simply a tool for specifying some of the transitions relating to these states.

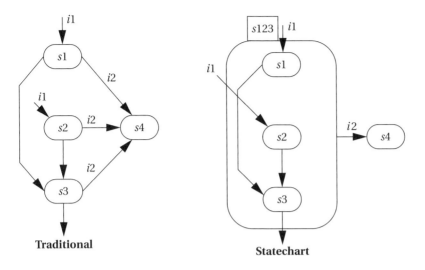

FIGURE 9.7

An OR state in Statecharts.

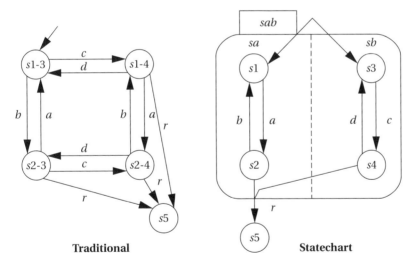

FIGURE 9.8

An AND state in Statecharts.

Figure 9.8 shows an example of an AND state specified in Statechart notation as compared to the equivalent in the traditional state machine model. In the traditional model, there are numerous transitions between the states; there is also one entry point into this cluster of states and one exit transition out of the cluster.

In the *Statechart*, the AND state *sab* is decomposed into two components, *sa* and *sb*. When the machine enters the AND state, it simultaneously inhabits the state *s*1 of component *sa* and the state *s*3 of component *sb*. We can think of the system's state as multidimensional. When it enters *sab*, knowing the complete state of the machine requires examining both *sa* and *sb*.

The names of the states in the traditional state machine reveal their relationship to the AND state components. Thus, state *s*1-3 corresponds to the Statechart machine having its *sa* component in *s*1 and its *sb* component in *s*3, and so forth. We can exit this cluster of states to go to state *s*5 only when, in the traditional specification, we are in state *s*2-4 and receive input *r*. In the AND state, this corresponds to *sa* in state *s*2, *sb* in state *s*4, and the machine receiving the *r* input while in this composite state. Although the *traditional* and *Statechart* models describe the same behavior, each component has only two states, and the relationships between these states are much simpler to see.

Leveson *et al.* [Lev94] used a different format, the **AND/OR table**, to describe similar relationships between states. An example AND/OR table and the Boolean expression it describes are shown in Figure 9.9. The rows in the AND/OR table are labeled with the basic variables in the expression. Each column corresponds to an AND term in the expression. For example, the AND term (*cond2* and not *cond3*) is represented in the second column with a *T* for *cond2*, an *F* for *cond3*, and a dash (don't-care) for *cond1*; this corresponds to the fact that *cond2* must be *T* and *cond3* *F* for the AND term to be true. We use the table to evaluate whether a given condition holds in the system. The current states of the variables are compared to the table elements. A column evaluates to true if all the current variable values correspond to the requirements given in the column. If any one of the columns evaluates to true, then the table's expression evaluates to true, as we would expect for an AND/OR expression. The most important difference between this notation

cond1 or (cond2 and !cond3)

Expression

OR

A	cond1	T	—
N	cond2	—	T
D	cond3	—	F

AND/OR table

FIGURE 9.9

An AND/OR table.

and Statecharts is that don't-cares are explicitly represented in the table, which was found to be of great help in identifying problems in a specification table.

9.3.2 Advanced Specifications

This section is devoted to a single example of a sophisticated system. Application Example 9.2 describes the specification of a real-world, safety-critical system used in aircraft. The specification techniques developed to ensure the correctness and safety of this system can also be used in many applications, particularly in systems where much of the complexity goes into the control structure.

Application Example 9.2

The TCAS II specification

TCAS II (Traffic Alert and Collision Avoidance System) is a collision avoidance system (CAS) for aircraft. Based on a variety of information, a TCAS unit in an aircraft keeps track of the position of other nearby aircraft. If TCAS decides that a mid-air collision may be likely, it uses audio commands to suggest evasive action—for example, a prerecorded voice may warn "DESCEND! DESCEND!" if TCAS believes that an aircraft above poses a threat and that there is room to maneuver below. TCAS makes sophisticated decisions in real time and is clearly safety critical. On the one hand, it must detect as many potential collision events as possible (within the limits of its sensors, etc.). On the other hand, it must generate as few false alarms as possible, since the extreme maneuvers it recommends are themselves potentially dangerous.

Leveson *et al.* [Lev94] developed a specification for the TCAS II system. We won't cover the entire specification here, but just enough to provide its flavor. The TCAS II specification was written in their RSML language. They use a modified version of State-chart notation for specifying states, in which the inputs to and outputs of the state are made explicit. The notation is illustrated below.

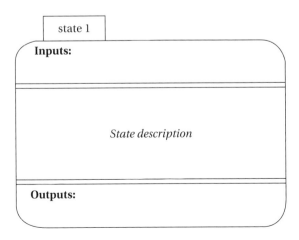

They also use a transition bus to show sets of states in which there are transitions between all (or almost all) states. In the following example, there are transitions from a, b, c, or d to any of the other states:

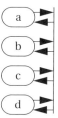

The top-level description of the CAS appears below.

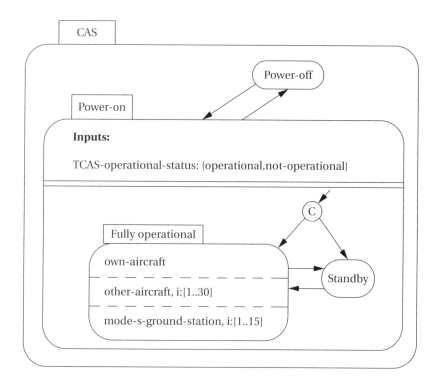

This diagram specifies that the system has *Power-off* and *Power-on* states. In the power-on state, the system may be in *Standby* or *Fully operational* mode. In the *Fully operational* mode, three components are operating in parallel, as specified by the AND State: the *own-aircraft* subsystem, a subsystem to keep track of up to 30 other aircraft,

and a subsystem to keep track of up to 15 Mode S ground stations, which provide radar information.

The next diagram shows a specification of the *Own-Aircraft* AND state. Once again, the behavior of *Own-Aircraft* is an AND composition of several subbehaviors. The *Effective-SL* and *Alt-SL* states are two ways to control the sensitivity level (SL) of the system, with each state representing a different sensitivity level. Differing sensitivities are required depending on distance from the ground and other factors. The *Alt-Layer* state divides the vertical airspace into layers, with this state keeping track of the current layer. *Climb-Inhibit* and *Descent-Inhibit* states are used to selectively inhibit climbs (which may be difficult at high altitudes) or descents (clearly dangerous near the ground), respectively. Similarly, the *Increase-Climb-Inhibit* and *Increase-Descend-Inhibit* states can inhibit high-rate climbs and descents. Because the *Advisory-Status* state is rather complicated, its details are not shown here.

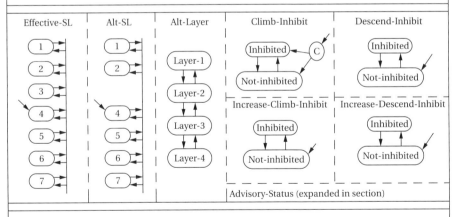

Own-Aircraft

Input:
own-alt-radio: integer
standby-discrete-input: {true, false}
own-alt-barometric: integer
mode-selector: {TA/RA, standby, TA-only,3,4,5,6,7}
radio-altimeter-status: {valid, not-valid}
own-air-status: {airborne, on-ground}
own-mode-s-address: integer
barometric-altimeter-status: {fine-coarse}

traffic-display-permitted: {true, false}
aircraft-altitude-limit: integer
prox-traffic-display: {true, false}
own-alt-rate: integer
config-climb-inhibit: {true, false}
altitude-climb-inhib-active: {true, false}
increase-climb-inhibit-discrete: {true, false}

Effective-SL Alt-SL Alt-Layer Climb-Inhibit Descend-Inhibit

Increase-Climb-Inhibit Increase-Descend-Inhibit

Advisory-Status (expanded in section)

Output:
sound-aural-alarm: {true, false}
aural-alarm-inhibit: {true, false}
combined-control-out: enumerated
vertical-control-out: enumerated

climb-RA: enumerated
descent-RA: enumerated
own-goal-alt-rate: integer
vertical-RAC: enumerated
horizontal-RAC: enumerated

9.4 SYSTEM ANALYSIS AND ARCHITECTURE DESIGN

In this section we consider how to turn a specification into an architecture design. We already have a number of techniques for making specific decisions; in this section we look at how to get a handle on the overall system architecture. The **CRC card** methodology is a well-known and useful way to help analyze a system's structure. It is particularly well suited to object-oriented design since it encourages the encapsulation of data and functions. The acronym *CRC* stands for the following three major items that the methodology tries to identify:

- *Classes* define the logical groupings of data and functionality.

- *Responsibilities* describe what the classes do.

- *Collaborators* are the other classes with which a given class works.

The name *CRC card* comes from the fact that the methodology is practiced by having people write on index cards. (In the United States, the standard size for index cards is $3'' \times 5''$, so these cards are often called 3×5 cards.) An example card is shown in Figure 9.10; it has space to write down the class name, its responsibilities and collaborators, and other information. The essence of the CRC card methodology is to have people write on these cards, talk about them, and update the cards until they are satisfied with the results.

This technique may seem like a primitive way to design computer systems. However, it has several important advantages. First, it is easy to get noncomputer people to create CRC cards. Getting the advice of domain experts (automobile designers for automotive electronics or human factors experts for PDA design, for example) is very important in system design. The CRC card methodology is informal enough that it will not intimidate non-computer specialists and will allow you to capture their input. Second, it aids even computer specialists by encouraging

Class name: Superclasses: Subclasses: Responsibilities: Collaborators:	Class name: Class's function: Attributes:
Front	**Back**

FIGURE 9.10

Layout of a CRC card.

them to work in a group and analyze scenarios. The walkthrough process used with CRC cards is very useful in scoping out a design and determining what parts of a system are poorly understood. This informal technique is valuable to tool-based design and coding. If you still feel a need to use tools to help you practice the CRC methodology, software engineering tools are available that automate the creation of CRC cards.

Before going through the methodology, let's review the CRC concepts in a little more detail. We are familiar with classes—they encapsulate functionality. A class may represent a real-world object or it may describe an object that has been created solely to help architect the system. A class has both an internal state and a functional interface; the functional interface describes the class's capabilities. The responsibility set is an informal way of describing that functional interface. The responsibilities provide the class's interface, not its internal implementation. Unlike describing a class in a programming language, however, the responsibilities may be described informally in English (or your favorite language). The collaborators of a class are simply the classes that it talks to, that is, classes that use its capabilities or that it calls upon to help it do its work.

The class terminology is a little misleading when an object-oriented programmer looks at CRC cards. In the methodology, a class is actually used more like an object in an OO programming language—the CRC card class is used to represent a real actor in the system. However, the CRC card class is easily transformable into a class definition in an object-oriented design.

CRC card analysis is performed by a team of people. It is possible to use it by yourself, but a lot of the benefit of the method comes from talking about the developing classes with others. Before becoming the process, you should create a large number of CRC cards using the basic format shown in Figure 9.10. As you are working in your group, you will be writing on these cards; you will probably discard many of them and rewrite them as the system evolves. The CRC card methodology is informal, but you should go through the following steps when using it to analyze a system:

1. *Develop an initial list of classes:* Write down the class name and perhaps a few words on what it does. A class may represent a real-world object or an architectural object. Identifying which category the class falls into (perhaps by putting a star next to the name of a real-world object) is helpful. Each person can be responsible for handling a part of the system, but team members should talk during this process to be sure that no classes are missed and that duplicate classes are not created.

2. *Write an initial list of responsibilities and collaborators:* The responsibilities list helps describe in a little more detail what the class does. The collaborators list should be built from obvious relationships between classes. Both the responsibilities and collaborators will be refined in the later stages.

3. *Create some usage scenarios:* These scenarios describe what the system does. Scenarios probably begin with some type of outside stimulus, which is one important reason for identifying the relevant real-world objects.

4. *Walk through the scenarios:* This is the heart of the methodology. During the walk-through, each person on the team represents one or more classes. The scenario should be simulated by acting: people can call out what their class is doing, ask other classes to perform operations, and so on. Moving around, for example, to show the transfer of data, may help you visualize the system's operation. During the walk-through, all of the information created so far is targeted for updating and refinement, including the classes, their responsibilities and collaborators, and the usage scenarios. Classes may be created, destroyed, or modified during this process. You will also probably find many holes in the scenario itself.

5. *Refine the classes, responsibilities, and collaborators:* Some of this will be done during the course of the walkthrough, but making a second pass after the scenarios is a good idea. The longer perspective will help you make more global changes to the CRC cards.

6. *Add class relationships:* Once the CRC cards have been refined, subclass and superclass relationships should become clearer and can be added to the cards.

Once you have the CRC cards, you need to somehow use them to help drive the implementation. In some cases, it may work best to use the CRC cards as direct source material for the implementors; this is particularly true if you can get the designers involved in the CRC card process. In other cases, you may want to write a more formal description, in UML or another language, of the information that was captured during the CRC card analysis, and then use that formal description as the design document for the system implementors.

Example 9.2 illustrates the use of the CRC card methodology.

Example 9.2

CRC card analysis

Let's perform a CRC card analysis of the elevator system of Section 8.7. First, we need the following basic set of classes:

- *Real-world classes:* elevator car, passenger, floor control, car control, and car sensor.

- *Architectural classes:* car state, floor control reader, car control reader, car control sender, and scheduler.

For each class, we need the following initial set of responsibilities and collaborators. (An asterisk, *, is used to remind ourselves which classes represent real-world objects.)

Class	Responsibilities	Collaborators
Elevator car*	Moves up and down	Car control, car sensor, car control sender
Passenger*	Pushes floor control and car control buttons	Floor control, car control
Floor control*	Transmits floor requests	Passenger, floor control reader
Car control*	Transmits car requests	Passenger, car control reader
Car sensor*	Senses car position	Scheduler
Car state	Records current position of car	Scheduler, car sensor
Floor control reader	Interface between floor control and rest of system	Floor control, scheduler
Car control reader	Interface between car control and rest of system	Car control, scheduler
Car control sender	Interface between scheduler and car	Scheduler, elevator car
Scheduler	Sends commands to cars based upon requests	Floor control reader, car control reader, car control sender, car state

Several usage scenarios define the basic operation of the elevator system as well as some unusual scenarios:

1. One passenger requests a car on a floor, gets in the car when it arrives, requests another floor, and gets out when the car reaches that floor.

2. One passenger requests a car on a floor, gets in the car when it arrives, and requests the floor that the car is currently on.

3. A second passenger requests a car while another passenger is riding in the elevator.

4. Two people push floor buttons on different floors at the same time.

5. Two people push car control buttons in different cars at the same time.

At this point, we need to walk through the scenarios and make sure they are reasonable. Find a set of people and walk through these scenarios. Do the classes, responsibilities, collaborators, and scenarios make sense? How would you modify them to improve the system specification?

9.5 QUALITY ASSURANCE

The quality of a product or service can be judged by how well it satisfies its intended function. A product can be of low quality for several reasons, such as it was

shoddily manufactured, its components were improperly designed, its architecture was poorly conceived, and the product's requirements were poorly understood. Quality must be designed in. You can't test out enough bugs to deliver a high-quality product. The ***quality assurance (QA)*** process is vital for the delivery of a satisfactory system. In this section we will concentrate on portions of the methodology particularly aimed at improving the quality of the resulting system.

The software testing techniques described earlier in the book constitute one component of quality assurance, but the pursuit of quality extends throughout the design flow. For example, settling on the proper requirements and specification cannot be overlooked as an important determinant of quality. If the system is too difficult to design, it will probably be difficult to keep it working properly. Customers may desire features that sound nice but in fact don't add much to the overall usefulness of the system. In many cases, having too many features only makes the design more complicated and the final device more prone to breakage.

To help us understand the importance of QA, Application Example 9.3 describes serious safety problems in one computer-controlled medical system. Medical equipment, like aviation electronics, is a safety-critical application; unfortunately, this medical equipment caused deaths before its design errors were properly understood. This example also allows us to use specification techniques to understand software design problems. In the rest of the section, we look at several ways of improving quality: design reviews, measurement-based QA, and techniques for debugging large systems.

Application Example 9.3

The Therac-25 medical imaging system

The Therac-25 medical imaging system caused what Leveson and Turner called "the most serious computer-related accidents to date (at least nonmilitary and admitted)" [Lev93]. In the course of six known accidents, these machines delivered massive radiation overdoses, causing deaths and serious injuries. Leveson and Turner analyzed the Therac-25 system and the causes for these accidents.

The Therac-25 was controlled by a PDP-11 minicomputer. The computer was responsible for controlling a radiation gun that delivered a dose of radiation to the patient. It also runs a terminal that presents the main user interface. The machine's software was developed by a single programmer in PDP-11 assembly language over several years. The software includes four major components: stored data, a scheduler, a set of tasks, and interrupt services. The three major critical tasks in the system follow:

- A treatment monitor controls and monitors the setup and delivery of the treatment in eight phases.

- A servo task controls the radiation gun, machine motions, and so on.

- A housekeeper task takes care of system status interlocks and limit checks. (A limit check determines whether some system parameter has gone beyond preset limits.)

The code was relatively crude—the software allowed several processes access to shared memory, there was no synchronization mechanism aside from shared variables, and test-and-set for shared variables were not indivisible operations.

Let's examine the software problems responsible for one series of accidents. Leveson and Turner reverse-engineered a specification for the relevant software as follows:

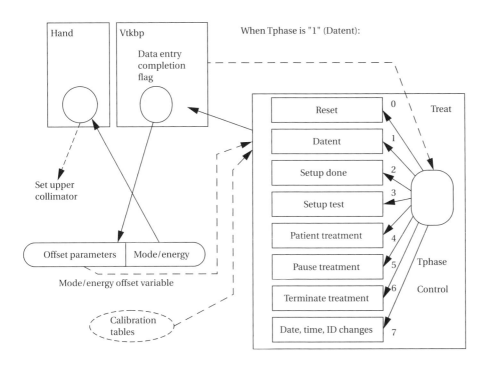

Treat is the treatment monitor task, divided into eight subroutines (*Reset, Datent,* and so on). *Tphase* is a variable that controls which of these subroutines is currently executing. *Treat* reschedules itself after the execution of each subroutine. The *Datent* subroutine communicates with the keyboard entry task via the data entry completion flag, which is a shared variable. *Datent* looks at this flag to determine when it should leave the data entry mode and go to the *Setup test* mode. The *Mode/energy offset* variable is a shared variable: The top byte holds offset parameters used by the *Datent* subroutine, and the low-order byte holds mode and energy offset used by the *Hand* task.

When the machine is run, the operator is forced to enter the mode and energy (there is one mode in which the energy is set to a default), but the operator can later edit the mode and energy separately. The software's behavior is timing dependent. If the keyboard handler sets the completion variable before the operator changes the *Mode/energy* data, the *Datent* task will not detect the change—once *Treat* leaves *Datent*, it will not enter that subroutine again during the treatment. However, the *Hand* task, which runs concurrently, will see the

new *Mode/energy* information. Apparently, the software included no checks to detect the incompatible data.

After the *Mode/energy* data are set, the software sends parameters to a digital/analog converter and then calls a Magnet subroutine to set the bending magnets. Setting the magnets takes about 8 seconds and a subroutine called Ptime is used to introduce a time delay. Due to the way that *Datent*, Magnet, and Ptime are written, it is possible that changes to the parameters made by the user can be shown on the screen but will not be sensed by *Datent*. One accident occurred when the operator initially entered *Mode/energy*, went to the command line, changed *Mode/energy*, and returned to the command line within 8 s. The error therefore depended on the typing speed of the operator. Since operators become faster and more skillful with the machine over time, this error is more likely to occur with experienced operators.

Leveson and Turner emphasize that the following poor design methodologies and flawed architectures were at the root of the particular bugs that led to the accidents:

- The designers performed a very limited safety analysis. For example, low probabilities were assigned to certain errors with no apparent justification.

- Mechanical backups were not used to check the operation of the machine (such as testing beam energy), even though such backups were employed in earlier models of the machine.

- Programmers created overly complex programs based on unreliable coding styles.

In summary, the designers of the Therac-25 relied on system testing with insufficient module testing or formal analysis.

In this section, we review the QA process in more detail. Section 9.5.1 introduces some QA techniques, Section 9.5.2 focuses on verifying requirements and specifications, and Section 9.5.3 discusses design reviews.

9.5.1 Quality Assurance Techniques

The International Standards Organization (ISO) has created a set of quality standards known as *ISO 9000*. ISO 9000 was created to apply to a broad range of industries, including but not limited to embedded hardware and software. A standard developed for a particular product, such as wooden construction beams, could specify criteria particular to that product, such as the load that a beam must be able to carry. However, a wide-ranging standard such as ISO 9000 cannot specify the detailed standards for every industry. Consequently, ISO 9000 concentrates on processes used to create the product or service. The processes used to satisfy ISO 9000 affect the entire organization as well as the individual steps taken during design and manufacturing.

A detailed description of ISO 9000 is beyond the scope of this book; several books [Sch94, Jen95] describe ISO 9000's applicability to software development. We can, however, make the following observations about quality management based on ISO 9000:

- *Process is crucial:* Haphazard development leads to haphazard products and low quality. Knowing what steps are to be followed to create a high-quality product is essential to ensuring that all the necessary steps are in fact followed.

- *Documentation is important:* Documentation has several roles: The creation of the documents describing processes helps those involved understand the processes; documentation helps internal quality monitoring groups to ensure that the required processes are actually being followed; and documentation also helps outside groups (customers, auditors, etc.) understand the processes and how they are being implemented.

- *Communication is important:* Quality ultimately relies on people. Good documentation is an aid for helping people understand the total quality process. The people in the organization should understand not only their specific tasks but also how their jobs can affect overall system quality.

Many types of techniques can be used to verify system designs and ensure quality. Techniques can be either *manual* or *tool based*. Manual techniques are surprisingly effective in practice. In Section 9.5.3 we discuss design reviews, which are simply meetings at which the design is discussed and which are very successful in identifying bugs. Many of the software testing techniques described in Section 5.10 can be applied manually by tracing through the program to determine the required tests. Tool-based verification helps considerably in managing large quantities of information that may be generated in a complex design. Test generation programs can automate much of the drudgery of creating test sets for programs. Tracking tools can help ensure that various steps have been performed. Design flow tools automate the process of running design data through other tools.

Metrics are important to the quality control process. To know whether we have achieved high levels of quality, we must be able to measure aspects of the system and our design process. We can measure certain aspects of the system itself, such as the execution speed of programs or the coverage of test patterns. We can also measure aspects of the design process, such as the rate at which bugs are found. Section describes ways in which measurements can be used in the QA process.

Tool and manual techniques must fit into an overall process. The details of that process will be determined by several factors, including the type of product being designed (e.g., video game, laser printer, air traffic control system), the number of units to be manufactured and the time allowed for design, the existing practices in the company into which any new processes must be integrated, and many other factors. An important role of ISO 9000 is to help organizations study their total process, not just particular segments that may appear to be important at a particular time.

One well-known way of measuring the quality of an organization's software development process is the ***Capability Maturity Model (CMM)*** developed by Carnegie Mellon University's Software Engineering Institute [SEI99]. The CMM

provides a model for judging an organization. It defines the following five levels of maturity:

1. *Initial:* A poorly organized process, with very few well-defined processes. Success of a project depends on the efforts of individuals, not the organization itself.

2. *Repeatable:* This level provides basic tracking mechanisms that allow management to understand cost, scheduling, and how well the systems under development meet their goals.

3. *Defined:* The management and engineering processes are documented and standardized. All projects make use of documented and approved standard methods.

4. *Managed:* This phase makes detailed measurements of the development process and product quality.

5. *Optimizing:* At the highest level, feedback from detailed measurements is used to continually improve the organization's processes.

The Software Engineering Institute has found very few organizations anywhere in the world that meet the highest level of continuous improvement and quite a few organizations that operate under the chaotic processes of the initial level. However, the CMM provides a benchmark by which organizations can judge themselves and use that information for improvement.

9.5.2 Verifying the Specification

The requirements and specification are generated very early in the design process. Verifying the requirements and specification is very important for the simple reason that bugs in the requirements or specification can be extremely expensive to fix later on. Figure 9.11 shows how the cost of fixing bugs grows over the course of the design process (we use the waterfall model as a simple example, but the same holds for any design flow). The longer a bug survives in the system, the more expensive it will be to fix. A coding bug, if not found until after system deployment, will cost money to recall and reprogram existing systems, among other things. But a bug introduced earlier in the flow and not discovered until the same point will accrue all those costs and more costs as well. A bug introduced in the requirements or specification and left until maintenance could force an entire redesign of the product, not just the replacement of a ROM. Discovering bugs early is crucial because it prevents bugs from being released to customers, minimizes design costs, and reduces design time. While some requirements and specification bugs will become apparent in the detailed design stages—for example, as the consequences of certain requirements are better understood—it is possible and desirable to weed out many bugs during the generation of the requirements and spec.

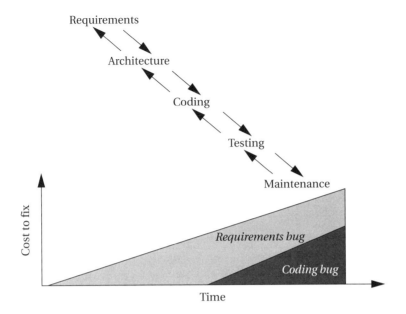

FIGURE 9.11

Long-lived bugs are more expensive to fix.

The goal of validating the requirements and specification is to ensure that they satisfy the criteria we originally applied in Section 9.2 to create the specification, including correctness, completeness, consistency, and so on. Validation is in fact part of the effort of generating the requirements and specification. Some techniques can be applied while they are being created to help you understand the requirements and specifications, while others are applied on a draft, with results used to modify the specs.

Since requirements come from the customer and are inherently somewhat informal, it may seem like a challenge to validate them. However, there are many things that can be done to ensure that the customer and the person actually writing the requirements are communicating. ***Prototypes*** are a very useful tool when dealing with end users—rather than simply describe the system to them in broad, technical terms, a prototype can let them see, hear, and touch at least some of the important aspects of the system. Of course, the prototype will not be fully functional since the design work has not yet been done. However, user interfaces in particular are well suited to prototyping and user testing. Canned or randomly generated data can be used to simulate the internal operation of the system. A prototype can help the end user critique numerous functional and nonfunctional requirements, such as data displays, speed of operation, size, weight, and so forth. Certain programming languages, sometimes called ***prototyping languages***

or *specification languages*, are especially well suited to prototyping. Very high-level languages (such as Matlab in the signal processing domain) may be able to perform functional attributes, such as the mathematical function to be performed, but not nonfunctional attributes such as the speed of execution. ***Preexisting systems*** can also be used to help the end user articulate his or her needs. Specifying what someone does or doesn't like about an existing machine is much easier than having them talk about the new system in the abstract. In some cases, it may be possible to construct a prototype of the new system from the preexisting system. Particularly when designing **cyber-physical systems** that use real-time computers for physical control, simulation is an important technique for validating requirements. Requirements for cyber-physical systems depend in part on the physical properties of the plant being controlled. Simulators that model the physical plant can help system designers understand the requirements on the cyber side of the system.

The techniques used to validate requirements are also useful in verifying that the specifications are correct. Building prototypes, specification languages, and comparisons to preexisting systems are as useful to system analysis and designers as they are to end users. Auditing tools may be useful in verifying consistency, completeness, and so forth. Working through ***usage scenarios*** often helps designers fill out the details of a specification and ensure its completeness and correctness. In some cases, ***formal techniques*** (that is, design techniques that make use of mathematical proofs) may be useful. Proofs may be done either manually or automatically. In some cases, proving that a particular condition can or cannot occur according to the specification is important. Automated proofs are particularly useful in certain types of complex systems that can be specified succinctly but whose behavior over time is complex. For example, complex protocols have been successfully formally verified.

9.5.3 Design Reviews

The ***design review*** [Fag76] is a critical component of any QA process. The design review is a simple, low-cost way to catch bugs early in the design process. A design review is simply a meeting in which team members discuss a design, reviewing how a component of the system works. Some bugs are caught simply by preparing for the meeting, as the designer is forced to think through the design in detail. Other bugs are caught by people attending the meeting, who will notice problems that may not be caught by the unit's designer. By catching bugs early and not allowing them to propagate into the implementation, we reduce the time required to get a working system. We can also use the design review to improve the quality of the implementation and make future changes easier to implement.

A design review is held to review a particular component of the system. A design review team has the following members:

- The *designers* of the component being reviewed are, of course, central to the design process. They present their design to the rest of the team for review and analysis.

- The *review leader* coordinates the pre-meeting activities, the design review itself, and the post-meeting follow-up.

- The *review scribe* records the minutes of the meeting so that designers and others know which problems need to be fixed.

- The *review audience* studies the component. Audience members will naturally include other members of the project for which this component is being designed. Audience members from other projects often add valuable perspective and may notice problems that team members have missed.

The design review process begins before the meeting itself. The design team prepares a set of documents (code listings, flowcharts, specifications, etc.) that will be used to describe the component. These documents are distributed to other members of the review team in advance of the meeting, so that everyone has time to become familiar with the material. The review leader coordinates the meeting time, distribution of handouts, and so forth.

During the meeting, the leader is responsible for ensuring that the meeting runs smoothly, while the scribe takes notes about what happens. The designers are responsible for presenting the component design. A top–down presentation often works well, beginning with the requirements and interface description, followed by the overall structure of the component, the details, and then the testing strategy. The audience should look for all types of problems at every level of detail, including the problems listed below.

- Is the design team's view of the component's specification consistent with the overall system specification, or has the team misinterpreted something?

- Is the interface specification correct?

- Does the component's internal architecture work well?

- Are there coding errors in the component?

- Is the testing strategy adequate?

The notes taken by the scribe are used in meeting follow-up. The design team should correct bugs and address concerns raised at the meeting. While doing so, the team should keep notes describing what they did. The design review leader coordinates with the design team, both to make sure that the changes are made and to distribute the change results to the audience. If the changes are straightforward, a written report of them is probably adequate. If the errors found during the review caused a major reworking of the component, a new design review meeting for the new implementation, using as many of the original team members as possible, may be useful.

SUMMARY

System design takes a comprehensive view of the application and the system under design. To ensure that we design an acceptable system, we must understand the application and its requirements. Numerous techniques, such as object-oriented design, can be used to create useful architectures from the system's original requirements. Along the way, by measuring our design processes, we can gain a clearer understanding of where bugs are introduced, how to fix them, and how to avoid introducing them in the future.

What We Learned

- Design methodologies and design flows can be organized in many different ways.

- A poor understanding of requirements means that the final system won't do what it is supposed to do, even if you use the best possible implementation techniques.

- CRC cards help us understand the system architecture in the initial phases of architecture design.

- We want to catch bugs as early as possible to minimize the cost of fixing those bugs.

FURTHER READING

Pressman [Pre97] provides a thorough introduction to software engineering. Davis [Dav90] gives a good survey of software requirements. Beizer [Bei84] surveys system-level testing techniques. Leveson [Lev86] provides a good introduction to software safety. Schmauch [Sch94] and Jenner [Jen95] both describe ISO 9000 for software development. A tutorial edited by Chow [Cho85] includes a number of important early papers on software quality assurance. Cusumano [Cus91] provides a fascinating account of software factories in both the United States and Japan.

QUESTIONS

Q9-1 Briefly describe the differences between the waterfall and spiral development models.

Q9-2 What skills might be useful in a cross-functional team that is responsible for designing a set-top box?

Q9-3 Provide realistic examples of how a requirements document may be:

 a. ambiguous,

 b. incorrect,

 c. incomplete,

 d. unverifiable.

Q9-4 How can poor specifications lead to poor quality code—do aspects of a poorly-constructed specification necessarily lead to bad software?

Q9-5 Estimate the cost of finding and fixing a single software bug.

Q9-6 What are the main phases of a design review?

LAB EXERCISES

L9-1 Draw a diagram showing the developmental steps of one of the projects you recently designed. Which development model did you follow (waterfall, spiral, etc.)?

L9-2 Find a detailed description of a system of interest to you. Write your own description of what it does and how it works.

UML Notations

INTRODUCTION

In this appendix we review the basics of UML notation for easy reference. We do not cover all the aspects of UML. For a more thorough treatment of UML, see references such as Booch *et al.* [Boo99]. This appendix includes only a basic summary of UML diagrams; for a more detailed introduction to what these symbols mean, see Section 1.3.

A.1 PRIMITIVE ELEMENTS

The most fundamental primitives of UML are the object and the class; an object is an instance of a class. In addition, various types of relations between objects and classes are possible. Other types of elements have also been defined. Primitives are summarized in Figure A.1. A *class* has attributes and behaviors. An *object* may have its attributes assigned particular values. An anonymous object belongs to a class but has no name, probably because it does not play a major role in the system. A *package* is an organizational unit of the system that may contain class definitions, objects, and so on. A *state* is used in state diagrams to describe *behavior*. A physical processor is a hardware element. A component is a physical part of a system that implements a set of interfaces.

We often find that we use a certain combination of elements in an object or class many times. We can give these combinations names; such a definition is called a **stereotype** in UML. The ≪*signal*≫ shown in Figure A.2 is an example of a stereotype.

An active class is a class that will implement a separate thread of control. As shown in Figure A.3, an active class is identified by its heavy borders.

A.2 DIAGRAM TYPES

The UML primitives can be put together in a number of ways. This section provides examples of several of the basic UML diagram types that we use in this book.

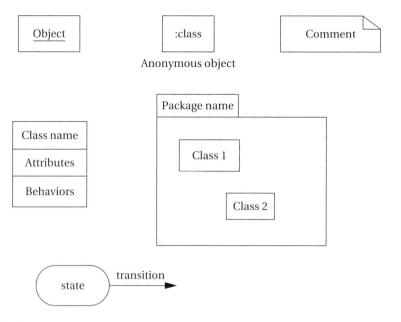

FIGURE A.1

Primitive elements for UML diagrams.

FIGURE A.2

A UML stereotype.

FIGURE A.3

An active class.

A.2.1 **Class Diagram**

The class diagram defines classes and describes the relationships between them. One type of relationship between classes is subtype/supertype, which is shown in Figure A.4; note that the derivation arrows go from derived to base. In the upper part of the figure, *Class 2* is derived from *Class 1*. In the lower part of the figure, *Class b* is derived from both *Class a1* and *Class a2*, an example of multiple inheritance.

Many relationships other than inheritance can be represented in UML. A class diagram with associations and their multiplicities is shown in Figure A.5. This diagram shows how many objects of one class interact with a given number of objects of another class.

A.2.2 **State Diagram**

The state diagram shows the structure of states and transitions for a behavior. A basic state diagram is shown in Figure A.6. A transition may be labeled with the event that causes entry onto that transition and the actions taken on the transition.

UML allows you to describe Statechart-style substates. Examples of sequential and concurrent substates are shown in Figure A.7. The sequential substates

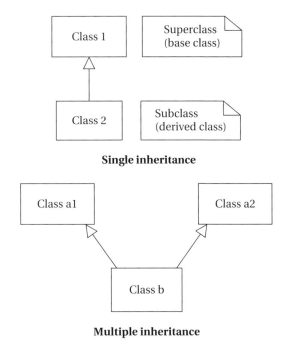

FIGURE A.4

Class derivation in a UML class diagram.

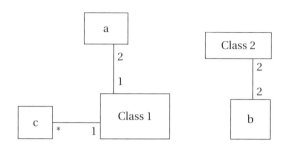

FIGURE A.5

A UML class diagram showing associations.

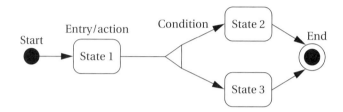

FIGURE A.6

A state diagram in UML.

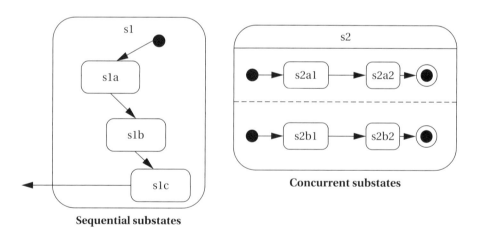

FIGURE A.7

Substates in UML.

(similar to the Statechart OR state) describe detailed behavior within an over-all system state; the concurrent substates (similar to the Statechart AND state) describe two distinct activities going on concurrently within the same system state.

A.2.3 Sequence and Collaboration Diagrams

Sequence and collaboration diagrams both illustrate scenarios, but in different ways. Figure A.8 shows a UML sequence diagram that includes a timeline. The bars show when different objects are active. Figure A.9 shows a UML collaboration diagram for the same sequence of events. In this diagram, messages are given sequence numbers to indicate time.

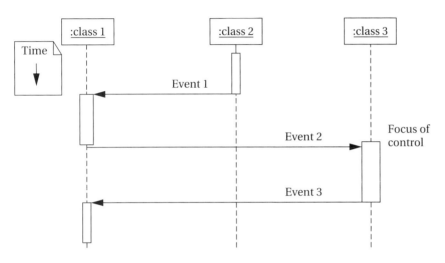

FIGURE A.8

A sequence diagram in UML.

FIGURE A.9

A UML collaboration diagram.

Glossary

Absolute address An address of an exact location in memory (Section 2.3.2, Section 5.3).

AC0–AC3 The four accumulators available in the C55x (Section 2.3.1).

Accumulator A register that is used as both the source and destination for arithmetic operations, as in accumulating a sum (Section 2.3).

Ack Short for *acknowledge*, a signal used in handshaking protocols (Section 4.1.1).

ACPI Advanced Configuration and Power Interface, an industry standard for power management interfaces (Section 6.6).

Activation record A data structure that describes the information required by a currently active procedure call (Section 2.2.3).

Active class A UML class that can create its own thread of control (Section 6.2.4).

Active low A logic 0 that denotes activity for a device, as compared to the normal logic 1 (Section 4.1.1).

Actuator A physical output device (Section 8.1).

A/D converter See *analog/digital converter*.

ADPCM Adaptive differential pulse code modulation (Section 6.7.1).

Allocation The assignment of responsibility for a computation to a processing element (Section 7.3.2).

Analog/digital converter A device that converts an analog signal into digital form (Section 4.3.2).

AND/OR table A technique for specifying control-oriented functionality (Section 9.3).

Application layer In the OSI model, the end-user interface (Section 8.1.2).

ASIC Application-specific integrated circuit (Section 4.5.2, Section 7.2).

Aspect ratio In a memory, the ratio of the number of addressable units to the number of bits read per request (Section 4.2.1).

Assembler A program that creates object code from a symbolic description of instructions (Section 5.3).

Asynchronous An event not coordinated with a clock (Section 4.2.2).

Atomic operation An operation that cannot be interrupted (Section 6.4.1).

Auto-indexing Automatically incrementing or decrementing a value before or after using it (Section 2.2.2).

Average-case execution time A typical execution time for typical inputs (Section 5.6).

Bank A block of memory in a memory system or cache.

Base-plus-offset addressing Calculating the address by adding a base address to an offset (usually contained in a register) (Section 2.2.2).

Basis paths A set of execution paths that cover the possible execution paths (Section 5.10.1).

Best-case execution time The shortest execution time for any possible set of inputs (Section 5.6).

Best-effort routing The Internet routing methodology, which does not guarantee completion (Section 8.4.1).

Big-endian A data format in which the low-order byte is stored in the highest bits of the word (Section 2.2.1).

BIOS Basic Input/Output System. Originally, low-level IBM PC software; today, low-level operating software in any computer system (Section 4.5.3).

Black-box testing Testing a program without knowledge of its implementation (Section 5.10.2).

Blocking communication Communication that requires a process to wait after sending a message (Section 6.4).

Boot-block flash A type of flash memory that protects some of its contents (Section 4.2.3).

Bottom–up design Using information from lower levels of abstraction to modify the design at higher levels of abstraction (Section 1.2).

Bounce Repeated make-break contacts upon of a switch (Section 4.3.3).

Branch table A multiway branching mechanism that uses a value to index into a table of branch targets (Example 2.5).

Branch target The destination address of a branch (Section 2.2.3).

Branch testing A technique to generate a set of tests for conditionals (Section 5.10.1).

Breakpoint A stopping point for system execution (Section 4.6.2).

Bridge A logic unit that acts as an interface between two buses (Section 4.1.3).

Bundle A collection of logically related signals (Section 4.1.1).

Burst transfer A bus transfer that transfers several contiguous locations without separate addresses for each (Section 4.1.1).

Bus Generally, a shared connection. CPUs use buses to connect themselves to external devices and memory (Section 4.1).

Bus grant The granting of ownership of the bus to a device (Section 4.1.2).

Bus master The current owner of the bus (Section 4.1.2).

Bus request A request to obtain ownership of the bus (Section 4.1.2).

Busy-wait I/O Servicing an I/O device by executing instructions that test the device's state (Section 3.1.3).

Cache A small memory that holds copies of certain main memory locations for fast access (Section 3.4.1).

Cache hit A memory reference to a location currently held in the cache (Section 3.4.1).

Cache miss A memory reference to a location not currently in the cache (Section 3.4.1).

Cache miss penalty The extra time incurred for a memory reference that is a cache miss (Section 3.5.2).

CAN bus A serial bus for networked embedded systems, originally designed for automobiles (Section 8.5.1).

Capability Maturity Model A method developed at the Software Engineering Institute of Carnegie Mellon University for assessing the quality of software development processes (Section 9.5.1).

Capacity miss A cache miss that occurs because the program's working set is too large for the cache (Section 3.4.1).

CAS See *column address select*.

CDFG See *control/data flow graph*.

Central processing unit The part of the computer system responsible for executing instructions fetched from memory (Section 2.1).

Changing In logic timing analysis, a signal whose value is changing at a particular moment in time (Section 4.1.1).

Circular buffer An array used to hold a window of a stream of data (Section 5.1.2).

CISC Complex instruction set computer. Typically uses a number of instruction formats of varying length and provides complex operations in some instructions (Section 2.1).

Class A type description in an object-oriented language (Section 1.3).

Class diagram A UML diagram that defines classes and shows derivation relationships among them (Section 1.4.3).

Clear-box testing Generating tests for a program with knowledge of its structure (Section 5.10.1).

CMM See *Capability Maturity Model*.

CMOS Complementary metal oxide semiconductor, the dominant VLSI technology today (Section 3.6).

Code motion A technique for moving operations in a program without affecting its behavior (Section 5.7.1).

Cold miss See *compulsory miss*.

Collaboration diagram A UML diagram that shows communication among classes without the use of a timeline (Section 1.4.3, Section A.2.3). See also *sequence diagram*.

Column address select A DRAM signal that indicates the column part of the address is being presented to the memory (Section 4.2.2).

Communication link A connection between processing elements (Section 8.1).

Completion time The time at which a process finishes executing (Section 6.1.4).

Compulsory miss A cache miss that occurs the first time a location is used (Section 3.4.1).

Computational kernel A small portion of an algorithm that performs a long function (Introduction of Chapter 7).

Computing platform A hardware system used for embedded computing (Introduction of Chapter 4).

Concurrent engineering Simultaneous design of several different system components (Section 9.1.2).

Conflict graph A graph that represents incompatibilities between entities; used in register allocation (Section 5.5.5).

Conflict miss A cache miss caused by two locations in use mapping to the same cache location (Section 3.4.1).

Control/data flow graph A graph that models both the data and control operations in a program (Section 5.2).

Context The state of a process (Section 6.2.1).

Controllability The ability to set a value in system state during testing.

Co-processor An optional unit added to a CPU that is responsible for executing some of the CPU's instructions (Section 3.3).

Co-routine A manual method of programming concurrency (Section 6.1.7).

Counter A device that counts asynchronous external events (Section 4.3.1).

CPSR Current program status register in the ARM processor (Section 2.2.2).

CPU See *central processing unit*.

CPU time The total execution time of a process (Section 6.1.4).

CRC card A technique for capturing design information (Section 9.4).

Critical instant In RMA, the worst-case combination of processes (Section 6.3.1).

Cycle-accurate simulator A CPU simulation that is accurate to the clock-cycle level (Section 5.6.2).

Cyclomatic complexity A measure of the control complexity of a program (Section 5.10.1).

D/A converter See *digital/analog converter*.

Data flow graph A graph that models data operations without conditionals (Section 5.2.1).

Data flow testing A technique for generating tests by examining the data flow representation of a program (Section 5.10.1).

Data link layer In the OSI model, the layer responsible for reliable data transport (Section 8.1.2).

Dead code elimination Eliminating code that can never be executed (Section 5.5.2).

Deadline The time at which a process must finish (Section 6.1.3).

Debouncing Eliminating the bouncing of a switch (Section 4.3.3).

Decision node A node in a CDFG that models a conditional (Section 5.2.2).

Def-use analysis Analyzing the relationships between reads and writes of variables in a program (Section 5.10.1).

Delayed branch A branch instruction that always executes one or more instructions after the branch, independent of whether the branch is taken (Section 3.5.1).

Dense instruction set An instruction set designed to provide compact code (Section 5.9).

Design flow A series of steps used to implement a system (Section 9.1.2).

Design methodology A method of proceeding through levels of abstraction to complete a design (Section 9.1).

Design process See *design methodology*.

Digital/analog converter A device that converts a sequence of digital values into an analog waveform (Section 4.3.2).

Digital signal processor A microprocessor whose architecture is optimized for digital signal processing applications (Introduction of Chapter 2).

Direct-mapped cache A cache with a single set (Section 3.4.1).

Direct memory access A bus transfer performed by a device without executing instructions on the CPU (Section 4.1.2).

Distributed embedded system An embedded system built around a network or one in which communication between processing elements is explicit (Introduction of Chapter 8).

DMA See *direct memory access*.

DMA controller A logic unit designed to execute DMA transfers (Section 4.1.2).

DNS See *Domain Name Server*.

Domain Name Server An Internet service that translates names to Internet addresses (Section 8.4.1).

DRAM See *dynamic random access memory*.

DSP See *digital signal processor*.

Dynamic power management A power management technique that looks at the CPU activity (Section 3.6).

Dynamic priority Process priorities that change during execution (Section 6.3).

Dynamic random access memory A memory that relies on stored charge (Section 4.2.2).

Dynamically linked library A code library that is linked into the program at the start of execution (Section 5.3.2).

Earliest deadline first A variable priority scheduling scheme (Section 6.3.2).

EDF See *earliest deadline first*.

Embedded computer system A computer used to implement some of the functionality of something other than a general-purpose computer (Section 1.1).

Encoded keyboard A keyboard that produces codes for key depressions (Section 4.3.3).

Energy The ability to do work (Section 3.6).

Enq Short for *enquiry*, a signal used in handshaking protocols (Section 4.1.1).

Entry point A label in an assembly language module that can be referred to by other program modules (Section 5.3.2).

Error injection Evaluating test coverage by inserting errors into a program and using your tests to try to find those errors (Section 5.10.3).

Ethernet A local area network (Section 8.2.2).

Evaluation board A printed circuit board designed to provide a typical platform (Section 4.5.2).

Executable binary An object program that is ready for execution (Section 5.3).

Exception Any unusual condition in the CPU that is recognized during execution (Section 3.2.2).

Expression simplification Rewriting an arithmetic expression (Section 5.5.1).

External reference A reference in an assembly language program to another module's entry point (Section 5.3.2).

Factory-programmed ROM A ROM that is programmed during manufacture (Section 4.2.3).

Fast return In the C55x, a procedure return that uses some registers rather than the stack to store certain values (Section 2.3.4).

Federated architecture An architecture for networked embedded systems that is constructed from several networks, each corresponding to an operational subsystem (Section 8.5.2).

Field-programmable gate array An integrated circuit that can be programmed by the user and that provides multilevel logic (Section 4.5.2, Section 7.2).

First-level cache The cache closest to the CPU (Section 3.4.1).

Flash memory An electrically-erasable programmable read-only memory (Section 4.2.3).

FlexRay A network designed for real-time systems (Section 8.5.1).

Four-cycle handshake A handshaking protocol that goes through four states (Section 4.1.1).

FPGA See *field-programmable gate array*.

Frame pointer Points to the end of a procedure stack frame (Section 5.4.2).

Function In a programming language, a procedure that can return a value to the caller (Section 2.2.3).

Functional requirements Requirements that describe the logical behavior of the system (Section 9.2).

Glue logic Interface logic (Section 4.4.2).

Glueless interface An interface between components that requires no glue logic (Section 4.4.2).

Handshake A protocol designed to confirm the arrival of data (Section 4.1.1).

Hardware/software co-design The simultaneous design of hardware and software components to meet system requirements (Section 7.2).

Harvard architecture A computer architecture that provides separate memories for instructions and data (Section 2.1.1).

Hit rate The probability of a memory access being a cache hit (Section 3.4.1).

Host system Any system that is used as an interface to another system (Section 4.6.1, Section 7.2).

Huffman coding A method of data compression (Section 3.7.1).

Hyperperiod The least common multiple of the periods in a system (Section 6.1.6).

I^2C bus A serial bus for distributed embedded systems (Section 8.2.1).

IEEE 1394 A high-speed serial network for peripherals (Section 4.5.3).

Immediate operand An operand embedded in an instruction rather than fetched from another location (Section 2.2.2).

Induction variable elimination A loop optimization technique that eliminates references to variables derived from the loop control variable (Section 5.7.1).

Initiation time The time at which a process actually starts to execute (Section 6.1.4).

Instruction-level simulator A CPU simulator that is accurate to the level of the programming model but not to timing (Section 4.6.2).

Instruction set The definition of the operations performed by a CPU (Introduction of Chapter 2).

Internet A worldwide network based on the Internet Protocol (Section 8.4.1).

Internet appliance An information system that makes use of the Internet (Section 8.4).

Internet-enabled embedded system Any embedded system that includes an Internet interface (Section 8.4).

Internet Protocol A packet-based protocol (Section 8.4.1).

Interpreter A program that executes a given program by analyzing a high-level description of the program at execution time (Section 5.5.9).

Interprocess communication A mechanism for communication between processes (Section 6.4).

Interrupt A mechanism that allows a device to request service from the CPU (Section 3.1.4).

Interrupt handler A routine called upon an interrupt to service the interrupting device (Section 3.1.4).

Interrupt priority Priorities used to determine which of several interrupts gets attention first (Section 3.1.4).

Interrupt vector Information used to select which segment of the program should be used to handle the interrupt request (Section 3.1.4).

I/O Input/output (Section 3.1).

IP See *Internet Protocol*.

ISO 9000 A series of international standards for quality process management (Section 9.5.1).

Iteration vector A specification of the loop iteration variable values that describe a particular iteration of a set of nested loops (Section 5.6.2).

JIT compiler A just-in-time compiler; compiles program sections on demand during execution (Section 5.5.9).

L1 cache See *first-level cache*.

L2 cache See *second-level cache*.

label In assembly language, a symbolic name for a memory location (Section 2.1.2).

LCD Liquid-crystal display (Section 4.3.5).

LED Light emitting diode (Section 4.3.4).

Lightweight process A process that shares its memory spaces with other processes.

Line replaceable unit In avionics, an electronic unit that corresponds to a functional unit, such as a flight instrument (Section 8.5.2).

Linker A program that combines multiple object program units, resolving references between them (Section 5.3.2).

Linux A well-known, open-source version of Unix.

Little-endian A data format in which the low-order byte is stored in the lowest bits of the word (Section 2.2.1).

Load balancing Adjusting scheduling and allocation to even out system load in a network (Section 8.3.1).

Loader A program that loads a given program into memory for execution (Section 5.3).

Load map A description of where object modules should be placed in memory (Section 5.3.2).

Load-store architecture An architecture in which only load and store operations can be used to access data and ALU, and other instructions cannot directly access memory (Section 2.2.2).

Logic analyzer A machine that captures multiple channels of digital signals to produce a timing diagram view of execution (Section 4.6.2).

Longest path The path through a weighted graph that gives the largest total sum of weights (Section 7.3.1).

Loop nest A set of loops, one inside the other (Section 5.7.2).

Loop unrolling Rewriting a loop so that several instances of the loop body are included in a single iteration of the modified loop (Section 5.5.4).

LRU See *line replaceable unit*.

Masking In interrupts, causing lower-priority interrupts to be held in order to service higher-priority interrupts (Section 3.1.4).

Memory controller A logic unit designed as an interface between DRAM and other logic (Section 4.2.2).

Memory management unit A unit responsible for translating logical addresses into physical addresses (Section 3.4.2).

Memory-mapped I/O Performing I/O by reading and writing memory locations that correspond to device registers (Section 3.1.2).

Memory mapping Translating addresses from logical to physical form (Section 3.4.2).

Message delay The delay required to send a message on a network with no interference (Section 8.3).

Message passing A style of interprocess communication (Section 6.4, Section 8.1.4).

Methodology Used to describe an overall design process (Section 1.2, Section 9.1).

Microcontroller A microprocessor that includes memory and I/O devices, often including timers, on a single chip (Section 1.1).

Miss rate The probability that a memory access will be a cache miss (Section 3.4.1).

MMU See *memory management unit*.

Motion vector A vector describing the displacement between two units of an image (Section 7.9.1).

Multihop network A network in which messages may go through an intermediate PE when traveling from source to destinations.

Multiprocessor A computer system that includes more than one processing element (Introduction of Chapter 7).

Multirate Operations that have different deadlines, causing the operations to be performed at different rates (Section 1.1, Section 6.1.2).

Network A system for communicating between components (Introduction of Chapter 8).

Network layer In the OSI model, the layer that provides end-to-end service (Section 8.1.2).

n-key rollover Reading the correct sequence of key depressions when keys are depressed simultaneously (Section 4.3.3).

NMI See *nonmaskable interrupt*.

Nonblocking communication Interprocess communication that allows the sender to continue execution after sending a message (Section 6.4).

Nonfunctional requirements Requirements that do not describe the logical behavior of the system; examples include size, weight, and power consumption (Section 1.2.1, Section 9.2).

Nonmaskable interrupt An interrupt that must always be handled, independent of other system activity (Section 3.1.4).

Object A program unit that includes both internal data and methods that provide an interface to the data (Section 1.3).

Object code A program in binary form (Section 5.3).

Object oriented Any use of objects and classes in design; can be applied at many different levels of abstraction (Section 1.3).

Observability The ability to determine a portion of system state during testing.

Operating system A program responsible for scheduling the CPU and controlling access to devices (Introduction of Chapter 6).

Origin The starting address of an assembly language module.

OSI model A model for levels of abstraction in networks (Section 8.1.2).

Overhead In operating systems, the CPU time required for the operating system to switch contexts (Section 6.1.6).

P() Traditional name for the procedure that takes a semaphore (Section 6.4.1).

Page fault A reference to a memory page not currently in physical memory (Section 3.4.2).

Page mode An addressing mechanism for RAMs (Section 4.2.2).

Paged addressing Division of memory into equal-sized pages (Section 3.4.2).

Partitioning Dividing a functional description into smaller modules that can be separately implemented.

PC 1. In computer architecture, see *program counter*. 2. Personal computer (Section 4.5.3).

PC sampling Generating a program trace by periodically sampling the PC during execution (Section 5.6.2).

PCI A high-performance bus for PCs and other applications (Section 4.5.3).

PC-relative addressing An addressing mode that adds a value to the current PC (Section 2.2.3).

PE See *processing element*.

Peek A high-level language routine that reads an arbitrary memory location (Section 3.1.2).

Performance The speed at which operations occur (Section 1.2).

Period In real-time scheduling, a periodic interval of execution (Section 6.1.3).

Physical layer In the OSI model, the layer that defines electrical and mechanical properties (Section 8.1.2).

Pipeline A logic structure that allows several operations of the same type to be performed simultaneously on multiple values, with each value having a different part of the operation performed at any one time (Section 3.5.1).

PLC See *program location counter*.

Platform Hardware and associated software that is designed to serve as the basis for a number of different systems to be implemented.

Poke A high-level language routine that writes an arbitrary location (Section 3.1.2).

Polling Testing one or more devices to determine whether they are ready (Section 3.1.3).

POSIX A standardized version of Unix.

Post-indexing An addressing mode in which an index is added to the base address after the fetch (Section 2.1.2).

Power Energy per unit time (Section 3.6).

Power-down mode A mode invoked in a CPU that causes the CPU to reduce its power consumption (Section 3.6).

Power management policy A scheme for making power management decisions (Section 6.6).

Power state machine A finite-state machine model for the behavior of a component under power management (Section 3.6).

Predictive shutdown A power management technique that predicts appropriate times for system shutdown (Section 6.6).

Preemptive multitasking A scheme for sharing the CPU in which the operating system can interrupt the execution of processes (Section 6.2).

Presentation layer In the OSI model, the layer responsible for data formats (Section 8.1.2).

Priority-driven scheduling Any scheduling technique that uses priorities of processes to determine the running process (Section 6.3).

Priority inversion A situation in which a lower-priority process prevents a higher-priority process from executing (Section 6.3.4).

Procedure A programming language construct that allows a single piece of code to be called at multiple points in the program (Section 2.2.3). Generally, a synonym for *subroutine;* see also *function*.

Procedure call stack A stack of records for currently active processes (Section 2.2.3).

Procedure linkage A convention for passing parameters and other actions required to call a procedure (Section 2.2.3).

Process A unique execution of a program (Introduction of Chapter 6).

Process control block A record that holds the context or state of a process (Section 6.2.1).

Processing element A component that performs a computation under the coordination of the system (Section 7.2, Introduction of Chapter 8).

Profiling A procedure for counting the relative execution times of different parts of a program (Section 5.7.3).

Program counter A common name for the register that holds the address of the currently executing instruction (Section 2.1.1).

Program location counter A variable used by an assembler to assign memory addresses to instructions and data in the assembled program (Section 5.3.1).

Programming model The CPU registers visible to the programmer (Section 2.1.1).

Pseudo-op An assembly language statement that does not generate code or data (Section 2.1.2, Section 5.3.1).

Quality assurance A process for ensuring that systems are designed and built to high quality standards (Section 9.5).

RAM See *random-access memory*.

Random-access memory A memory that can be addressed in arbitrary order (Section 4.2.2).

Random testing Testing a program using randomly generated inputs (Section 5.10.2).

RAS See *row address select*.

Raster scan (or order) display A display that writes pixels by rows and columns (Section 4.3.5).

Rate Inverse of period (Section 6.1.3).

Rate-monotonic scheduling A fixed-priority scheduling scheme (Section 6.3.1).

Reactive system A system designed to react to external events (Section 5.4.2).

Read-only memory A memory with fixed contents (Section 4.2.3).

Real time A system that must perform operations by a certain time (Section 1.1).

Real-time operating system An operating system designed to be able to satisfy real-time constraints (Introduction of Chapter 6).

Re-entrancy The ability of a program to be executed multiple times, using the same memory image without error.

Refresh Restoring the values kept in a DRAM (Section 4.2.2).

Register Generally, an electronic component that holds state. In the context of computer programming, storage internal to the CPU that is part of the programming model (Section 2.1.1).

Register allocation Assigning variables to registers (Section 5.5.5).

Register-indirect addressing Fetching from a first memory location to find the address of the memory location that contains the operand (Section 2.2.2).

Regression testing Testing hardware or software by applying previously used tests (Section 5.10.2).

Relative address An address measured relative to some other location, such as the start of an object module (Section 5.3).

Release time The time at which a process becomes ready to execute (Section 6.1.3).

Repeat In instruction sets, an instruction that allows another instruction or set of instructions to be repeated in order to create low-overhead loops (Section 2.3.4).

Requirements An informal description of what a system should do (Section 1.2). A precursor to a specification.

Reservation table A hardware technique for scheduling instructions (Section 5.5.6).

Response time The time span between the initial request for a process and its completion (Section 6.3.1).

RISC Reduced instruction set computer (Section 2.1).

RMA Rate-monotonic analysis, another term for *rate-monotonic scheduling*.

Rollover Reading multiple keys when two keys are pressed at once (Section 4.3.3).

ROM See *read-only memory*.

Row address select A DRAM signal that indicates the row part of the address is being presented (Section 4.2.2).

RTOS See *real-time operating system*.

Saturation arithmetic An arithmetic system that provides a result at the maximum/minimum value on overflow/underflow.

Scheduling Determining the time at which an operation will occur (Section 7.3.2).

Scheduling overhead The execution time required to make a scheduling decision (Section 6.3.1, Section 6.3.4).

Scheduling policy A methodology for making scheduling decisions (Section 6.3.1).

SDL A software specification language (Section 9.3.1).

Second-level cache A cache after the first-level cache but before main memory (Section 3.4.1).

Segmented addressing Dividing memory into large, unequal-sized segments (Section 3.4.2).

Semaphore A mechanism for coordinating communicating processes (Section 6.4.1).

Sensor An input device that reads a physical value (Section 8.1).

Sequence diagram A UML diagram type that shows how objects communicate over time using a timeline (Section 1.3.2). See also *collaboration diagram*.

Session layer In the OSI model, the layer responsible for application dialog control (Section 8.1.2).

Set-associative cache A cache with multiple sets (Section 3.4.1).

Set-top box A system used for cable or satellite television reception.

Shared memory A communication style that allows multiple processes to access the same memory locations (Section 6.4).

Signal 1. A Unix interprocess communication method (Section 6.4.3). 2. A UML stereotype for communication (Section 6.4.3).

Single-assignment form A program that writes to each variable once at most (Section 5.2.1).

Single-hop network A network in which messages can travel from one PE to any other PE without going through a third PE.

Slow return In the C55x, a procedure return that the stack to store certain values, providing a procedure return than is provided by the *fast return* (Section 2.3.4).

Software interrupt See *trap*.

Software pipelining A technique for scheduling instructions in loops (Section 5.5.6).

Specification A formal description of what a system should do (Section 1.2). More precise than a requirements document.

Speedup The ratio of system performance before and after a design modification (Section 7.3.1).

Spill Writing a register value to main memory so that the register can be used for another purpose (Section 5.5.5).

Spiral model A design methodology in which the design iterates through specification, design, and test at increasingly detailed levels of abstraction (Section 9.1.2).

SRAM See *static random-access memory*.

Stable In logic timing analysis, a signal whose value is not changing at a particular moment in time (Section 4.1.1).

Stack pointer Points to the top of a procedure call stack (Section 5.4.2).

Statecharts A specification technique that uses compound states (Section 9.3.1).

State machine Generally, a machine that goes through a sequence of states over time. May be implemented in software (Section 1.3.2, Section 5.1.1).

State mode A logic analyzer mode that provides reduced timing resolution in return for longer time spans (Section 4.6.2).

Static power management A power management technique that does not consider the current CPU behavior (Section 3.6).

Static random-access memory A RAM that consumes power to continuously maintain its stored values (Section 4.2.2).

Static priority A scheduling policy in which process priorities are fixed (Section 6.3).

Streaming data A sequence of data values that is received periodically, such as in digital signal processing (Section 2.1.1).

Strength reduction Replacing an operation with another equivalent operation that is less expensive (Section 5.7.1).

Subroutine A synonym for *procedure* (Section 2.2.3).

Successive refinement A design methodology in which the design goes through the levels of abstraction several times, adding detail in each refinement phase (Section 9.1.2).

Superscalar An execution method that can perform several different instructions simultaneously.

Supervisor mode A CPU execution mode with unlimited privileges (Section 3.2.1). See also *user mode*.

Symbol table Generally, a table relating symbols in a program to their meaning; in an assembler, a table giving the locations specified by labels (Section 5.3.1).

Synchronous DRAM A memory that uses a clock (Section 4.2.2).

System-on-silicon A single-chip system that includes computation, memory, and I/O.

Tag The part of a cache block that gives the address bits from which the cache entry came (Section 3.4.1).

Target system A system being debugged with the aid of a host (Section 4.6.1).

Task graph A graph that shows processes and data dependencies among them (Section 6.4.2).

TCP See *Transmission Control Protocol*.

TDMA See *Time Divison Multiple Access*.

Test-and-set A hardware primitive, commonly used to implement semaphores, that reads a memory location and changes it without allowing another intervening access to the location (Section 6.4.1).

Testbench A setup used to test a design; may be implemented in software to test other software (Section 4.6.1).

Testbench program A program running on a host used to interface to a debugger that runs on an embedded processor (Section 4.6.1).

Thread See *lightweight process*.

Time Division Multiple Access A scheduling policy that divides the schedule into time slots (Section 6.1.6).

Timer A device that measures time from a clock input (Section 4.3.1).

Timing constraint A relationship among two or more events on signals in a logic network (Section 4.1.1).

Timing diagram A diagram that shows the relationships between signal transitions, possibly with arrows showing timing constraints (Section 4.1.1).

Timing mode A logic analyzer mode that provides increased timing resolution (Section 4.6.2).

TLB See *translation lookaside buffer*.

Top–down design Designing from higher levels of abstraction to lower levels of abstraction (Section 1.2).

Touchscreen A combination display and input device that allows pointing (Section 4.3.6).

Trace A record of the execution path of a program (Section 5.6.2).

Trace-driven analysis Analyzing a trace of a program's execution (Section 5.6.2).

Translation lookaside buffer A cache used to speed up virtual-to-physical address translation (Section 3.4.2).

Transmission Control Protocol A connection-oriented protocol built upon the IP (Section 8.4.1).

Transport layer In the OSI model, the layer responsible for connections (Section 8.1.2).

Trap An instruction that causes the CPU to execute a predetermined handler (Section 3.2.3).

UART Universal Asynchronous Receiver/Transmitter, a serial I/O device (Section 3.1.1).

UML See *Unified Modeling Language*.

Unified cache A cache that holds both instructions and data (Section 3.4.1).

Unified Modeling Language A widely used graphical language that can be used to describe designs at many levels of abstraction (Section 1.3).

Unrolled schedule A schedule whose length is the hyperperiod (Section 6.1.6).

Usage scenario A description of how a system will be used (Section 9.5.2).

USB Universal Serial Bus, a high-performance serial bus for PCs and other systems.

User mode A CPU execution mode with limited privileges (Section 3.2.1). See also *supervisor mode*.

Utilization In general, the fractional or percentage time that we can effectively use a resource; the term is most often applied to how processes make use of a CPU (Section 6.1.4).

V() Traditional name for the procedure that releases a semaphore (Section 6.4.1).

Virtual addressing Translating an address from a logical to a physical location (Section 3.4.2).

VLSI Acronym for *very large scale integration*; generally means any modern integrated circuit fabrication process (Section 1.1).

Von Neumann architecture A computer architecture that stores instructions and data in the same memory (Section 2.1).

Wait state A state in a bus transaction that waits for the response of a memory or device (Section 4.1.1).

Watchdog timer A timer that resets the system when the system fails to periodically reset the timer (Section 4.3.1).

Waterfall model A design methodology in which the design proceeds from higher to lower levels of abstraction (Section 9.1.2).

Way A bank in a cache (Section 3.4.1).

White-box testing See *clear-box testing*.

Word The basic unit of memory access in a computer (Section 2.2.1).

Working set The set of memory locations used during a chosen interval of a program's execution (Section 3.4.1).

Worst-case execution time The longest execution time for any possible set of inputs (Section 5.6).

Write-back Writing to main memory only when a line is removed from the cache (Section 3.4.1).

Write-through Writing to main memory for every write into the cache (Section 3.4.1).

References

[Aho06] Alfred V. Aho, Monica S. Lam, Ravi Sethi, and Jeffrey D. Ullman, *Compilers: Principles, Techniques, and Tools, second edition*. Reading, MA: Addison-Wesley, 2006.

[Ald73] Robin Alder, Mark Baker, and Howard D. Marshall, "The logic analyzer: a new instrument for observing logic signals," *Hewlett-Packard Journal* 25(2) (1973): 2–16.

[Ale05] Aleph One, 2005 http://www.aleph1.co.uk/yaffs.

[ARM99A] ARM Limited, *AMBA (TM) Specification (Rev 2.0)*, 1999, http://www.arm.com.

[ARM99B] ARM Limited, *ARM7TDMI-S Technical Reference Manual*, 1999, http://www.arm.com.

[Aus04] Todd Austin, David Blaauw, Scott Mahlke, Trevor Mudge, Chaitali Chakrabarti, and Wayne Wolf, "Mobile supercomputers," *IEEE Computer* 37(5) (2004): 81–83.

[Ban93] Uptal Banerjee, *Loop Transformations for Restructuring Compilers: The Foundations*. Boston: Kluwer Academic Publishers, 1993.

[Ban94] Uptal Banerjee, *Loop Parallelization*. Boston: Kluwer Academic Publishers, 1994.

[Ban95] Amir Ban, "Flash file system," U.S. Patent 5,404,485, April 4, 1995.

[Bar07] Richard Barry, 2007 http://www.freertos.org.

[Bay76] Bryce E. Bayer, "Color imaging array," U.S. Patent 3,971,065, July 20, 1976.

[Bei84] Boris Beizer, *Software System Testing and Quality Assurance*. New York: Van Nostrand Reinhold, 1984.

[Bei90] Boris Beizer, *Software Testing Techniques*, 2nd edition. New York: Van Nostrand Reinhold, 1990.

[Ben00] L. Benini, A. Bogliolo, and G. De Micheli, "A survey of design techniques for system-level dynamic power management," *IEEE Transactions on VLSI Systems* 8(3) (2000): 299–316.

[Bod95] Nanette J. Boden, Danny Cohen, Robert E. Felderman, Alan E. Kulawik, Charles L. Seitz, Jakov N. Seizovic, and Wen-King Su, "Myrinet—a gigabit-per-second local-area network," *IEEE Micro* February (1995): 29–36.

[Boe87] Barry W. Boehm, "A spiral model of software development and enhancement," in *Software Engineering Project Management*, 1987, pp. 128–142. Reprinted in Richard H. Thayer and Merlin Dorfman, eds., *System and Software Requirements Engineering*, Los Alamitos, CA: IEEE Computer Society Press, 1990.

[Boo91] Grady Booch, *Object-Oriented Design*. Redwood City, CA: Benjamin/Cummings, 1991.

[Boo99] Grady Booch, James Rumbaugh, and Ivar Jacobson, *The Unified Modeling Language User Guide*. Reading, MA: Addison-Wesley, 1999.

[Bos07] Robert Bosch GMBH, *Automotive Electrics Automotive Electronics*, 5th edition. Cambridge, MA: Bentley Publishers, 2007.

[Cal00] Timothy J. Callahan, John R. Hauser, and John Wawrzynek, "The Garp architecture and C compiler," *IEEE Computer* 33(4) (2000): 62–69.

[Cat98] Francky Catthoor, Sven Wuytack, Eddy De Greef, Florin Balasa, Lode Nachtergaele, and Arnout Vandecappelle, *Custom Memory Management Methodology: Exploration of Memory*

Organization for Embedded Multimedia System Design. Norwell, MA: Kluwer Academic Publishers, 1998.

[Cha92] Anantha P. Chandrakasan, Samuel Sheng, and Robert W. Brodersen, "Low-power CMOS digital design," *IEEE Journal of Solid-State Circuits* 27(4) (1992): 473–484.

[Chi94] M. Chiodo, P. Giusto, H. Hsieh, A. Jurecska, L. Lavagno, and A. Sangiovanni-Vincentelli, "Hardware/software co-design of embedded systems," *IEEE Micro* 14(4) (1994): 26–36.

[Cho85] Tsun S. Chow, *Tutorial: Software Quality Assurance: A Practical Approach*. Silver Spring, MD: IEEE Computer Society Press, 1985.

[Cir04A] Cirrus Logic, "CRD7410-CD18 User's Guide," DS620UMC2, February 2004, http://www.cirrus.com.

[Cir04B] Cirrus Logic, "CS7410 CD/MP3/WMA/AAC Audio Controller," DS553PP3, June 2004, http://www.cirrus.com.

[Coh81] Danny Cohen, "On holy wars and a plea for peace," *Computer* 14(10) (1981): 48–54.

[Col97] Robert R. Collins, "In-circuit emulation," *Dr. Dobb's Journal* September (1997): 111–113.

[Cra97] Timothy Cramer, Richard Friedman, Terrence Miller, David Seberger, Robert Wilson, and Mario Wolczko, "Compiling Java just in time," *IEEE Micro* May/June (1997): 36–43.

[Cus91] Michael A. Cusumano, *Japan's Software Factories*. New York: Oxford University Press, 1991.

[Dah00] Tom Dahlin, "Reach out and touch: designing a resistive touch screen," *Circuit Cellar*, 114 (2000): 20–25.

[Dav90] Alan M. Davis, *Software Requirements: Analysis and Specification*. Englewood Cliffs, NJ: Prentice Hall, 1990.

[DeM01] Giovanni De Micheli, Rolf Ernst, and Wayne Wolf, eds., *Readings in Hardware/Software Co-Design*, Morgan Kaufmann, 2001.

[Dou98] Bruce Powel Douglass, *Real-Time UML: Developing Efficient Objects for Embedded Systems*. Reading, MA: Addison-Wesley Longman, 1998.

[Dut96] Santanu Dutta and Wayne Wolf, "A flexible parallel architecture adapted to block-matching motion-estimation algorithms," *IEEE Transactions on Circuits and Systems for Video Technology* 6(1) (1996): 74–86.

[Dzu05] Dacfey Dzung, Martin Naedele, Thomas P. von Hoff, and Mario Crevatin, "Security for industrial communication systems," *Proceedings of the IEEE* 93(6) (2005): 1152–1177.

[Dec05] Jean-Dominique Decotignie, "Ethernet-based real-time and industrial communications," *Proceedings of the IEEE* 93(6) (2005): 1102–1117.

[Ear97] Richard W. Earnshaw, Lee D. Smith, and Kevin Welton, "Challenges in cross-development," *IEEE Micro* 17(4) July/August (1997): 28–36.

[Ern93] Rolf Ernst, Joerg Henkel, and Thomas Benner, "Hardware-software cosynthesis for microcontrollers," *IEEE Design and Test of Computers* 10(4) (1993): 64–75.

[Fag76] M. E. Fagan, "Design and code inspections to reduce errors in program development," *IBM Systems Journal* 15(3) (1976): 219–248.

[Fal92] Ohad Falik and Gideon Intrater, "NSC's Digital Answering Machine Solution," In *Proceedings, ICCD '92*, Los Alamitos, CA: IEEE Computer Society Press, 1992.

[Fel05] Max Felser, "Real-time Ethernet—industry perspective," *Proceedings of the IEEE* 93(6) (2005): 1118–1129.

[Fra88] Phyllis G. Frankl and Elaine J. Weyuker, "An applicable family of data flow testing criteria," *IEEE Transactions on Software Engineering*, 14(10) (1988): 1483–1498.

[Fur96] Steve Furber, *ARM System Architecture*. Harlow, England: Addison-Wesley, 1996.

[Gar81] John R. Garman, "The 'bug' heard 'round the world," *Software Engineering Notes* 6(5) (1981): 3–10.

[Gar94] Sonya Gary, Pete Ippolito, Gianfranco Gerosa, Carl Dietz, Jim Eno, and Hector Sanchez, "PowerPC 603, a microprocessor for portable computers," *IEEE Design and Test of Computers* 11(4) (1994): 14–23.

[Gat94] David A. Gatenby, Paul M. Lee, Randall E. Howard, Kaveh Hushyar, Rich Layendecker, and John Wesner, "Concurrent engineering: an enabler for fast, high-quality product realization," *AT & T Technical Journal* January/February (1994): 34–47.

[Gup93] Rajesh K. Gupta and Giovanni De Micheli, "Hardware-software cosynthesis for digital systems," *IEEE Design and Test of Computers* 10(3) (1993): 29–40.

[Har87] D. Harel, "Statecharts: a visual formalism for complex systems," *Science of Computer Programming* 8 (1987): 231–274.

[Hel04] Albert Helfrick, *Principles of Avionics*, 3rd edition. Avionics Communications Inc., 2004.

[Hen94] J. Henkel, R. Ernst, U. Holtmann, and T. Benner, "Adaptation of partitioning and high-level synthesis in hardware/software co-synthesis." In *Proceedings, ICCAD-94*. Los Alamitos, CA: IEEE Computer Society Press, 1994, pp. 96–100.

[Hen06] John L. Hennessy and David A. Patterson, *Computer Architecture: A Quantitative Approach*, 4th edition. San Francisco: Morgan Kaufmann, 2006.

[Hor96] Joseph R. Horgan and Aditya P. Mathur, "Software testing and reliability," Chapter 13. In *Handbook of Software Reliability Engineering*, ed. Michael R. Lyu, 531–566. Los Alamitos, CA: IEEE Computer Society Press/McGraw-Hill, 1996.

[How82] W. E. Howden, "Weak mutation testing and the completeness of test cases," *IEEE Transactions on Software Engineering* SE-8(4) (1982): 371–379.

[Huf52] David A. Huffman, "A method for the construction of minimum-redundancy codes," *Proceedings of the IRE* September (1952): 1098–1101.

[Int99] Intel, *Intel StrongARM SA-1100 Microprocessor Technical Reference Manual*, March 1999, http://www.intel.com.

[Jag95] Dave Jaggar, ed., *Advanced RISC Machines Architectural Reference Manual*. London: Prentice Hall, 1995.

[Jen95] Michael G. Jenner, *Software Quality Management and ISO 9001: How to Make Them Work for You*. New York: John Wiley and Sons, 1995.

[Kar06] Holger Karl and Andreas Willig, *Protocols and Architectures for Wireless Sensor Networks*. New York: John Wiley and Sons, 2006.

[Kas79] J. M. Kasson, "The ROLM Computerized Branch Exchange: an advanced digital PBX," *IEEE Computer* June (1979): 24–31.

[Kem98] T. M. Kemp, R. K. Montoye, J. D. Harper, J. D. Palmer, and D. J. Auerbach, "A decompression core for PowerPC," *IBM Journal of Research and Development* 42(6) (1998): 807–812.

[Ker88] Brian W. Kernighan and Dennis M. Ritchie, *The C Programming Language*, 2nd edition. New York: Prentice Hall, 1988.

[Kog81] Peter M. Kogge, *The Architecture of Pipelined Computers*. New York: McGraw-Hill, 1981.

[Kop97] Hermann Kopetz, *Real-Time Systems: Design Principles for Distributed Embedded Applications*. Boston: Kluwer Academic Publishers, 1997.

[Lev86] Nancy G. Leveson, "Software safety: Why, what, and how," *Computing Surveys* 18(2) (1986): 125–163.

[Lev93] Nancy G. Leveson and Clark S. Turner, "An investigation of the Therac-25 accidents," *IEEE Computer* July (1993): 18–41.

[Lev94] Nancy G. Leveson, Mats Per Erik Heimdahl, Holly Hildreth, and Jon Damon Reese, "Requirements specification for process-control systems," *IEEE Transactions on Software Engineering* 20(9) (1994): 684–707.

[Li97] Yau-Tsun Steven Li and Sharad Malik, "Performance analysis of embedded software using implicit path enumeration," *IEEE Transactions on CAD/ICAS* 16(12) (1997): 1477–1487.

[Li98] Yanbing Li and Joerg Henkel, "A framework for estimating and minimizing energy dissipation of embedded HW/SW systems." In *Proceedings, DAC '98*. New York: ACM Press, 1998, pp. 188–193.

[Li99] Yanbing Li and Wayne Wolf, "A task-level hierarchical memory model for system synthesis of multiprocessors," *IEEE Transactions on CAD* 18(10) (1999): 1405–1417.

[Lie98] Clifford Liem and Pierre Paulin, "Compilation techniques and tools for embedded processor architectures," Chapter 5. In *Hardware/Software Co-Design: Principles and Practice*, eds. J. Staunstrup and W. Wolf, Boston: Kluwer Academic Publishers, 1998.

[Liu73] C. L. Liu and James W. Layland, "Scheduling algorithms for multiprogramming in a hard-real-time environment," *Journal of the ACM* 20(1) (1973): 46–61.

[Liu00] Jane W. S. Liu, *Real-Time Systems*. Prentice Hall, 2000.

[Los97] Pete Loshin, *TCP/IP Clearly Explained*, 2nd edition. New York: Academic Press, 1997.

[Lyu96] Michael R. Lyu, ed., *Handbook of Software Reliability Engineering*. Los Alamitos, CA: IEEE Computer Society Press/McGraw-Hill, 1996.

[Mal96] Sharad Malik, Wayne Wolf, Andrew Wolfe, Yao-Tsun Steven Li, and Ti-Yen Yen, "Performance analysis of embedded systems." In *Hardware-Software Co-Design*, eds. G. De Micheli and M. Sami, Boston: Kluwer Academic Publishers, 1996.

[Mar78] John Marley, "Evolving microprocessors which better meet the needs of automotive electronics," *Proceedings of the IEEE* 66(2) (1978): 142–150.

[McC76] T. J. McCabe, "A complexity measure," *IEEE Transactions on Software Engineering* 2 (1976): 308–320.

[McD98] Charles E. McDowell, Bruce R. Montague, Michael R. Allen, Elizabeth A. Baldwin, and Marcelo E. Montoreano, "Javacam: Trimming Java down to size," *IEEE Internet Computing* May/June (1998): 53–59.

[Met97] Hufeza Metha, Robert Michael Owens, Mary Jane Irwin, Rita Chen, and Debashree Ghosh, "Techniques for low energy software." In *Proceedings, 1997 International Symposium on Low Power Electronics and Design*. New York: ACM Press, 1997, pp. 72–75.

[Mic00] Micron Technology, Inc., "512 Mb Synchronous SDRAM," 2000 http://www.micron.com/products/dram/sdram.

[Min95] Mindshare, Inc., Tom Shanley, and Don Anderson, *PCI System Architecture*, 3rd edition. Reading, MA: Addison-Wesley, 1995.

[Mor07] Michael J. Morgan, "Boeing B-777," Chapter 9. In *Digital Avionics Handbook, Second Edition: Avionics Development and Implementation*, ed. Cary R. Spitzer, Boca Raton, FL: CRC Press, 2007.

[Muc97] Steven S. Muchnick, *Advanced Compiler Design and Implementation*. San Francisco: Morgan Kaufmann, 1997.

[Mye79] G. Myers, *The Art of Software Testing*. New York: John Wiley and Sons, 1979.

[Obe99] James Oberg, "Why the Mars probe went off course," *IEEE Spectrum* 36(12) December (1999): 34–39.

[Pat07] David A. Patterson and John L. Hennessy, *Computer Organization and Design: The Hardware/Software Interface*, revised third edition. San Francisco: Morgan Kaufmann, 2007.

[Phi89] Philips Semiconductors, "Using the 8XC751 microcontroller as an I^2C bus master," Philips Application Note AN422, September 1989, revised June 1993. In *Application Notes and Development Tools for 80C51 Microcontrollers*, Philips Semiconductors, 1995.

[Phi92] "The I^2C bus and how to use it (including specification)," January 1992. In *Application Notes and Development Tools for 80C51 Microcontrollers*, Philips Semiconductors, 1995.

[Pil05] Dan Pilone and Neil Pitman, *UML 2.0 in a Nutshell*. Sebastopol, CA: O'Reilly Media, 2005.

[Pre97] Roger S. Pressman, *Software Engineering: A Practitioner's Approach*. New York: McGraw-Hill, 1997.

[Roc82] Anders Rockström and Roberto Saracco, "SDL–CCITT specification and description language," *IEEE Transactions on Communication* 30(6) (1982): 1310–1318.

[Rum91] James Rumbaugh, Michael Blaha, William Premerlani, Frederick Eddy, and William Lorensen, *Object-Oriented Modeling and Design*. Englewood Cliffs, NJ: Prentice Hall, 1991.

[Sas91] Steven J. Sasson and Robert G. Hills, "Electronic still camera utilizing image compression and digital storage," U.S. Patent 5,016,107, May 14, 1991.

[Sch94] Charles H. Schmauch, *ISO 9000 for Software Developers*. Milwaukee: ASQC Quality Press, 1994.

[SEI99] Software Engineering Institute, "Capability Maturity Model (SW-CMM) for Software," 1999, www.sei.cmu.edu/cmm/cmm.html.

[Sel94] Bran Selic, Garth Gullekson, and Paul T. Ward, *Real-Time Object-Oriented Modeling*. New York: John Wiley and Sons, 1994.

[Sha89] Alan C. Shaw, "Reasoning about time in higher-level language software," *IEEE Transactions on Software Engineering*, 15 (1989): 875–889.

[Shl92] Sally Shlaer and Stephen J. Mellor, *Object Lifecycles: Modeling the World in States*. New York: Yourdon Press Computing Series, 1992.

[Spi07] Cary R. Spitzer, ed., *Digital Avionics Handbook, Second Edition: Avionics Development and Implementation*. Boca Raton, FL: CRC Press, 2007.

[Sri94] Amitabh Srivastava and Alan Eustace, "ATOM: A system for building customized program analysis tools," Digital Equipment Corp., WRL Research Report 94/2, March 1994, http://www.research.digital.com.

[Sta97A] William Stallings, *Data and Computer Communication*, 5th edition. Upper Saddle River, NJ: Prentice Hall, 1997.

[Sta97B] Joergen Staunstrup and Wayne Wolf, eds., *Hardware/Software Co-Design: Principles and Practice*. Boston: Kluwer Academic Publishers, 1997.

[Sto95] Thomas M. Stout and Theodore J. Williams, "Pioneering work in the field of computer process control," *IEEE Annals of the History of Computing* 17(1) (1995): 6–18.

[Str97] Bjarne Stroustrup, *The C++ Programming Language*, 3rd edition, Reading, MA: Addison-Wesley Professional, 1997.

[Tay06] Jim Taylor, *DVD Demystified*, 3rd edition. New York: McGraw Hill, 2006.

[Tex00B] Texas Instruments, *TMS320C55x DSP Functional Overview*, SPRU312, June 2000.

[Tex01] Texas Instruments, *TMS320C55x DSP Programmer's Guide, Preliminary Draft*, document SPRU376A, August 2001.

[Tex02] Texas Instruments, *TMS320C55x DSP Mnemonic Instruction Set Reference Guide*, document SPRU374G, October 2002.

[Tex04] Texas Instruments, *TMS320C55x DSP CPU Reference Guide*, document SPRU371F, February 2004.

[Tiw94] Vivek Tiwari, Sharad Malik, and Andrew Wolfe, "Power analysis of embedded software: a first step toward software power minimization," *IEEE Transactions on VLSI Systems* 2(4) (1994): 437–445.

[van97] Albert van der Werf, Font Brüls, Richard Kleinhorst, Erwin Waterlander, Matt Verstraeler, and Thomas Friedrich, "I.McIC: a single-chip MPEG2 video encoder for storage," In *ISSCC '97 Digest of Technical Papers*. Castine, ME: John W. Wuorinen, 1997, pp. 254–255.

[Wal97] Dave Walsh, "Reducing system cost with software modems," *IEEE Micro* July/August (1997): 37–55.

[Wal07] Randy Walter and Chris Watkins, "Genesis Platform," Chapter 12. In *Digital Avionics Handbook, Second Edition: Avionics Development and Implementation*, ed. Cary R. Spitzer. Boca Raton, FL: CRC Press, 2007.

[Whi72] Thomas M. Whitney, France Rode, and Chung C. Tung, "The 'powerful pocketful': an electronic calculator challenges the slide rule," *Hewlett-Packard Journal* (1972): 2–9.

[Wol92] Wayne Wolf, "Expert opinion: in search of simpler software integration," *IEEE Spectrum* 29(1) (1992): 31.

[Wol08] Wayne Wolf, *Modern VLSI Design: IP-Based System Design*, 4th edition. Upper Saddle River, NJ: Prentice Hall, 1998.

[Yan89] Kun-Min Yang, Ming-Ting Sun, and Lancelot Wu, "A family of VLSI designs for the motion compensation block-matching algorithm," *IEEE Transactions on Circuits and Systems* 36(10) (1989): 1317–1325.

[Zha04] Feng Zhao and Leonidas Guibas, *Wireless Sensor Networks: An Information Processing Approach*. San Francisco: Morgan Kaufmann, 2004.

Index